Advances in Mechanics and Mathematics

Volume 26

T0190460

For further volumes:
http://www.springer.com/series/5613

Stanisław Migórski • Anna Ochal • Mircea Sofonea

Nonlinear Inclusions and Hemivariational Inequalities

Models and Analysis of Contact Problems

Springer

Stanisław Migórski
Faculty of Mathematics
and Computer Science
Institute of Computer Science
Jagiellonian University
Kraków, Poland

Anna Ochal
Faculty of Mathematics
and Computer Science
Institute of Computer Science
Jagiellonian University
Kraków, Poland

Mircea Sofonea
Laboratoire de Mathématiques et Physique
(LAMPS)
Université de Perpignan Via Domitia
Perpignan, France

ISSN 1571-8689 ISSN 1876-9896 (electronic)
ISBN 978-1-4899-9561-2 ISBN 978-1-4614-4232-5 (eBook)
DOI 10.1007/978-1-4614-4232-5
Springer New York Heidelberg Dordrecht London

Printed on acid-free paper

Springer is part of Springer Science+Business Media (www.springer.com)

To My MM
(Stanisław Migórski)

To My Parents
(Anna Ochal)

To Carmen and Mircea, with Love
(Mircea Sofonea)

Preface

An important number of problems arising in mechanics, physics, and engineering science lead to mathematical models expressed in terms of nonlinear inclusions. For this reason the mathematical literature dedicated to this field is extensive and the progress made in the last four decades is impressive. It concerns both results on the existence, uniqueness, regularity, and behavior of the solution for various classes of nonlinear inclusions and results on the numerical approaches to the solution of the corresponding problems. Hemivariational inequalities represent a class of nonlinear inclusions that are associated with the Clarke subdifferential operator.

Contact between deformable bodies abound in industry and everyday life. A few simple examples are brake pads in contact with wheels, tires on roads, and pistons with skirts. Because of the importance of contact processes in structural and mechanical systems, considerable effort has been put into their modeling, analysis, and numerical simulations.

The purpose of this book is to introduce to the reader the theory of nonlinear inclusions and hemivariational inequalities with emphasis on contact mechanics. The content covers both abstract results for nonlinear inclusions and hemivariational inequalities and the study of specific contact problems, including their modeling and variational analysis. In carrying out the variational analysis of various contact models we systematically use results on hemivariational inequalities and, in this way, we illustrate the applications of nonlinear analysis in contact mechanics. Our intention is to introduce new mathematical results and to apply them in the study of nonlinear problems which describe the contact between deformable bodies and foundations.

Our book is divided into three parts with eight chapters and is addressed to mathematicians, applied mathematicians, engineers, and scientists. Advanced graduate students can also benefit from the material presented in this book. It is organized with two different aims, so that readers who are not interested in modeling and applications can skip Part III and will find an introduction to the study of nonlinear inclusions and hemivariational inequalities in Part II of the book; alternatively, readers who are interested only in modeling and applications can skip the technical proofs presented in Part II of the book and will find in Part III the variational analysis of various mathematical models which describe contact processes.

A brief description of the three parts of the book follows.

Part I is devoted to the basic results on functional analysis which are fundamental to the study of the problems treated in the rest of the book. We review some preliminary material on normed and Banach spaces, duality and weak topologies, and results on measure theory. Then, we continue with a description of the function spaces, including spaces of smooth functions, Lebesgue and Sobolev spaces as well as spaces of vector-valued functions. For each of these spaces we present the main properties which are relevant to the developments we present in the following chapters. Finally, we describe some elements of nonlinear analysis, including set-valued mappings, nonsmooth analysis and operators of monotone type. We pay a particular attention to the subdifferentiability of the functionals and the properties of subdifferential mappings. The material presented in this introductory part is standard, although some of it is very recent, and can be found in various textbooks and monographs. For this reason we present only very few details of the proofs.

Part II includes original results; some of which have not been published before. We present various classes of nonlinear inclusions, both in the stationary and in the evolutionary cases. We prove existence results and, in some cases, we prove uniqueness results. To this end we use methods based on monotonicity, compactness, and fixed-point arguments. We then use our results in the study of various classes of stationary and evolutionary hemivariational inequalities, including inequalities with Volterra integral terms.

Part III is also based on our original research. It deals with the study of static and dynamic frictional contact problems. We model the material's behavior with elastic or viscoelastic constitutive laws and, in the case of viscoelastic materials, we consider both short- and long-term memory laws. We pay particular attention to piezoelectric materials. The contact is modeled with normal compliance or normal damped response. Friction is modeled with versions of Coulomb's law and its regularizations. For each of the problems we provide a variational formulation, which usually is in a form of a stationary or evolutionary hemivariational inequality. Then, we use the abstract results in Part II and establish existence and, sometimes, uniqueness results.

Each part ends with a section entitled *Bibliographical Notes* that discusses references on the principal results treated, as well as information on important topics related to, but not included, in the body of the text. The list of the references at the end of the book is by no means exhaustive. It only includes works that are closely related to the subjects treated in this monograph.

The present manuscript is a result of the cooperation between the authors over the last four years. It was written within the project Polonium "Nonsmooth Analysis with Applications to Contact Mechanics" under contract no. 7817/R09/R10 between the Jagiellonian University and the University of Perpignan. This research was supported by a Marie Curie International Research Staff Exchange Scheme Fellowship within the seventh European Community Framework Programme under

Grant Agreement no. PIRSES-GA-2011-295118 as well. The first two authors were also partially supported by the Ministry of Science and Higher Education of Poland under grants nos. N201 027 32/1449 and N N201 604640. Part of the material is related to our joint work with several collaborators to whom we express our thanks: Prof. Mikaël Barboteu (Perpignan), Prof. Zdzislaw Denkowski (Krakow), Prof. Weimin Han (Iowa City), Prof. Zhenhai Liu (Nanning, P.R. China), and Prof. Meir Shillor (Rochester, Michigan). We also thank Joëlle Sulian who prepared the figures for this book. We extend our gratitude to Professor David Y. Gao for inviting us to make this contribution to the Springer book series entitled *Advances in Mechanics and Mathematics*.

Kraków, Poland — Stanisław Migórski
Kraków, Poland — Anna Ochal
Perpignan, France — Mircea Sofonea

Contents

List of Symbols

Sets

$\mathbb{N}$: the set of positive integers

$\mathbb{N}_0$: the set of nonnegative integers, i.e., $\mathbb{N}_0 = \mathbb{N} \cup \{0\}$

$\mathbb{R}$: the real line

$\mathbb{R}_+$: the set of nonnegative real numbers, i.e., $\mathbb{R}_+ = \{ r \in \mathbb{R} \mid r \geq 0 \}$

$\overline{\mathbb{R}} = \mathbb{R} \cup \{-\infty, +\infty\}$

$\mathbb{R}^d$: the d-dimensional Euclidean space

$\mathbb{S}^d$: the space of second-order symmetric tensors on $\mathbb{R}^d$

2^X: all subsets of a set X

$\mathcal{P}(X)$: all nonempty subsets of a set X

Ω: a set in $\mathbb{R}^d$ with boundary $\Gamma = \partial\Omega$ usually, Ω is assumed to be bounded, connected and Γ is assumed to be Lipschitz

$\overline{\Omega}$: the closure of Ω, i.e., $\overline{\Omega} = \Omega \cup \Gamma$

$\Gamma = \overline{\Gamma}_D \cup \overline{\Gamma}_N \cup \overline{\Gamma}_C = \overline{\Gamma}_a \cup \overline{\Gamma}_b \cup \overline{\Gamma}_C$: two partitions on Γ such that $\overline{\Gamma}_D$, $\overline{\Gamma}_N$, $\overline{\Gamma}_C$ on one hand, and $\overline{\Gamma}_a$, $\overline{\Gamma}_b$, $\overline{\Gamma}_C$ on the other hand, have mutually disjoint interiors

Γ_D: the part of the boundary where the displacement is prescribed; meas $(\Gamma_D) > 0$ is assumed throughout the book

Γ_N: the part of the boundary where tractions are prescribed

Γ_C: a measurable part of Γ; in Part III it respresents the part of the boundary where contact takes place

Γ_a: the part of the boundary where the electric potential is prescribed; meas $(\Gamma_a) > 0$ is assumed throughout the book

Γ_b: the part of the boundary where the electric charges are specified

$[0, T]$: time interval of interest, $0 < T < \infty$

$Q = \Omega \times (0, T)$, $\Sigma_D = \Gamma_D \times (0, T)$, $\Sigma_N = \Gamma_N \times (0, T)$, $\Sigma_C = \Gamma_C \times (0, T)$
$\Sigma_a = \Gamma_a \times (0, T)$, $\Sigma_b = \Gamma_b \times (0, T)$

Abstract Spaces

X: a Hilbert space with inner product $\langle \cdot, \cdot \rangle_X$, or a normed space with norm $\| \cdot \|_X$

X^*: the dual of X

$\langle \cdot, \cdot \rangle_{X^* \times X}$: the duality paring

$(w\text{–}X)$, X_w: the space X endowed with the weak convergence

$(w^*\text{–}X^*)$, $X^*_{w^*}$: the space X^* endowed with the weak* convergence

$\mathcal{L}(X, Y)$: the space of linear continuous operators from a normed space X to a normed space Y

$\mathcal{L}(X) = \mathcal{L}(X, X)$

$X \times Y$: the product of Hilbert spaces X and Y, with inner product $\langle \cdot, \cdot \rangle_{X \times Y}$

$\mathcal{P}_{f(c)}(X) = \{ A \subseteq X \mid A \text{ is nonempty, closed, (convex)} \}$

$\mathcal{P}_{(w)k(c)}(X) = \{ A \subseteq X \mid A \text{ is nonempty, (weakly) compact, (convex)} \}$

(V, H, V^*): an evolution triple

$\mathcal{V} = L^2(0, T; V)$, $\mathcal{V}^* = L^2(0, T; V^*)$, $\mathcal{W} = \{ v \in \mathcal{V} \mid v' \in \mathcal{V}^* \}$

Operators

$A \colon X \to 2^Y$: a multivalued operator

$D(A) = \{ x \in X \mid A(x) \neq \emptyset \}$: the domain of A

$R(A) = \cup_{x \in X} A(x)$: the range of A

$\mathrm{Gr}\,(A) = \{ (x, y) \in X \times Y \mid y \in A(x) \}$: the graph of A

$A^{-1} \colon Y \to X$, $A^{-1}(y) = \{ x \in X \mid y \in A(x) \}$: the inverse of A

$\mathcal{P}_K$: the projection operator onto a set K

∇: the gradient operator

Div: the divergence operator for tensor fields

div: the divergence operator for vector fields

γ: the trace operator

$\boldsymbol{I}_d$: the identity operator on $\mathbb{R}^d$

Function Spaces

$C^m(\overline{\Omega})$: the space of functions whose derivatives up to and including order m are continuous up to the boundary Γ

$C_0^\infty(\Omega)$: the space of infinitely differentiable functions with compact support in Ω

$L^p(\Omega)$: the Lebesgue space of p-integrable functions, with the usual modification if $p = \infty$

$W^{k,p}(\Omega)$: the Sobolev space of functions whose weak derivatives of order less than or equal to k are p-integrable on Ω

$W_0^{k,p}(\Omega)$: the closure of $C_0^\infty(\Omega)$ in $W^{k,p}(\Omega)$

$W^{-k,q}(\Omega)$: the dual of $W_0^{k,p}(\Omega)$, $\frac{1}{p} + \frac{1}{q} = 1$

$H^k(\Omega) \equiv W^{k,2}(\Omega)$

$H_0^k(\Omega) \equiv W_0^{k,2}(\Omega)$

$H^{-1}(\Omega) \equiv W^{-1,2}(\Omega)$

$L^p(\Omega;\mathbb{R}^d) = \left(L^p(\Omega)\right)^d$

$W^{k,p}(\Omega;\mathbb{R}^d) = \left(W^{k,p}(\Omega)\right)^d$

$H^k(\Omega;\mathbb{R}^d) = \left(H^k(\Omega)\right)^d$

$Z = H^\delta(\Omega;\mathbb{R}^d),\ \delta \in \left(\frac{1}{2}, 1\right)$

$V = \{\, \boldsymbol{v} \in H^1(\Omega;\mathbb{R}^d) \mid \boldsymbol{v} = \boldsymbol{0} \text{ a.e. on } \Gamma_D \,\}$

$H = \{\, \boldsymbol{u} = (u_i) \mid u_i \in L^2(\Omega) \,\} = L^2(\Omega;\mathbb{R}^d)$

$\mathcal{H} = \{\, \boldsymbol{\sigma} = (\sigma_{ij}) \mid \sigma_{ij} = \sigma_{ji} \in L^2(\Omega) \,\} = L^2(\Omega;\mathbb{S}^d)$

$\Phi = \{\, \psi \in H^1(\Omega) \mid \psi = 0 \text{ a.e. on } \Gamma_a \,\}$

$C(0,T;X) = \{\, v\colon [0,T] \to X \mid v \text{ continuous} \,\}$

$C^m(0,T;X) = \{\, v \in C(0,T;X) \mid v^{(i)} \in C(0,T;X),\ i = 1,\dots,m \,\}$

$L^p(0,T;X) = \{\, v\colon (0,T) \to X \text{ measurable} \mid \|v\|_{L^p(0,T;X)} < \infty \,\}$

$$W^{k,p}(0,T;X) = \{\, v \in L^p(0,T;X) \mid \|v^{(i)}\|_{L^p(0,T;X)} < \infty \text{ for all } i \le k \,\}$$

$$H^k(0,T;X) \equiv W^{k,2}(0,T;X)$$

Other Symbols

c: a generic positive constant

$r_+ = \max\{0, r\}$: the positive part of r

$r_- = \max\{0, -r\}$: the negative part of r

$[r]$: integer part of r

$\forall$: for all

$\exists$: there exist(s)

$\Longrightarrow$: implies

$\Longleftrightarrow$: equivalence

$\overline{K}$: the closure of the set K

∂K: the boundary of the set K

δ_{ij}: the Kronecker delta

a.e.: almost everywhere

lsc: lower semicontinuous

usc: upper semicontinuous

ψ_K: the indicator function of the set K

χ_K: the characteristic function of the set K

$\partial\varphi$: the generalized subdifferential of the function φ

f', f'', $f^{(i)}$: the time derivatives of the function f

$\boldsymbol{u} \cdot \boldsymbol{v}$: the canonical inner product of the vectors $\boldsymbol{u}, \boldsymbol{v} \in \mathbb{R}^d$

$\boldsymbol{\sigma} : \boldsymbol{\tau}$: the canonical inner product of the tensors $\boldsymbol{\sigma}, \boldsymbol{\tau} \in \mathbb{S}^d$

$\boldsymbol{\nu}$: the unit outward normal vector to Γ

u_ν: the normal component of the vector $\boldsymbol{u}$, i.e. $u_\nu = \boldsymbol{u} \cdot \boldsymbol{\nu}$

$\boldsymbol{u}_\tau$: the tangential component of the vector $\boldsymbol{u}$, i.e. $\boldsymbol{u}_\tau = \boldsymbol{u} - u_\nu \boldsymbol{\nu}$

σ_ν: the normal component of the tensor $\boldsymbol{\sigma}$, i.e. $\sigma_\nu = \boldsymbol{\sigma}\boldsymbol{\nu} \cdot \boldsymbol{\nu}$

$\boldsymbol{\sigma}_\tau$: the tangential component of the tensor $\boldsymbol{\sigma}$, i.e. $\boldsymbol{\sigma}_\tau = \boldsymbol{\sigma}\boldsymbol{\nu} - \sigma_\nu \boldsymbol{\nu}$

Part I
Background on Functional Analysis

Chapter 1
Preliminaries

In this chapter we present preliminary material from functional analysis which will be used subsequently. The results are stated without proofs, since they are standard and can be found in many references. For the convenience of the reader we summarize definitions and results on normed spaces, Banach spaces, duality, and weak topologies which are mostly assumed to be known as a basic material from functional analysis. We then recall some standard results on measure theory that will be applied repeatedly in this book. We assume that the reader has some familiarity with the notions of linear algebra and general topology.

1.1 Normed Spaces

Background on normed spaces. The concept of linear or vector space is fundamental in linear algebra and functional analysis.

Definition 1.1. Let X be a set of elements, to be called vectors, and let $\mathbb{R}$ be the set of real scalars. Assume there are two operations: $(x, y) \mapsto x + y \in X$ and $(\alpha, x) \mapsto \alpha x \in X$, called addition and scalar multiplication, respectively, defined for any $x, y \in X$ and any $\alpha \in \mathbb{R}$. These operations are to satisfy the following rules:

(i) $x + y = y + x$ for all $x, y \in X$.
(ii) $(x + y) + z = x + (y + z)$ for all $x, y, z \in X$.
(iii) There exists an element $0 \in X$ such that $0 + x = x$ for all $x \in X$.
(iv) For each $x \in X$, there is an element $-x \in X$ such that $x + (-x) = 0$.
(v) $1x = x$ for all $x \in X$.
(vi) $\alpha(\beta x) = (\alpha\beta)x$ for all $x \in X$ and all $\alpha, \beta \in \mathbb{R}$.

S. Migórski et al., *Nonlinear Inclusions and Hemivariational Inequalities*, Advances in Mechanics and Mathematics 26, DOI 10.1007/978-1-4614-4232-5_1,

(vii) $\alpha(x+y) = \alpha x + \alpha y$ and $(\alpha + \beta)x = \alpha x + \beta x$ for all $x, y \in X$ and all α, $\beta \in \mathbb{R}$.

Then X is called a *real linear space* or a *real vector space*.

In this book we deal only with real linear spaces, so in what follows we call them linear spaces.

Definition 1.2. A *norm* on a linear space X is a map

$$\|\cdot\|_X : X \to [0, +\infty)$$

such that

(i) $\|x\|_X = 0$ if and only if $x = 0$.
(ii) $\|\alpha x\|_X = |\alpha| \, \|x\|_X$ for all $x \in X$ and $\alpha \in \mathbb{R}$.
(iii) $\|x + y\|_X \leq \|x\|_X + \|y\|_X$ for all $x, y \in X$.

The inequality (iii) above is known as the *triangle inequality* and it is in common use in mathematics. The linear space X equipped with the norm $\|\cdot\|_X$, denoted $(X, \|\cdot\|_X)$, is called a *linear normed space* or a *normed space*. For simplicity we say that X is a normed space when the definition of the norm is clear from the context.

Every normed space is a metric space and, consequently, a topological space. We can always define a metric on a normed space in terms of its norm, by taking

$$d(x, y) = \|x - y\|_X \quad \text{for all} \quad x, y \in X.$$

Definition 1.3. Let X be a normed space with the norm $\|\cdot\|_X$. A sequence $\{x_n\} \subset X$ is said to be *convergent* if there exists $x \in X$ such that

$$\lim_{n\to\infty} \|x_n - x\|_X = 0.$$

We also say that x is a *limit* of the sequence $\{x_n\}$, and write $x_n \to x$ in X, as $n \to \infty$ or $\lim\limits_{n\to\infty} x_n = x$ in X. When no confusion arises, for simplicity, we write $x_n \to x$ or $\lim\limits_{n\to\infty} x_n = x$ or $\lim x_n = x$.

It can be verified that any sequence can have at most one limit, i.e., when a limit exists, it is unique.

On a linear space various norms can be defined which may give rise to different forms of convergence. Nevertheless, this situation does not arise when two norms are equivalent.

Definition 1.4. Let X be a linear space. We say that two norms $\|\cdot\|_1$ and $\|\cdot\|_2$ on X are *equivalent* if there exists a positive constant c such that

$$c^{-1}\|x\|_1 \leq \|x\|_2 \leq c\|x\|_1 \quad \text{for all} \quad x \in X.$$

Obviously, if two norms are equivalent then a sequence $\{x_n\}$ converges in one norm if and only if it converges in the other norm. Conversely, if each sequence converging with respect to one norm also converges with respect to the other norm, then the two norms are equivalent. It is well known that over a finite-dimensional space, any two norms are equivalent and, therefore, on such spaces different norms lead to the same convergence notion.

We recall in what follows the concept of continuity of functions defined on normed spaces.

Definition 1.5. Let X be a normed space. A function $f: X \to \mathbb{R}$ is said to be *continuous* at $x \in X$ if for any sequence $\{x_n\}$ with $x_n \to x$, we have $f(x_n) \to f(x)$ as $n \to \infty$. The function f is said to be *continuous* on X if it is continuous at every $x \in X$.

Let $(X, \|\cdot\|_X)$ be a normed space. As a consequence of the triangle inequality (see Definition 1.2(iii)) it is easy to see that

$$\Big| \|x\|_X - \|y\|_X \Big| \leq \|x - y\|_X \quad \text{for all} \quad x, y \in X.$$

Therefore, if $x_n \to x$ then $\|x_n\|_X \to \|x\|_X$, which implies that the norm function is continuous.

We proceed with some definitions related to closedness and compactness which are sufficient for our purposes.

Definition 1.6. A subset D of a normed spaces X is called (normed or strongly) *closed* if any limit of any convergent sequence contained in D is itself in D.

We note that above we define "closedness" in terms of convergence of sequences which would be in topological spaces called rather "sequential closedness." The *closure* of a set D of X is the set obtained by adding to D the limits of all convergent sequences $\{x_n\} \subset D$. In other words, the closure of a set D is the smallest closed set containing D. The closure of a set D is denoted by $\overline{D}$.

Definition 1.7. Let $(X, \|\cdot\|_X)$ be a normed space. A subset D of X is called

(i) *dense* (in X), if $\overline{D} = X$.
(ii) *bounded*, if $\sup \{\|x\|_X \mid x \in D\} < \infty$.

If there exists a countable dense subset of X, we say that the normed space X is *separable*.

Definition 1.8. A subset D in a normed space X is *compact* if every sequence in D contains a convergent subsequence whose limit belongs to D. The subset $D \subset X$ is called *relatively compact* if its closure is compact.

We remark that in a general topological space a set D is called "compact" if every covering of D by open sets contains a finite subcovering, and a set D which satisfies Definition 1.8 is called "sequential compact." Nevertheless, we do not distinguish between these concepts because of the following result.

Theorem 1.9. *Let X be a normed space. Then, a set $D \subset X$ is compact if and only if it is sequentially compact.*

Cauchy sequences and Banach spaces. The following concept is important in the definition of a Banach space.

Definition 1.10. Let X be a normed space with the norm $\|\cdot\|_X$. A sequence $\{x_n\} \subset X$ is said to be a *Cauchy sequence* if for every $\varepsilon > 0$ there is $n_0 = n_0(\varepsilon) \in \mathbb{N}$ such that

$$\|x_m - x_n\|_X \leq \varepsilon \text{ for all } m, n \geq n_0.$$

It is clear that a convergent sequence is a Cauchy sequence. In the finite-dimensional space $\mathbb{R}^d$ any Cauchy sequence is convergent; however in a general infinite-dimensional space, a Cauchy sequence may fail to converge. This justify the following definition.

Definition 1.11. A normed space is said to be *complete* if every Cauchy sequence from the space converges to an element in the space. A complete normed space is called a *Banach space*.

The following are classical examples of infinite-dimensional Banach spaces.

1. $c_0 = \{\, x = \{x_n\} \subset \mathbb{R} \mid x_n \to 0 \,\}$ and
   $c = \{\, x = \{x_n\} \subset \mathbb{R} \mid x_n$ is a convergent sequence in $\mathbb{R}\,\}$,
   both with norm $\|x\| = \sup\limits_{n\in\mathbb{N}} |x_n|$.
2. $l^p = \left\{ x = \{x_n\} \subset \mathbb{R} \mid \sum_{n\in\mathbb{N}} |x_n|^p < \infty \right\}$
   with norm $\|x\| = \left(\sum_{n\in\mathbb{N}} |x_n|^p \right)^{1/p}$, $1 \leq p < \infty$.
3. The Lebesgue spaces $L^p(\Omega)$, $1 \leq p \leq \infty$ introduced in Sect. 2.2, with the norm $\|\cdot\|_{L^p(\Omega)}$ introduced on page 26.
4. $C_b(X) = \{\, v\colon X \to \mathbb{R} \mid v$ is continuous and bounded$\,\}$ and
   $B(X) = \{\, v\colon X \to \mathbb{R} \mid v$ is bounded $\}$,
   both with norm $\|v\| = \sup\limits_{x\in X} |v(x)|$, X being a metric space.

When X is a compact metric space, then every continuous function $v\colon X \to \mathbb{R}$ is bounded and then we write $C(X)$ instead of $C_b(X)$. On the other hand, given an open bounded set $\Omega \subset \mathbb{R}^d$ and $1 \leq p < \infty$, we define the p-norm by

$$\|v\|_p = \left(\int_\Omega |v(x)|^p \, dx \right)^{1/p} \quad \text{for all } v \in C(\overline{\Omega}).$$

It can be proved that the space $C(\overline{\Omega})$ endowed with the norm $\|\cdot\|_p$ is not a Banach space.

The following result is the celebrated "Banach contraction principle" or "Banach fixed point theorem," which guarantees the existence of a unique fixed point in a complete metric space and, therefore, in a Banach space.

Lemma 1.12 (Banach contraction principle). *Let (X, d) be a complete metric space and let $f: X \to X$ be a k-contraction (i.e., for all $x, y \in X$ we have $d(f(x), f(y)) \leq k\, d(x, y)$ with $k < 1$). Then f has a unique fixed point, i.e., there exists a unique element $x^* \in X$ such that $f(x^*) = x^*$.*

Although a Cauchy sequence is not necessarily convergent, it converges if it has a convergent subsequence, as shown in the following result.

Proposition 1.13. *If a Cauchy sequence $\{x_n\}$ in a normed space contains a subsequence $\{x_{n_k}\}$ such that $x_{n_k} \to x$ as $k \to \infty$, then $\lim_{n\to\infty} x_n = x$.*

Moreover, we recall the following convergence criterium in normed spaces.

Proposition 1.14. *If every subsequence of a sequence $\{x_n\}$ in a normed space contains a subsequence convergent to x, then $\lim_{n\to\infty} x_n = x$.*

Inner product spaces and Hilbert spaces. We introduce now the concept of inner product spaces. This class of spaces has more mathematical structure than the normed spaces discussed previously. Every inner product space is a normed space and, therefore, a metric space and a topological space. Furthermore, the notion of "orthogonality" of the elements is defined in an inner product space.

Definition 1.15. An *inner product* on a linear space X is a map

$$\langle \cdot, \cdot \rangle_X : X \times X \to \mathbb{R}$$

such that

(i) $\langle x, y\rangle_X = \langle y, x\rangle_X$ for all $x, y \in X$.
(ii) $\langle \alpha x + \beta y, z\rangle_X = \alpha\langle x, z\rangle_X + \beta\langle y, z\rangle_X$ for all $x, y, z \in X$, and $\alpha, \beta \in \mathbb{R}$.
(iii) $\langle x, x\rangle_X \geq 0$ for all $x \in X$, $\langle x, x\rangle_X = 0$ if and only if $x = 0$.

The linear space X equipped with an inner product $\langle \cdot, \cdot \rangle_X$, denoted $(X, \langle \cdot, \cdot \rangle_X)$, is called an *inner product space*. For simplicity we say that X is an inner product space when the definition of the inner product is clear from the context.

The inner product generates a norm on a linear space X given by

$$\|x\|_X = \sqrt{\langle x, x\rangle_X}. \tag{1.1}$$

Hence, all inner product spaces are normed spaces, but the converse is not true in general.

We use frequently the following properties of an inner product space: the Cauchy–Schwarz inequality, the continuity of the inner product, and the parallelogram law, which we summarized in the following.

Theorem 1.16. (i) (Cauchy–Schwarz inequality) *If X is an inner product space, then*

$$|\langle x, y\rangle_X| \leq \|x\|_X\, \|y\|_X \quad \text{for all } x, y \in X,$$

and the equality holds if and only if x and y are linearly dependent.

(ii) (Continuity of the inner product) *An inner product is continuous with respect to its induced norm, i.e., if the norm is defined by* (1.1)*, then* $\|x_n - x\|_X \to 0$ *and* $\|y_n - y\|_X \to 0$ *imply*

$$\langle x_n, y_n\rangle_X \to \langle x, y\rangle_X .$$

(iii) (Parallelogram law) *A norm* $\|\cdot\|_X$ *on a linear space X is induced by an inner product if and only if it satisfies the parallelogram law*

$$\|x+y\|_X^2 + \|x-y\|_X^2 = 2\left(\|x\|_X^2 + \|y\|_X^2\right) \quad \text{for all} \quad x, y \in X.$$

Among the inner product spaces, of particular importance are those which are complete in the norm generated by the inner product.

Definition 1.17. A *Hilbert space* is a complete inner product space.

The inner product allows to introduce the concept of orthogonality of elements in an inner product space.

Definition 1.18. Let X be an inner product space. Two elements $x,\ y \in X$ are said to be *orthogonal* if $\langle x, y\rangle_X = 0$. The *orthogonal complement* of a set $A \subseteq X$ is defined by

$$A^{\perp} = \{\, x \in X \mid \langle x, a\rangle_X = 0 \ \text{ for all } \ a \in A \,\}.$$

It is known that if A is an arbitrary set of a Hilbert space X, then its orthogonal complement $A^{\perp}$ is a closed linear subspace of X. An important role in the theory of Hilbert spaces is played by the following result.

Theorem 1.19. *Let* $A \subseteq X$ *be a closed linear subspace of a Hilbert space* X *and* $A^{\perp}$ *be its orthogonal complement. Then any element* $x \in X$ *can uniquely be represented as* $x = a + a'$*, where* $a \in A$ *and* $a' \in A^{\perp}$*. In this case we write* $X = A \oplus A^{\perp}$.

We conclude this section with the definition of the projection operator in a Hilbert space. To this end we start by recalling the following definition.

Definition 1.20. Let X be a linear space. A subset C of X is said to be *convex* if for all $x_1, x_2 \in C$ and $\lambda \in [0,1]$ we have

$$\lambda x_1 + (1-\lambda)x_2 \in C.$$

In other words, the set C is convex if it contains all the segments joining any two of its points.

An induction argument easily shows that if $x_i \in C$, $i = 1, \dots, n$, then $w = \sum_{i=1}^{n} \lambda_i x_i \in C$, whenever $\sum_{i=1}^{n} \lambda_i = 1$ and $\lambda_i \geq 0$. The vector w is called a *convex combination* of $x_1, \dots, x_n$.

We proceed with the following existence and uniqueness result.

Proposition 1.21 (Projection lemma). *Let K be a nonempty, closed, and convex subset of a Hilbert space X. Then for each $f \in X$ there exists a unique element $u \in K$ such that*

$$\|u - f\|_X = \min_{v \in K} \|v - f\|_X. \tag{1.2}$$

Proposition 1.21 allows to introduce the following definition.

Definition 1.22. Let K be a nonempty, closed, and convex subset of a Hilbert space X. Then, for each $f \in X$ the element u which satisfies (1.2) is called the *projection* of f on K and is usually denoted by $\mathcal{P}_K f$. Moreover, the operator $\mathcal{P}_K \colon X \to K$ is called the *projection operator* onto K.

A characterization of the projection, in terms of variational inequalities, is provided by the following result.

Proposition 1.23. *Let K be a nonempty, closed, and convex subset of a Hilbert space X and let $f \in X$. Then $u = \mathcal{P}_K f$ if and only if*

$$u \in K, \quad (u, v - u)_X \geq (f, v - u)_X \quad \text{for all } v \in K. \tag{1.3}$$

Using Proposition 1.23 it is easy to prove the following result, which will be useful in next chapters of the book.

Proposition 1.24. *Let K be a nonempty, closed, and convex subset of a Hilbert space X. Then the projection operator satisfies the following inequalities:*

$$(\mathcal{P}_K u - \mathcal{P}_K v, u - v)_X \geq 0 \quad \text{for all } u, v \in X, \tag{1.4}$$

$$\|\mathcal{P}_K u - \mathcal{P}_K v\|_X \leq \|u - v\|_X \quad \text{for all } u, v \in X. \tag{1.5}$$

Proof. Let $u, v \in X$. We use (1.3) to obtain

$$(\mathcal{P}_K u, \mathcal{P}_K v - \mathcal{P}_K u)_X \geq (u, \mathcal{P}_K v - \mathcal{P}_K u)_X,$$

$$(\mathcal{P}_K v, \mathcal{P}_K u - \mathcal{P}_K v)_X \geq (v, \mathcal{P}_K u - \mathcal{P}_K v)_X.$$

We add these inequalities to see that

$$(\mathcal{P}_K u - \mathcal{P}_K v, \mathcal{P}_K v - \mathcal{P}_K u)_X \geq (u - v, \mathcal{P}_K v - \mathcal{P}_K u)_X$$

and, therefore,

$$(\mathcal{P}_K u - \mathcal{P}_K v, u - v)_X \geq \|\mathcal{P}_K u - \mathcal{P}_K v\|_X^2. \tag{1.6}$$

Inequality (1.4) follows now from (1.6) whereas inequality (1.5) follows from (1.6) and the Cauchy–Schwarz inequality. □

1.2 Duality and Weak Topologies

In this section we deal with duality and weak topologies. We begin with some standard results on operators defined on normed spaces. Given two linear spaces X and Y, an *operator* $A: X \to Y$ is a rule that assigns to each element $x \in X$ a unique element $Ax \in Y$. A real-valued operator defined on a linear space is called a *functional.*

Linear and continuous operators. Some important classes of operators defined on linear normed spaces are introduced by the definitions below.

Definition 1.25. Let X and Y be two linear spaces. An operator $A: X \to Y$ is *linear* if

$$A(\alpha_1 x_1 + \alpha_2 x_2) = \alpha_1 A x_1 + \alpha_2 A x_2 \ \text{ for all } \ x_1, x_2 \in X, \alpha_1, \alpha_2 \in \mathbb{R}.$$

Definition 1.26. Let $(X, \|\cdot\|_X)$ and $(Y, \|\cdot\|_Y)$ be two normed spaces. An operator $A: X \to Y$ is said to be

(i) *continuous at* $x_0 \in X$ if for every $\varepsilon > 0$ there is $\delta > 0$ such that for every $x \in X$, $\|x - x_0\|_X \leq \delta$ entails $\|Ax - Ax_0\|_Y \leq \varepsilon$ or, equivalently, if $\{x_n\} \subset X$, $x_n \to x_0$ in X implies $Ax_n \to Ax_0$ in Y.
(ii) *continuous* (on X), if it is continuous at each point $x_0 \in X$.
(iii) *Lipschitz continuous* (on X) if there exists a constant $L_A > 0$ such that $\|Ax_1 - Ax_2\|_Y \leq L_A \|x_1 - x_2\|_X$ for all $x_1, x_2 \in X$.
(iv) *bounded* (on X) if for any $r > 0$, there exists $R > 0$ such that $\|x\|_X \leq r$ implies $\|Ax\|_Y \leq R$ or, alternatively, if for any set $B \subset X$ the inequality $\sup_{x \in B} \|x\|_X < \infty$ implies that $\sup_{x \in B} \|Ax\|_Y < \infty$.

The main properties of linear continuous operators on normed spaces are resumed in the following result.

Theorem 1.27. *Let $(X, \|\cdot\|_X)$ and $(Y, \|\cdot\|_Y)$ be two normed spaces and let $A: X \to Y$ be a linear operator. Then*

(i) *A is continuous over the whole space X if and only if A is continuous at any one point, say at $x = 0$.*
(ii) *A is bounded if and only if there exists a constant $M > 0$ such that*

$$\|Ax\|_Y \leq M \|x\|_X \ \text{ for all } \ x \in X.$$

(iii) *A is continuous if and only if it is bounded on X.*

It follows from Theorem 1.27 that a linear operator between normed spaces is continuous if and only if it is Lipschitz continuous. Moreover, under the assumptions of Proposition 1.24, it follows that the projection operator $\mathcal{P}_K: X \to K \subset X$ is

Lipschitz continuous with Lipschitz constant $L = 1$. Such kinds of operators are also called *nonexpansive* operators.

Everywhere in this book we use the notation $\mathcal{L}(X, Y)$ for the set of all linear bounded operators between the normed spaces X and Y. In the special case $X = Y$, we use $\mathcal{L}(X)$ instead of $\mathcal{L}(X, X)$. For $A \in \mathcal{L}(X, Y)$, the quantity

$$\|A\|_{\mathcal{L}(X,Y)} = \sup_{x \in X \setminus \{0\}} \frac{\|Ax\|_Y}{\|x\|_X} = \sup_{\|x\|_X \leq 1} \|Ax\|_Y = \sup_{\|x\|_X = 1} \|Ax\|_Y$$

is called the *operator norm* of A and, indeed, the mapping

$$\mathcal{L}(X, Y) \ni A \mapsto \|A\|_{\mathcal{L}(X,Y)} \in \mathbb{R}$$

defines a norm on the space $\mathcal{L}(X, Y)$. It is easy to see that for all $A \in \mathcal{L}(X, Y)$, we have

$$\|A\|_{\mathcal{L}(X,Y)} = \inf\{ M > 0 \mid \|Ax\|_Y \leq M \|x\|_X \text{ for all } x \in X \}.$$

Moreover, the operator norm enjoys the following compatibility condition:

$$\|Ax\|_Y \leq \|A\|_{\mathcal{L}(X,Y)} \|x\|_X \text{ for all } x \in X,$$

and, finally, the following completitude result holds.

Theorem 1.28. *Let X be a normed space and Y be a Banach space. Then $\mathcal{L}(X, Y)$ is also a Banach space.*

Dual and reflexive spaces. In the special case $Y = \mathbb{R}$, endowed with the usual topology, the space $\mathcal{L}(X, \mathbb{R})$ is called the (topological) *dual space* of X, and it is denoted by X^*. The elements of $X^* = \mathcal{L}(X, \mathbb{R})$ are also called linear continuous functionals. The bilinear (i.e., linear in both variables) mapping $\langle \cdot, \cdot \rangle_{X^* \times X} : X^* \times X \to \mathbb{R}$ defined by $\langle x^*, x \rangle_{X^* \times X} = x^*(x)$ is called the *duality pairing* or the *duality brackets* for the pair (X^*, X). This notation is convenient in many problems involving different dual spaces. The space X^* endowed with the dual norm

$$\|x^*\|_{X^*} = \sup_{\|x\|_X \leq 1} |\langle x^*, x \rangle_{X^* \times X}| = \sup_{\|x\|_X = 1} |\langle x^*, x \rangle_{X^* \times X}|$$

becomes a normed space. In fact, it is a Banach space since, by Theorem 1.28, the dual of any normed space (complete or not), is always complete, i.e., it is a Banach space. Obviously, we have the following useful inequality:

$$|\langle x^*, x \rangle_{X^* \times X}| \leq \|x\|_X \sup_{\|v\|_X \leq 1} |\langle x^*, v \rangle_{X^* \times X}| = \|x^*\|_{X^*} \|x\|_X$$

for all $x \in X$, $x^* \in X^*$. The dual space $\mathcal{L}(X^*, \mathbb{R})$ of X^* is called the *bidual space* or *second dual* of X and it is denoted by $X^{**} = (X^*)^*$. The bidual is also a Banach space. Each element $x \in X$ induces a linear continuous functional $l_x \in X^{**}$ by the relation $l_x(x^*) = \langle x^*, x\rangle_{X^*\times X}$ for all $x^* \in X^*$. The mapping $X \ni x \mapsto l_x \in X^{**}$ is linear and isometric, i.e., $\|l_x\|_{X^{**}} = \|x\|_X$ for all $x \in X$. Thus, the normed space X can be viewed as a linear subspace of the Banach space X^{**} under the mapping $x \mapsto l_x$. This mapping is called the *canonical injection* or *canonical embedding* of X into its bidual X^{**} and is denoted by $\mathcal{I}$. The surjectivity of $\mathcal{I}$ leads to the notion of reflexive space.

Definition 1.29. A normed space X is said to be *reflexive* if X may be identified with its bidual X^{**} by the canonical embedding $\mathcal{I}$, i.e., if $\mathcal{I}(X) = X^{**}$.

An immediate consequence of Definition 1.29 is that a reflexive normed space is always complete, i.e., it is a Banach space.

On Hilbert spaces, any linear continuous functional is of the form of inner product, as shown in the following result.

Theorem 1.30 (Riesz representation theorem). *Let X be a Hilbert space and $l \in X^*$. Then there exists a unique $u \in X$ such that $l(x) = (u, x)_X$ for all $x \in X$. Moreover, $\|l\|_{X^*} = \|u\|_X$.*

By the Riesz representation theorem, it is relatively straightforward to show that any Hilbert space is reflexive.

Weak convergence. We recall from Definition 1.3 that in a normed space X, a sequence $\{x_n\}$ is said to converge to an element $x \in X$, if $\lim_{n\to\infty} \|x_n - x\|_X = 0$. Such convergence is also called *convergence in norm* or *strong convergence*, and we write $x_n \to x$ in X. In a normed space it is possible to introduce another type of convergence, which is called *weak convergence*.

Definition 1.31. Let X be a normed space with X^* its dual space. A sequence $\{x_n\} \subset X$ *weakly converges* to $x \in X$, if

$$l(x_n) \to l(x) \quad \text{for all } l \in X^*.$$

In this case we say that x is a *weak limit* of the sequence $\{x_n\}$ and we write $x_n \to x$ weakly in X.

It can be proved that the weak limit $x \in X$, if it exists, is unique. The space X endowed with the weak convergence is denoted by $(w\text{–}X)$ or X_w. It is easy to see that any sequence converging in X is weakly convergent. The converse is not true, except in the case when X is a finite-dimensional space.

The weak convergence is used to define weakly closed sets in a normed space.

Definition 1.32. A subset D of a normed space X is called *weakly closed* if it contains the limits of all weakly convergent sequences $\{x_n\} \subset D$.

Evidently, every weakly closed subset of X is (normed or strongly) closed, the converse is not true, in general. An exception is provided by the class of convex sets, as shown in the following result.

Theorem 1.33 (Mazur theorem). *A convex subset of a Banach space is (strongly) closed if and only if it is weakly closed.*

We now review further results concerning the weak convergence. To this end, we recall that given a normed space X, a set $D \subset X$ is called *sequentially weakly compact*, if every sequence in D has a subsequence converging weakly to a point in D. Moreover, the *weak closure* of a set D of X is the set obtained by adding to D the limits of all weak convergent sequences $\{x_n\} \subset D$. In other words, the weak closure of a set D is the smallest weakly closed set containing D.

Theorem 1.34 (Eberlein–Smulian theorem). *Let D be a subset of a Banach space X. Then the weak closure of D is weakly compact if and only if for any sequence in D there exists a subsequence weakly convergent to some element of X.*

Theorem 1.35 (Kakutani theorem). *A Banach space X is reflexive if and only if the closed unit ball $\{x \in X \mid \|x\|_X \leq 1\}$ is weakly compact.*

Perhaps the most frequent use of the reflexive spaces is based on the following compactness result. It is a corollary of the previous two theorems.

Theorem 1.36. *Let X be a Banach space. Then X is reflexive if and only if every bounded sequence in X contains a weakly convergent subsequence.*

We also have the following result.

Proposition 1.37. *Let X be a Banach space. Then, the following statements hold:*

(i) *If a sequence $\{x_n\} \subset X$ converges weakly to $x \in X$, then it is bounded and*

$$\|x\|_X \leq \liminf \|x_n\|_X.$$

(ii) *If $\{x_n\} \subset X$, $\{l_n\} \subset X^*$, $x_n \to x$ weakly in X and $l_n \to l$ in X^*, then*

$$\langle l_n, x_n \rangle_{X^* \times X} \to \langle l, x \rangle_{X^* \times X}.$$

Proposition 1.37(i) implies that the norm function on the dual of a Banach space is weakly lower semicontinuous. Indeed, to be more precise, we recall that a function $f : X \to \overline{\mathbb{R}}$ on a normed space X is called *lower* (respectively, *upper*) *semicontinuous* (or sequentially lower (upper) semicontinuous), if for all $x \in X$ and any $\{x_n\} \subset X$ such that $x_n \to x$ in X, we have

$$f(x) \leq \liminf f(x_n) \ \text{(respectively, } \limsup f(x_n) \leq f(x)\text{)}. \tag{1.7}$$

If the convergence of $\{x_n\}$ refers to the weak one, the function f is called *weakly lower* (respectively, *upper*) *semicontinuous*.

As usual in the literature, we refer to a lower semicontinuous function as to a *lsc* function. Similarly, we refer to an upper semicontinuous function as to an *usc* function. Also, we recall that here and everywhere in this book the set $\overline{\mathbb{R}} = \mathbb{R} \cup \{\pm\infty\}$ is endowed with the usual operations of addition and scalar multiplication, with its usual structure of order as well as with its usual topology.

In what follows we recall some convexity and smoothness properties of the norm in Banach spaces that are important in the surjectivity result for pseudomonotone operators presented in Theorem 3.63.

Definition 1.38. A Banach space X is called *strictly convex* if

$$\|\lambda u + (1-\lambda)v\|_X < 1$$

provided $\|u\|_X = \|v\|_X = 1$, $u \neq v$, and $0 < \lambda < 1$. A Banach space X is called *locally uniformly convex* if for each $\varepsilon \in (0,2]$ and for each $u \in X$ with $\|u\|_X = 1$, there exists $\delta = \delta(\varepsilon, u) > 0$ such that for all $v \in X$ with $\|v\|_X = 1$ and $\|u - v\|_X \geq \varepsilon$, the following inequality holds

$$\frac{1}{2}\|u + v\|_X \leq 1 - \delta.$$

A Banach space X is called *uniformly convex* if and only if X is locally uniformly convex and δ can be chosen to be independent of u.

It is known that, given a Banach space X, the following implications hold:

$$X \text{ is uniformly convex} \implies X \text{ is locally uniformly convex},$$

$$X \text{ is locally uniformly convex} \implies X \text{ is strictly convex}.$$

It is also known that each Hilbert space is uniformly convex (as a consequence of the parallelogram law) and every uniformly convex Banach space is reflexive (result known as the Milman–Pettis theorem). Moreover, in every reflexive Banach space, an equivalent norm can be introduced so that both X and X^* are locally uniformly convex (result known as the Troyanski theorem). The latter result simplifies some proofs in the theory of monotone operators.

Subsequently, we recall the well-known notion of convex function.

Definition 1.39. Let X be a linear space. A function $f: X \to \overline{\mathbb{R}}$ is *convex* if

$$f(\lambda x_1 + (1-\lambda)x_2) \leq \lambda f(x_1) + (1-\lambda) f(x_2)$$

for all $x_1, x_2 \in X$ and all $\lambda \in [0,1]$.

From Theorem 1.33, we have the following result.

Corollary 1.40. *Let X be a Banach space and $f: X \to \overline{\mathbb{R}}$ be convex. Then f is lower semicontinuous in the strong topology on X if and only if f is lower semicontinuous in the weak topology on X.*

Another result concerns the weak* continuity of linear operators between normed spaces and it is stated as follows.

Proposition 1.41. *Let $(X, \|\cdot\|_X)$ and $(Y, \|\cdot\|_Y)$ be normed spaces and let $A: X \to Y$ be a linear operator. Then A is continuous if and only if A is weakly continuous, i.e., A is continuous from $(w\text{–}X)$ into $(w\text{–}Y)$.*

Weak* convergence. In the dual of a normed space X, denoted X^*, a third type of convergence can be introduced, the *weak* convergence*. It is defined as follows.

Definition 1.42. Let X be a normed space with X^* its dual space. A sequence of functionals $\{l_n\} \subset X^*$ is called *weakly* convergent* to $l \in X^*$, if

$$l_n(x) \to l(x) \quad \text{for all } x \in X.$$

In this case we say that l is the weak* limit of the sequence $\{l_n\}$ and we write $l_n \to l$ weakly* in X^*.

It can be proved that the weak* limit $l \in X^*$, if it exists, is unique. The space X^* endowed with the weak* convergence is denoted by $(w^*\text{–}X^*)$ or X^*_{w*}. It is easy to see that any sequence converging in X^* is weakly* convergent. Moreover, the most important feature of the weak* topology is contained in the following compactness result (see Theorem 3.4.44 of [66]).

Theorem 1.43 (Banach–Alaoglu theorem). *The closed unit ball of the dual space X^* of a normed space X is compact in the weak* topology.*

Theorem 1.43 is frequently used in nonlinear analysis and in the study of various boundary value problems, as well. Its following version, which holds under the separability assumption, may also find useful applications.

Theorem 1.44. *Let X be a separable normed space. Then, every bounded sequence in the dual space X^* contains a subsequence that is weakly* convergent to an element of X^*.*

We have also the *weak*–X^** version of Proposition 1.37.

Proposition 1.45. *Let X be a Banach space. Then, the following statements hold:*

(i) *If a sequence $\{l_n\} \subset X^*$ weakly* converges to $l \in X^*$, then it is bounded and*

$$\|l\|_{X^*} \leq \liminf \|l_n\|_{X^*}.$$

(ii) *If* $\{x_n\} \subset X$, $\{l_n\} \subset X^*$, $x_n \to x$ *in* X *and* $l_n \to l$ *weakly* in* X^*, *then*

$$\langle l_n, x_n \rangle_{X^* \times X} \to \langle l, x \rangle_{X^* \times X}.$$

We conclude from above that X^* has two quite natural weak topologies, that generated by X^{**} (which is the weak topology for X^*), and that generated by the elements of X (which is the weak* topology for X^*). If X is a reflexive space, then the weak and the weak* topologies on X^* are the same. If X is not reflexive, then the weak* topology of X^* is weaker than its weak topology. If X is finite dimensional all three topologies (the weak, the weak*, and the norm topology) coincide.

Evolution triples. The concept of evolution triple (or, equivalently, Gelfand triple) is widely used in the study of nonlinear evolutionary equations and inclusions. To introduce it we start by recalling some results on compact operator between Banach spaces.

Definition 1.46. Let X and Y be Banach spaces and $A: X \to Y$ be a continuous operator. We say that A is a *compact operator*, if for every nonempty bounded set $D \subset X$, the set $A(D) = \{ Ax \mid x \in D \}$ is relatively compact in Y. Equivalently, we say that a continuous operator $A: X \to Y$ is compact if the image $\{A(x_n)\}$ of any bounded sequence $\{x_n\}$ in X contains a strongly convergent subsequence in Y.

Definition 1.47. Let X and Y be Banach spaces and $A: X \to Y$ be an operator. We say that A is *completely continuous* or *totally continuous* if it maps weakly convergent sequences to strongly convergent ones. Equivalently, A is a *completely continuous*, if it is $(w\text{–}X)$ to Y continuous.

The relationship between compact and complete continuous operators is provided by the following result.

Theorem 1.48. *Let* X *be a reflexive Banach space and* Y *be a Banach space. Then*

(i) *Every completely continuous operator from* X *to* Y *is compact.*
(ii) *The converse does not hold, cf. e.g., Example* 1.1.5 *of* [67].
(iii) *For linear continuous operators from* X *to* Y, *the notions of compactness and complete continuity are equivalent.*

Next, we introduce the concept of compact embedding.

Definition 1.49. Let X and Y be normed spaces. We say that X is *embedded* in Y provided

(i) X is a vector subspace of Y.
(ii) The embedding operator $i: X \to Y$ defined by $i(x) = x$ for all $x \in X$ is continuous.

We say that X is *compactly embedded* in Y if the embedding operator i is compact.

Since the embedding operator is linear, the continuity condition (ii) is equivalent to the existence of a constant $c > 0$ such that

$$\|x\|_Y \leq c\,\|x\|_X \quad \text{for all } \; x \in X$$

or, equivalently, to the following condition: for any sequence $\{x_n\} \subset X$, $x_n \to x$ in X entails $x_n \to x$ in Y. Moreover, it can be proved that $X \subset Y$ compactly if and only if for any sequence $\{x_n\} \subset X$, $x_n \to x$ weakly in X entails $x_n \to x$ in Y.

An important class of linear continuous operators defined in terms of the notion of duality consists of *adjoint* operators, called also *transpose* or *dual* operators and defined as follows.

Definition 1.50. Let X and Y be normed spaces and let $A \in \mathcal{L}(X, Y)$. The *adjoint* $A^*\colon Y^* \to X^*$ is the (unique) operator defined via the relation

$$\langle A^* y^*, x\rangle_{X^* \times X} = \langle y^*, Ax\rangle_{Y^* \times Y}$$

for all $y^* \in Y^*$ and $x \in X$.

One of the main properties of the adjoint operators is the following.

Proposition 1.51. *Let X and Y be normed spaces and let $A \in \mathcal{L}(X, Y)$. Then $A^* \in \mathcal{L}(Y^*, X^*)$, $A^* \in \mathcal{L}(Y^*_{w^*}, X^*_{w^*})$, and*

$$\|A\|_{\mathcal{L}(X,Y)} = \|A^*\|_{\mathcal{L}(Y^*,X^*)}.$$

In the special situation when $X = Y$ is a Hilbert space and $A = A^*$, we say that A is a *self-adjoint* operator. If $A \in \mathcal{L}(X)$ is self-adjoint, we have a useful characterization of its norm

$$\|A\|_{\mathcal{L}(X)} = \sup_{\|x\|_X = 1} |\langle Ax, x\rangle_X|.$$

We are now in a position to introduce the concept of evolution triple of spaces, also known as Gelfand triple.

Definition 1.52. A triple of spaces (V, H, V^*) is said to be an *evolution triple* of spaces, if the following are true:

(i) V is a separable, reflexive Banach space.
(ii) H is a separable Hilbert space.
(iii) The embedding $V \subset H$ is continuous and V is dense in H.

For examples of evolution triples of spaces we refer to Example 2.20 on page 33. In an evolution triple we identify the Hilbert space H with its dual H^*. The main properties of the evolution triple, which will be frequently used in this book, are gathered in the following.

Proposition 1.53. *Let (V, H, V^*) be an evolution triple. Then*

(i) *For every $h \in H$, there exists a linear continuous functional $\overline{h} \in V^*$ defined by*

$$\langle \overline{h}, v \rangle_{V^* \times V} = \langle h, v \rangle_H \quad \text{for all } v \in V.$$

In addition, the mapping $H \ni h \mapsto \overline{h} \in V^*$ is linear, injective, and continuous. Moreover, identifying $\overline{h}$ with h, we have the continuous embedding $H \subset V^*$, and

$$\langle h, v \rangle_{V^* \times V} = \langle h, v \rangle_H \quad \text{for all } h \in H,\ v \in V,$$

$$\|h\|_{V^*} \le c\,\|h\|_H \quad \text{for all } h \in H \text{ with } c > 0.$$

(ii) *H is dense in V^*.*
(iii) *For any $f \in V^*$, there exists a sequence $\{f_n\} \subset H$ such that*

$$\langle f_n, v \rangle_H \to \langle f, v \rangle_{V^* \times V} \quad \text{for all } v \in V.$$

(iv) *If in addition the embedding $i: V \to H$ is compact, then so is the (adjoint) embedding $i^*: H \to V^*$.*

It follows from Proposition 1.53(i) that the duality pairing on $V^* \times V$ can be viewed as the extension by continuity of the inner product $\langle \cdot, \cdot \rangle_H$ acting on $H \times V$.

1.3 Elements of Measure Theory

The measure theory is a basis of integration theory and deals with set functions, called measures, defined on certain collection of sets. In the following we recall the bare minimum of measure and integration theory required for our purpose.

Definition 1.54. Given a set $\mathcal{O}$, a collection Σ of subsets of $\mathcal{O}$ is called *σ-algebra* (or *σ-field*) if

(i) $\emptyset \in \Sigma$.
(ii) If $A \in \Sigma$ then $\mathcal{O} \setminus A \in \Sigma$.
(iii) If $A_n \in \Sigma$, $n \in \mathbb{N}$ then $\cup_{n \ge 1} A_n \in \Sigma$.

The elements of Σ are called *measurable sets* or *Σ-measurable*. If $\mathcal{O}$ is a topological space, then the smallest σ-algebra containing all open sets is called the *Borel σ-algebra* and it is denoted by $\mathcal{B}(\mathcal{O})$.

Definition 1.55. A *measurable space* is a pair $(\mathcal{O}, \Sigma)$ where $\mathcal{O}$ is a set and Σ is a σ-algebra of subsets of $\mathcal{O}$.

Definition 1.56. (i) If $(\mathcal{O}_1, \Sigma_1)$ and $(\mathcal{O}_2, \Sigma_2)$ are measurable spaces, then a function $f: \mathcal{O}_1 \to \mathcal{O}_2$ is called *measurable* (or (Σ_1, Σ_2)-measurable) if $f^{-1}(\Sigma_2) \subseteq \Sigma_1$.
(ii) If $(\mathcal{O}, \Sigma)$ is a measurable space and Y is a Hausdorff topological space, then $f: \mathcal{O} \to Y$ is called *measurable* if $f^{-1}(\mathcal{B}(Y)) \subseteq \Sigma$, i.e., if it is $(\Sigma, \mathcal{B}(Y))$-measurable in the sense of part (i).

Remark 1.57. In the case (ii) of Definition 1.56 if $\mathcal{O} = X$ where X is a topological space and $f: X \to Y$, we say that f is *Borel measurable*, if it is $(\mathcal{B}(X), \mathcal{B}(Y))$-measurable. If $X = \mathbb{R}^d$, we say that f is *Lebesgue measurable*, if it is $(\mathcal{L}(\mathbb{R}^d), \mathcal{B}(Y))$-measurable, $\mathcal{L}(\mathbb{R}^d)$ being the Lebesgue σ-algebra of $\mathbb{R}^d$, see Definition 2.1.27 of [66]. Note that in all cases, when the range space is topological, we use the Borel σ-field. The reason is that the Lebesgue σ-field on the range space may be too large, see for instance Remark 2.1.49 of [66].

Definition 1.58. Let $\mathcal{O}$ be a set and Σ be a σ-field. A set function $\mu: \Sigma \to [0, \infty]$ is a *measure* (or *countably additive set function* or *σ-additive set function*) on Σ if $\mu(\emptyset) = 0$ and

$$\mu\Big(\bigcup_{n\geq 1} A_n\Big) = \sum_{n\geq 1} \mu(A_n)$$

for every infinite sequence $\{A_n\}_{n\geq 1}$ of pairwise disjoint sets from Σ. A measure on Σ is said to be *finite* if $\mu(\mathcal{O}) < \infty$. A measure on Σ is called *σ-finite* if $\mathcal{O} = \cup_{n\geq 1}\mathcal{O}_n$, $\mathcal{O}_n \in \Sigma$, and $\mu(\mathcal{O}_n) < \infty$ for all $n \geq 1$. If $(\mathcal{O}, \Sigma)$ is a measurable space and μ is a measure on Σ, then the triple $(\mathcal{O}, \Sigma, \mu)$ is called a *measure space*.

In the study of measurability properties of set-valued mappings we need also the following notion of complete measure space.

Definition 1.59. Let $(\mathcal{O}, \Sigma, \mu)$ be a measure space. The measure μ is said to be *complete*, if for every $A \in \Sigma$ with $\mu(A) = 0$ it follows that every $B \subset A$ belongs to Σ. Then $(\mathcal{O}, \Sigma, \mu)$ is said to be a *complete measure space*.

Roughly speaking, completeness is a property of the σ-algebra Σ, but it is common practice to use the term complete for the measure. It is well known that every measure space can be "completed." For this and other results in this direction, we refer to Chap. 2 of [66].

The strategy for defining the integral of a function defined on a measure space consists of two steps. In the first step the integral of simple functions is defined and, in the second step, the integral is extended to limits of simple functions.

Definition 1.60. Let $(\mathcal{O}, \Sigma)$ be a measurable space. A function $s: \mathcal{O} \to \mathbb{R}$ which assumes only a finite number of values $\{\alpha_i\}_{i=1}^n$ is said to be a *simple* (or a *step*, or a *finitely-valued*) function, if $A_i = s^{-1}(\{\alpha_i\}) \in \Sigma$ for every $i \in \{1, \ldots, n\}$. In other words, a simple function is a finite linear combination of characteristic functions of measurable sets, i.e., $s(\omega) = \sum_{i=1}^n \alpha_i \chi_{A_i}(\omega)$ for $\omega \in \mathcal{O}$ where, recall, the *characteristic function* of a set A is defined by

$$\chi_A(\omega) = \begin{cases} 1 & \text{if } \omega \in A, \\ 0 & \text{if } \omega \notin A. \end{cases}$$

The integral of a nonnegative, simple function is defined in an intuitive way.

Definition 1.61. Let $(\mathcal{O}, \Sigma, \mu)$ be a measure space and $s : \mathcal{O} \to [0, +\infty)$ be a simple function having the representation $s = \sum_{i=1}^{n} \alpha_i \chi_{A_i}$ with $\alpha_i \geq 0$, $A_i \in \Sigma$ for $i \in \{1, \ldots, n\}$. Then the *integral* of s is defined by

$$\int_{\mathcal{O}} s(\omega)\, d\mu(\omega) = \sum_{i=1}^{n} \alpha_i \mu(A_i).$$

If for some $i \in \{1, \ldots, n\}$, $\alpha_i = 0$, and $\mu(A_i) = +\infty$, we set $\alpha_i \mu(A_i) = 0$ (according to the usual arithmetic on extened real line).

It can be observed that the integral of a simple function is independent of its particular representation. Next, the integral can be defined for a nonnegative measurable function.

Definition 1.62. Let $(\mathcal{O}, \Sigma, \mu)$ be a measure space and $f : \mathcal{O} \to [0, +\infty]$ be a Σ-measurable function. The *integral* of f with respect to μ is defined by

$$\int_{\mathcal{O}} f(\omega)\, d\mu(\omega) = \sup \left\{ \int_{\mathcal{O}} s(\omega)\, d\mu(\omega) \mid s \text{ is a simple function, } 0 \leq s \leq f \right\}.$$

We say that f is *integrable* if $\int_{\mathcal{O}} f d\mu < +\infty$. Finally, if $A \in \Sigma$, we can define the integral of f over A by

$$\int_A f\, d\mu = \int_{\mathcal{O}} f(\omega)\, \chi_A(\omega)\, d\mu(\omega).$$

The definition of the integral is completed by defining the integral of a measurable $\overline{\mathbb{R}}$-valued function. Given a Σ-measurable function $f : \mathcal{O} \to \overline{\mathbb{R}}$, we define its *positive* and *negative* parts by $f^+ = \max\{f, 0\}$ and $f^- = \max\{-f, 0\}$ and note that they are both Σ-measurable nonnegative functions. Moreover, $f = f^+ - f^-$ and $|f| = f^+ + f^-$.

Definition 1.63. Let $(\mathcal{O}, \Sigma, \mu)$ be a measure space and $f : \mathcal{O} \to \overline{\mathbb{R}}$ be a Σ-measurable function. We say that f is *μ-integrable*, if both f^+ and f^- are integrable in the sense of Definition 1.62. In that case we define

$$\int_{\mathcal{O}} f\, d\mu = \int_{\mathcal{O}} f^+\, d\mu - \int_{\mathcal{O}} f^-\, d\mu.$$

We denote the class of μ-integrable functions on $\mathcal{O}$ by $\mathcal{L}^1(\mathcal{O})$.

Furthermore, a statement about $\omega \in \Omega$ is said to hold "μ-almost everywhere" (μ-a.e. for short) if and only if it holds for all $\omega \notin A$ for some $A \in \Sigma$ with $\mu(A) = 0$. And, when the measure μ is known from the context, we simply say that a statement holds almost everywhere (a.e. for short) if it holds μ-a.e.

The following two results are useful is the sequel.

Theorem 1.64 (Fatou lemma). *Let $(\mathcal{O}, \Sigma, \mu)$ be a measure space and $f_n: \mathcal{O} \to \overline{\mathbb{R}}$ be a sequence of Σ-measurable functions such that there is $h \in \mathcal{L}^1(\mathcal{O})$ with $f_n \leq h$ μ-a.e. on $\mathcal{O}$ for all $n \in \mathbb{N}$. Then*

$$\limsup \int_{\mathcal{O}} f_n \, d\mu \leq \int_{\mathcal{O}} \limsup f_n \, d\mu.$$

If there is a function $h_1 \in \mathcal{L}^1(\mathcal{O})$ such that $f_n \geq h_1$ μ-a.e. on $\mathcal{O}$ for all $n \in \mathbb{N}$, then

$$\int_{\mathcal{O}} \liminf f_n \, d\mu \leq \liminf \int_{\mathcal{O}} f_n \, d\mu.$$

Theorem 1.65 (Lebesgue-dominated convergence theorem). *Let $(\mathcal{O}, \Sigma, \mu)$ be a measure space and $f_n: \mathcal{O} \to \overline{\mathbb{R}}$ be a sequence of Σ-measurable functions such that $f_n(\omega) \to f(\omega)$ μ-a.e. on $\mathcal{O}$ and $|f_n(\omega)| \leq h(\omega)$ μ-a.e. on $\mathcal{O}$ for all $n \in \mathbb{N}$ with $h \in \mathcal{L}^1(\mathcal{O})$. Then $f \in \mathcal{L}^1(\mathcal{O})$ and*

$$\int_{\mathcal{O}} f \, d\mu = \lim \int_{\mathcal{O}} f_n \, d\mu.$$

We can generalize the linear space $\mathcal{L}^1(\mathcal{O})$ as follows.

Let $(\mathcal{O}, \Sigma, \mu)$ be a measure space and $0 < p < \infty$. By $\mathcal{L}^p(\mathcal{O})$ we denote the set of all Σ-measurable functions $f: \mathcal{O} \to \overline{\mathbb{R}}$ such that $\int_{\mathcal{O}} |f|^p \, d\mu < +\infty$, i.e., $|f|^p \in \mathcal{L}^1(\mathcal{O})$. If $1 \leq p < \infty$, the quantity

$$\|f\|_p = \left(\int_{\mathcal{O}} |f|^p \, d\mu \right)^{1/p}$$

is called the L^p-seminorm. On $\mathcal{L}^p(\mathcal{O})$ we consider the equivalence relation $\sim$ defined by $f \sim g$ if and only if $f = g$ μ-a.e. on $\mathcal{O}$. Then we set $L^p(\mathcal{O}) = \mathcal{L}^p(\mathcal{O})/\sim$. On the quotient space $L^p(\mathcal{O})$ the quantity $\|\cdot\|_p$ is a norm, and, in fact, $(L^p(\mathcal{O}), \|\cdot\|_p)$ becomes a Banach space, see Sect. 2.2 for details.

We conclude this section by recalling some notion and results used in the next chapters of the book, related to integration with respect to product measures.

Lemma 1.66. *Let $(\mathcal{O}_1, \Sigma_1)$ and $(\mathcal{O}_2, \Sigma_2)$ be measurable spaces and $f: \mathcal{O}_1 \times \mathcal{O}_2 \to \mathbb{R}$ be a $\Sigma_1 \times \Sigma_2$-measurable function. Then $f(\omega_1, \cdot)$ is Σ_2-measurable for each $\omega_1 \in \mathcal{O}_1$ and $f(\cdot, \omega_2)$ is Σ_1-measurable for each $\omega_2 \in \mathcal{O}_2$.*

Definition 1.67. Let $(\mathcal{O}, \Sigma)$ be a measurable space and Y_1, Y_2 be topological spaces. A function $f: \mathcal{O} \times Y_1 \to Y_2$ is said to be a *Carathéodory function* if $f(\cdot, y)$ is $(\Sigma, \mathcal{B}(Y_2))$-measurable for every $y \in Y_1$ and $f(\omega, \cdot)$ is continuous for every $\omega \in \mathcal{O}$.

Lemma 1.68. *If $(\mathcal{O}, \Sigma)$ is a measurable space, Y_1 is a separable metric space, Y_2 is a metric space, $f : \mathcal{O} \times Y_1 \to Y_2$ is a Carathéodory function and $x : \mathcal{O} \to Y_1$ is Σ-measurable, then $\mathcal{O} \ni \omega \mapsto f(\omega, x(\omega)) \in Y_2$ is Σ-measurable.*

Theorem 1.69 (Fubini theorem). *Let $(\mathcal{O}_1, \Sigma_1, \mu_1)$, $(\mathcal{O}_2, \Sigma_2, \mu_2)$ be σ-finite measure spaces, and let $f : \mathcal{O}_1 \times \mathcal{O}_2 \to \overline{\mathbb{R}}$ be $\mu_1 \times \mu_2$ integrable function. Then, we have*

$$\text{the function } f(\omega_1, \cdot) \text{ is } \mu_2\text{-integrable, for } \mu_1\text{-almost all } \omega_1 \in \mathcal{O}_1,$$

$$\text{the function } \int_{\mathcal{O}_2} f(\cdot, \omega_2)\, d\mu_2(\omega_2) \text{ is } \mu_1\text{-integrable.}$$

Similarly, we have

$$\text{the function } f(\cdot, \omega_2) \text{ is } \mu_1\text{-integrable, for } \mu_2\text{-almost all } \omega_2 \in \mathcal{O}_2,$$

$$\text{the function } \int_{\mathcal{O}_1} f(\omega_1, \cdot)\, d\mu_1(\omega_1) \text{ is } \mu_2\text{-integrable.}$$

Moreover,

$$\begin{aligned} \int_{\mathcal{O}_1 \times \mathcal{O}_2} f(\omega_1, \omega_2)\, d(\mu_1 \times \mu_2) &= \int_{\mathcal{O}_1} \left(\int_{\mathcal{O}_2} f(\omega_1, \omega_2)\, d\mu_2(\omega_2) \right) d\mu_1(\omega_1) \\ &= \int_{\mathcal{O}_2} \left(\int_{\mathcal{O}_1} f(\omega_1, \omega_2)\, d\mu_1(\omega_1) \right) d\mu_2(\omega_2). \end{aligned}$$

The above result enables to evaluate integrals with respect to product measures in terms of iterated integrals. For definitions and basic properties related to product measure spaces we send the reader to Sect. 2.4 of [66].

Chapter 2
Function Spaces

In this chapter we introduce function spaces that will be relevant to the subsequent developments in this monograph. The function spaces to be discussed include spaces of continuous and continuously differentiable functions, smooth functions, Lebesgue and Sobolev spaces, associated with an open bounded domain in $\mathbb{R}^d$. In order to treat time-dependent problems, we also introduce spaces of vector-valued functions, i.e., spaces of mappings defined on a time interval $[0, T]$ with values in a Banach or a Hilbert space.

2.1 Spaces of Smooth Functions

Spaces of continuously differentiable functions. Everywhere in this section Ω denotes an open bounded subset of $\mathbb{R}^d$ and $x = (x_1, \ldots, x_d)$ will represent a generic point of Ω. We define $C(\Omega)$ to be the set of all continuous functions from Ω to $\mathbb{R}$. The set $C(\Omega)$ forms a linear space under the usual addition and scalar multiplication. Similarly, the notation $C(\overline{\Omega})$ is used for the space of real-valued functions continuous on $\overline{\Omega}$. Since Ω is a bounded set, the space $C(\overline{\Omega})$ consists of functions which are *uniformly continuous* on Ω and it is a Banach space with the norm

$$\|v\|_{C(\overline{\Omega})} = \sup_{x \in \overline{\Omega}} |v(x)| = \max_{x \in \overline{\Omega}} |v(x)|.$$

It is clear that $C(\overline{\Omega}) \subset C(\Omega)$ with the proper inclusion. Indeed a simple one-dimensional example of function which belongs to $C(\Omega)$ and does not belong to $C(\overline{\Omega})$ is given by taking $v(x) = 1/x$ on $\Omega = (0, 1) \subset \mathbb{R}$.

We introduce some space of continuously differentiable functions which can be endowed with a Banach space structure. To this end we adopt the following notion of multi-indices which is useful as a compact notation for partial derivatives.

S. Migórski et al., *Nonlinear Inclusions and Hemivariational Inequalities*, Advances in Mechanics and Mathematics 26, DOI 10.1007/978-1-4614-4232-5_2,

A *multi-index* m is an ordered d-tuple of integers $m = (m_1, \dots, m_d)$, $m_i \geq 0$ for $i \in \{1, \dots, d\}$. The length $|m|$ of the multi-index is the sum of the components of m, $|m| = \sum_{i=1}^{d} m_i$. Let $v\colon \Omega \to \mathbb{R}$, $D_i = \partial/\partial x_i$ and

$$D^m v = D_1^{m_1} \cdots D_d^{m_d} v = \frac{\partial^{|m|} v}{\partial x^m} = \frac{\partial^{m_1 + \cdots + m_d} v}{\partial x_1^{m_1} \partial x_2^{m_2} \cdots \partial x_d^{m_d}}$$

with $D^0 v = v$.

Let $k \in \mathbb{N}$. We denote by $C^k(\Omega)$ the vector space of all functions v which, together with all their partial derivatives $D^m v$ of orders $|m| \leq k$, are continuous on Ω. Here and everywhere in this book, for $k = 0$, we set $C^0(\Omega) = C(\Omega)$ and $C^0(\overline{\Omega}) = C(\overline{\Omega})$. The set of k-times continuously differentiable functions on $\overline{\Omega}$ is recursively defined by

$$C^k(\overline{\Omega}) = \{v \in C^{k-1}(\overline{\Omega}) \mid D^m v \in C(\overline{\Omega}) \text{ for all } m \text{ such that } |m| = k\}.$$

The set $C^k(\overline{\Omega})$ is a Banach space with the norm

$$\|v\|_{C^k(\overline{\Omega})} = \sum_{|m| \leq k} \|D^m v\|_{C(\overline{\Omega})}.$$

We also have a proper inclusion $C^k(\overline{\Omega}) \subset C^k(\Omega)$. The spaces of infinitely differentiable functions are defined by

$$C^\infty(\Omega) = \bigcap_{k \in \mathbb{N}_0} C^k(\Omega), \quad C^\infty(\overline{\Omega}) = \bigcap_{k \in \mathbb{N}_0} C^k(\overline{\Omega}).$$

Given $\Omega_1, \Omega_2 \subset \mathbb{R}^d$, we recall that $\Omega_1 \subset\subset \Omega_2$ means that $\overline{\Omega}_1 \subset \Omega_2$ and $\overline{\Omega}_1$ is compact in $\mathbb{R}^d$. For a function $v\colon \Omega \to \mathbb{R}$, its *support* is defined by

$$\operatorname{supp} v = \overline{\{x \in \Omega \mid v(x) \neq 0\}}.$$

We say that v has a *compact support* if $\operatorname{supp} v \subset\subset \Omega$. We set

$$C_0^\infty(\Omega) = \{v \in C^\infty(\Omega) \mid \operatorname{supp} v \subset\subset \Omega\}.$$

It is clear that $C_0^\infty(\Omega) \subset C^\infty(\overline{\Omega})$.

Space of Hölder continuous functions. A function $v\colon \overline{\Omega} \to \mathbb{R}$ is said to be *Hölder continuous* in $\overline{\Omega}$ if there exists two constants $c > 0$ and $\lambda \in (0, 1]$ such that

$$|v(x) - v(y)| \leq c\, \|x - y\|_{\mathbb{R}^d}^{\lambda} \text{ for all } x, y \in \overline{\Omega}.$$

The constant λ is called the *Hölder exponent* of v. If $\lambda = 1$ the function is said to be *Lipschitz continuous*. The space of Lipschitz continuous functions in $\overline{\Omega}$, denoted by $C^{0,1}(\overline{\Omega})$, is a Banach space with the norm

$$\|v\|_{C^{0,1}(\overline{\Omega})} = \|v\|_{C(\overline{\Omega})} + \sup_{\substack{x,y \in \overline{\Omega} \\ x \neq y}} \frac{|v(x) - v(y)|}{\|x - y\|_{\mathbb{R}^d}}.$$

Let $k \in \mathbb{N}_0$ and $\lambda \in (0, 1]$. A function $v \in C^k(\overline{\Omega})$ is said to be (k, λ)-Hölder continuous in $\overline{\Omega}$ if there exists a constant $c > 0$ such that

$$|D^m v(x) - D^m v(y)| \leq c\, \|x - y\|_{\mathbb{R}^d}^{\lambda} \quad \text{for all } x, y \in \overline{\Omega}$$

for all multi-indices $|m| \leq k$. The set of (k, λ)-Hölder continuous functions in $\overline{\Omega}$ is denoted by $C^{k,\lambda}(\overline{\Omega})$ and it is a Banach space endowed with the norm

$$\|v\|_{C^{k,\lambda}(\overline{\Omega})} = \|v\|_{C^k(\overline{\Omega})} + \max_{0 \leq |m| \leq k} \sup_{\substack{x,y \in \overline{\Omega} \\ x \neq y}} \frac{|D^m v(x) - D^m v(y)|}{\|x - y\|_{\mathbb{R}^d}^{\lambda}}.$$

For a nonnegative integer k and $0 < \mu < \lambda < 1$, we have the following strict inclusions:

$$C^{k,1}(\overline{\Omega}) \subset C^{k,\lambda}(\overline{\Omega}) \subset C^{k,\mu}(\overline{\Omega}) \subset C^k(\overline{\Omega})$$

which hold for any open subset Ω of $\mathbb{R}^d$. It is also clear that $C^{k,1}(\overline{\Omega}) \not\subseteq C^{k+1}(\overline{\Omega})$.

The following result is a direct consequence of the definitions of the spaces introduced above, see for instance Theorem 1.31 of [1].

Theorem 2.1. *Let Ω be an open subset of $\mathbb{R}^d$, let k be a nonnegative integer and $0 < \mu < \lambda \leq 1$. Then we have the following embeddings:*

$$C^{k+1}(\overline{\Omega}) \subset C^k(\overline{\Omega}), \tag{2.1}$$

$$C^{k,\lambda}(\overline{\Omega}) \subset C^k(\overline{\Omega}), \tag{2.2}$$

$$C^{k,\lambda}(\overline{\Omega}) \subset C^{k,\mu}(\overline{\Omega}). \tag{2.3}$$

Moreover, if Ω is bounded, then the embeddings (2.2) *and* (2.3) *are compact. In addition, if Ω is convex, we have the embeddings*

$$C^{k+1}(\overline{\Omega}) \subset C^{k,1}(\overline{\Omega}),$$

$$C^{k+1}(\overline{\Omega}) \subset C^{k,\lambda}(\overline{\Omega}). \tag{2.4}$$

And, finally, if Ω is bounded and convex, then the embeddings (2.1) *and* (2.4) *are compact.*

2.2 Lebesgue Spaces

In this section we recall the definition of L^p spaces as well as their main properties. Proofs of standard results are omitted. We restrict ourselves to the spaces defined on an open subset of $\mathbb{R}^d$ although the theory is well developed in an abstract measure space setting, as shown in Sects. 2.2 and 3.8 of [66].

Let Ω be an open subset of $\mathbb{R}^d$ and consider the Lebesgue measure on $\mathbb{R}^d$, denoted meas $(\cdot)$. As on page 21, given two measurable functions $u, v: \Omega \to \mathbb{R}$, we say that u is equivalent to v, and we write $u \sim v$ if $u(x) = v(x)$ for a.e. $x \in \Omega$. We note that $\sim$ is an equivalence relation in the class of measurable functions. For convenience, with an abuse of notation, we identify a measurable function with its equivalence class.

For $1 \leq p < \infty$, we define

$$L^p(\Omega) = \{u: \Omega \to \mathbb{R} \mid u \text{ is a measurable function such that } \|u\|_{L^p(\Omega)} < \infty\},$$

where

$$\|u\|_{L^p(\Omega)} = \left(\int_\Omega |u(x)|^p \, dx \right)^{1/p}.$$

If $p = \infty$, then

$$L^\infty(\Omega) = \{u: \Omega \to \mathbb{R} \mid u \text{ is measurable function such that } \|u\|_{L^\infty(\Omega)} < \infty\},$$

where the *essential supremum* norm is given by

$$\|u\|_{L^\infty(\Omega)} = \text{ess sup}\,|u| = \inf\{\alpha \in \mathbb{R} \mid |u(x)| \leq \alpha \text{ for a.e. } x \in \Omega\}.$$

The elements of $L^p(\Omega)$ are thus equivalence classes of measurable functions. For simplicity, and when there is no possibility of confusion, we denote the $L^p(\Omega)$ spaces simply by L^p and the norms $\|\cdot\|_{L^p(\Omega)}$ by $\|\cdot\|_{L^p}$ or $\|\cdot\|_p$.

Given a natural number $s \geq 1$ and a real number $1 \leq p < \infty$ we denote by $L^p(\Omega; \mathbb{R}^s)$ the space of functions $u: \Omega \to \mathbb{R}^s$ whose components are in $L^p(\Omega)$, i.e., $L^p(\Omega; \mathbb{R}^s) = (L^p(\Omega))^s$. We endow $L^p(\Omega; \mathbb{R}^s)$ with the norm

$$\|u\|_{L^p(\Omega;\mathbb{R}^s)} = \left(\int_\Omega \|u(x)\|^p_{\mathbb{R}^s} \, dx \right)^{1/p}.$$

For a measurable function $u: \Omega \to \mathbb{R}$, if $u \in L^p(\Omega')$ for any $\Omega' \subset\subset \Omega$, then we say that u is *locally p-integrable* (or locally in $L^p(\Omega)$) and write $u \in L^p_{loc}(\Omega)$.

The basic properties of L^p spaces are provided in the following theorem.

Theorem 2.2. *Let Ω be an open bounded subset of $\mathbb{R}^d$. Then*

(a) *For $1 \leq p \leq \infty$, $L^p(\Omega)$ is a Banach space.*
(b) *For $1 < p < \infty$, $L^p(\Omega)$ is reflexive and uniformly convex.*
(c) *For $1 \leq p < \infty$, $L^p(\Omega)$ is separable.*

(d) *$L^2(\Omega)$ is a Hilbert space with respect to the inner product*

$$\langle u, v\rangle_{L^2(\Omega)} = \int_\Omega u(x)\, v(x)\, dx.$$

(e) *If $1 \le p \le r \le \infty$, then $L^r(\Omega) \subset L^p(\Omega)$ and*

$$\|v\|_p \le \mathrm{meas}(\Omega)^{1/p-1/r} \|v\|_r \quad \text{for all } v \in L^r(\Omega).$$

(f) *For $1 \le p \le \infty$, if $\{u_n\}$ converges to u in $L^p(\Omega)$, then there exist a subsequence $\{u_{n_k}\}$ of $\{u_n\}$ and a function $h \in L^p(\Omega)$ such that $\{u_{n_k}(x)\}$ converges to $u(x)$ a.e. $x \in \Omega$ and $|u_{n_k}(x)| \le h(x)$ a.e. $x \in \Omega$ and for all $k \in \mathbb{N}$.*

(g) *For $1 \le p < \infty$, both the space $C_0^\infty(\Omega)$ and the family of all simple functions in $L^p(\Omega)$ are dense in $L^p(\Omega)$.*

The triangle inequality for $L^p(\Omega)$ norm is called the Minkowski inequality.

Theorem 2.3 (Minkowski inequality). *Let $1 \le p \le \infty$ and $u, v \in L^p(\Omega)$. Then $u + v \in L^p(\Omega)$ and*

$$\|u + v\|_p \le \|u\|_p + \|v\|_p.$$

For $1 \le p \le \infty$, we define its (Hölder) *conjugate* q by the relation $1/p + 1/q = 1$ and adopt the convention $1/\infty = 0$. It is easy to see that $1 \le q \le \infty$. Moreover, $1 < q < \infty$ if $1 < p < \infty$, $q = 1$ if $p = \infty$, and $q = \infty$ if $p = 1$. With this notation we have the following result.

Theorem 2.4 (Hölder inequality). *Let $1 \le p \le \infty$ and let q be its conjugate exponent. If $u, v\colon \Omega \to \mathbb{R}$ are measurable functions, then*

$$\|uv\|_1 \le \|u\|_p\, \|v\|_q.$$

In particular, if $u \in L^p(\Omega)$ and $v \in L^q(\Omega)$, then $uv \in L^1(\Omega)$.

The Hölder inequality for $p = 2$ is actually the well-known *Cauchy–Bunyakovsky–Schwarz inequality*. A third basic inequality associated with integrable functions is the so-called *Jensen inequality*.

Theorem 2.5 (Jensen inequality). *Let I be an open interval in $\mathbb{R}$, $f\colon I \to \mathbb{R}$ be a convex function, $u \in L^1(\Omega)$ with $u(\Omega) \subset I$ and $f \circ u \in L^1(\Omega)$. Then*

$$f\left(\frac{1}{\mathrm{meas}(\Omega)} \int_\Omega u\, dx\right) \le \frac{1}{\mathrm{meas}(\Omega)} \int_\Omega f \circ u\, dx.$$

Next, we recall two elementary inequalities of Young and Gronwall which are frequently used in the book.

Lemma 2.6 (Young inequality). *Let $1 < p < \infty$, $1/p + 1/q = 1$, and $\varepsilon > 0$. Then*

$$a\, b \le \frac{\varepsilon^p}{p} |a|^p + \frac{1}{\varepsilon^q q} |b|^q \quad \text{for all } a, b \in \mathbb{R}.$$

Lemma 2.7 (Gronwall inequality). *Assume that $f, g: [0,T] \to \mathbb{R}$ are continuous functions, $h \in L^1(0,T)$, $h \geq 0$ and*

$$f(t) \leq g(t) + \int_0^t h(s)\, f(s)\, ds \quad \text{for all } \ t \in [0,T].$$

Then

$$f(t) \leq g(t) + \int_0^t \exp\Big(\int_s^t h(r) dr\Big) h(s)\, g(s)\, ds \quad \text{for all } \ t \in [0,T].$$

A characterization of the dual of the L^p spaces follows from the following well-known result.

Theorem 2.8 (Riesz representation theorem for L^p). *Let $1 \leq p < \infty$, $1/p + 1/q = 1$ and let $l \in (L^p(\Omega))^*$. Then there exists $v \in L^q(\Omega)$ such that for all $u \in L^p(\Omega)$, we have*

$$l(u) = \int_\Omega u\, v\, dx$$

and, moreover, $\|v\|_{L^q(\Omega)} = \|l\|_{(L^p(\Omega))^*}$. Thus $(L^p(\Omega))^* = L^q(\Omega)$.

To complete the statement in Theorem 2.8 we recall that the space $L^\infty(\Omega)$ is not separable and $(L^\infty(\Omega))^*$ is much larger than $L^1(\Omega)$. Details can be found in Theorem 3.8.6 of [66]. Moreover, $L^1(\Omega)$ and $L^\infty(\Omega)$ are not reflexive spaces.

The following result is useful in the study of weak solutions to partial differential equations.

Lemma 2.9 (Variational lemma or **Lagrange lemma).** *Let Ω be an open subset of $\mathbb{R}^d$, $u \in L^1_{loc}(\Omega)$ and assume that*

$$\int_\Omega u \varphi\, dx = 0 \ \text{ for all } \ \varphi \in C_0^\infty(\Omega).$$

Then $u = 0$ a.e. on Ω.

We conclude this section with a collection of results on weak convergence in L^p. Let $1 \leq p \leq \infty$ and q be the conjugate exponent of p. If $p = 1$ or $p = \infty$, then assume in addition that Ω is bounded. By Theorem 2.8, it follows that a sequence $\{u_n\} \subset L^p(\Omega)$ converges weakly (weakly* if $p = \infty$) to a function $u \in L^p(\Omega)$ if and only if

$$\int_\Omega u_n\, v\, dx \to \int_\Omega u\, v\, dx \quad \text{for all } \ v \in L^q(\Omega).$$

Proposition 2.10. *Let Ω be an open subset of $\mathbb{R}^d$, $1 \leq p \leq \infty$, and assume that Ω is bounded when $p = \infty$. Let $\{u_n\} \subset L^p(\Omega)$.*

(a) *If $u_n \to u$ weakly in $L^p(\Omega)$ (weakly* if $p = \infty$), then*

$$\|u\|_p \leq \liminf \|u_n\|_p \leq \sup_{n\in\mathbb{N}} \|u_n\|_p < \infty.$$

(b) *If* $1 < p < \infty$ *and* $\sup_{n\in\mathbb{N}} \|u_n\|_p < \infty$, *then there exists a subsequence* $\{u_{n_k}\}$ *such that* $u_{n_k} \to u$ *weakly in* $L^p(\Omega)$ *for some* $u \in L^p(\Omega)$. *This property also holds in* $L^\infty(\Omega)$ *with respect to the weak* convergence.*
(c) *If* $1 < p < \infty$, $u_n \to u$ *weakly in* $L^p(\Omega)$ *and* $\|u\|_p = \lim \|u_n\|_p$, *then* $u_n \to u$ *in* $L^p(\Omega)$.

2.3 Sobolev Spaces

The theory of Sobolev spaces has been developed by generalizing the notion of classical derivatives and introducing the idea of weak or generalized derivatives. These spaces are among the most common function spaces used both in the study of partial differential equations and in diverse fields of mechanics. In this section we summarize the main properties of Sobolev spaces.

Definition and basic properties. Let Ω be an open subset of $\mathbb{R}^d$ and $d \in \mathbb{N}$. Below, we adopt the multi-index notation introduced in Sect. 2.1. Let $m = (m_1, \ldots, m_d)$ with $m_i \geq 0$ for any $i \in \{1, \ldots, d\}$, $|m| = \sum_{i=1}^{d} m_i$ and $D^m = D_1^{m_1} \cdots D_d^{m_d}$ with $D_i = \partial/\partial x_i$.

Let $u \in L^1_{loc}(\Omega)$. A function $w \in L^1_{loc}(\Omega)$ is called the mth *weak derivative* or *generalized derivative* of u, if

$$\int_\Omega u\, D^m \varphi \, dx = (-1)^{|m|} \int_\Omega w \varphi \, dx \quad \text{for all } \varphi \in C_0^\infty(\Omega).$$

The weak derivative, if it exists, is uniquely defined up to a set of measure zero. It will be denoted by $w = D^m u$.

We recall that $C_0^\infty(\Omega)$ stands for the space of infinitely differentiable functions with compact support in Ω and it is called the space of *test functions*. It can be furnished with a convergence structure (see, e.g., Definition 3.9.1 of [66]). The definition of weak derivative can be extended to distributions (linear and continuous functionals on $C_0^\infty(\Omega)$). Note that the weak derivative of function, as well as the distributional derivative introduced in Definition 2.45, has a global feature, i.e., it can not be defined pointwise. This represents one of the differences with respect the classical derivative which, in contrast, can be defined in each point of the domain Ω.

Let $1 \leq p \leq \infty$ and $k \in \mathbb{N}$. The Sobolev space $W^{k,p}(\Omega)$ is the space of functions $u \in L^p(\Omega)$ which have generalized derivatives up to order k such that $D^m u \in L^p(\Omega)$ for all $|m| \leq k$. For $k = 0$, we set $W^{0,p}(\Omega) = L^p(\Omega)$.

The space $W^{k,p}(\Omega)$ becomes a Banach space with the norm

$$\|u\|_{W^{k,p}(\Omega)} = \begin{cases} \left(\sum_{|m| \le k} \|D^m u\|^p_{L^p(\Omega)} \right)^{1/p} & \text{if } 1 \le p < \infty, \\ \max_{|m| \le k} \|D^m u\|_{L^\infty(\Omega)} & \text{if } p = \infty. \end{cases}$$

The Sobolev space $W^{k,2}(\Omega)$ is denoted by $H^k(\Omega)$. By using Rademacher's theorem (Theorem 5.6.16 in [66] or Corollary 4.19 in [49]), it can be shown that a real-valued Lipschitz continuous function defined on Ω is almost everywhere differentiable on Ω and, moreover, $W^{1,\infty}(\Omega) = C^{0,1}(\Omega)$. Details can be found in Sect. 5.8 of [80].

We now proceed with the following definition.

Definition 2.11. Let $1 \le p \le \infty$ and $k \in \mathbb{N}$. Then, the Sobolev space $W_0^{k,p}(\Omega)$ is the closure of the space $C_0^\infty(\Omega)$ in the norm of the space $W^{k,p}(\Omega)$.

It follows from the definition above that the space $W_0^{k,p}(\Omega)$ is a Banach space with the norm $\|\cdot\|_{W^{k,p}(\Omega)}$. We write $W_0^{k,2}(\Omega) = H_0^k(\Omega)$. For $1 \le p < \infty$, the dual space of $W_0^{k,p}(\Omega)$ is denoted by $W^{-k,q}(\Omega)$ where q is the conjugate exponent of p. We usually use the notation $W^{-1,2}(\Omega) = H^{-1}(\Omega)$. Moreover, for $k, l \in \mathbb{N}$, $k \le l$ we have the inclusions

$$C_0^\infty(\Omega) \subset H_0^l(\Omega) \subset H_0^k(\Omega) \subset L^2(\Omega) \subset H^{-k}(\Omega) \subset H^{-l}(\Omega) \subset (C_0^\infty(\Omega))^*,$$

each of these spaces being dense in the following one.

Additional properties of the Sobolev spaces are provided by the following result.

Theorem 2.12. *Let Ω be an open bounded subset of $\mathbb{R}^d$, $d \ge 1$ and $k \in \mathbb{N}$. Then $W^{k,p}(\Omega)$ is a uniformly convex Banach space (and hence reflexive) for $1 < p < \infty$, and separable for $1 \le p < \infty$. The Sobolev space $H^k(\Omega)$ is a Hilbert space with the inner product*

$$\langle u, v \rangle_{H^k(\Omega)} = \sum_{|m| \le k} \langle D^m u, D^m v \rangle_{L^2(\Omega)}.$$

In the case of vector-valued functions $v \colon \Omega \to \mathbb{R}^s$, $s \ge 1$, $v = (v_1, \ldots, v_s)$, we use the notation $W^{k,p}(\Omega; \mathbb{R}^s) = (W^{k,p}(\Omega))^s$ and

$$\|v\|_{W^{k,p}(\Omega;\mathbb{R}^s)} = \left(\sum_{i=1}^s \|v_i\|^p_{W^{k,p}(\Omega)} \right)^{1/p}.$$

Regularity of the boundary. Open sets in $\mathbb{R}^d$ could have very bad boundaries. Some properties of Sobolev spaces require a certain degree of regularity of the boundary and, for this reason, many theorems concerning Sobolev spaces hold under additional conditions on the boundary of the open set Ω. To present them, below, we restrict ourselves to the case of bounded sets.

Definition 2.13. Let Ω be an open bounded subset of $\mathbb{R}^d$ with a boundary Γ. We say that the *boundary* Γ *is of class* $C^{k,\lambda}$ or is (k, λ)-*Hölderian*, $k \in \mathbb{N}_0$, $\lambda \in (0, 1]$, if there exists l Cartesian coordinate systems S_j, $j = 1, \dots, l$, $S_j = (x_{j,1}, \dots, x_{j,d-1}, x_{j,d}) = (x'_j, x_{j,d})$, two real numbers α, $\beta > 0$, and l functions a_j with

$$a_j \in C^{k,\lambda}([-\alpha, \alpha]^{d-1}), \; j = 1, \dots, l$$

such that the sets defined by

$$\Lambda^j = \left\{(x'_j, x_{j,d}) \in \mathbb{R}^d \mid \|x'_j\|_{\mathbb{R}^{d-1}} \le \alpha, \; x_{j,d} = a_j(x'_j)\right\},$$

$$V^j_+ = \left\{(x'_j, x_{j,d}) \in \mathbb{R}^d \mid \|x'_j\|_{\mathbb{R}^{d-1}} \le \alpha, \; a_j(x'_j) < x_{j,d} < a_j(x'_j) + \beta\right\},$$

$$V^j_- = \left\{(x'_j, x_{j,d}) \in \mathbb{R}^d \mid \|x'_j\|_{\mathbb{R}^{d-1}} \le \alpha, \; a_j(x'_j) - \beta < x_{j,d} < a_j(x'_j)\right\},$$

possess the following properties:

$$\Lambda^j \subset \Gamma, \qquad V^j_+ \subset \Omega, \qquad V^j_- \subset \mathbb{R}^d \setminus \Omega$$

for all $j = 1, \dots, l$ and

$$\bigcup_{j=1}^{l} \Lambda^j = \Gamma.$$

If the mappings a_j belong to $C^k([-\alpha, \alpha]^{d-1})$, $j = 1, \dots, l$, we say that the boundary Γ is of class C^k. A boundary of $C^{0,1}$-class is then naturally called a *Lipschitz boundary*, which means that Γ is locally the graph of a Lipschitz continuous function. An open bounded subset of $\mathbb{R}^d$ with Lipschitz boundary is called a *Lipschitz domain*.

We remark that the smooth domains (the sphere, parallelpiped, pyramid, etc.) have a Lipschitz boundary and, in engineering applications, most domains are Lipschitz domains. Well-known domains which are not Lipschitz domains are the circles with a radius removed and the smooth domains with cracks. In the rest of the book we always assume that Ω has a Lipschitz boundary, which is a quite natural requirement. Then the functions a_j are Lipschitz continuous and, therefore, as mentioned on page 30, they possess derivatives almost everywhere on Λ^j.

Definition 2.14. Consider now the function a_j from Definition 2.13. Then the vector $\nu \in \mathbb{R}^d$ given by

$$\nu = \frac{1}{s}\left(\frac{\partial a_j}{\partial x_{j,1}}, \dots, \frac{\partial a_j}{\partial x_{j,d-1}}, -1\right), \quad \text{where } s = \left(1 + \sum_{i=1}^{d-1}\left(\frac{\partial a_j}{\partial x_{j,i}}\right)^2\right)^{1/2}$$

is called the *unit outward normal* to the boundary Γ.

The unit outward normal is defined uniquely almost everywhere on Γ and its components ν_i are bounded measurable functions, i.e., $\nu \in L^\infty(\Gamma;\mathbb{R}^d)$. It can be proved that the vector ν does not depend on the concrete coordinate system. In this case the $(d-1)$-dimensional surface measure is well defined and it allows to define $L^p(\Gamma)$ spaces and to introduce the trace operator. Also, it allows to extend the integration by parts formula to Sobolev functions. More details in this matter can be find in Chap. 3.9 of [66], for instance. Finally, it is known that if Ω is open bounded and has a Lipschitz continuous boundary, then the Sobolev space $W^{k,p}(\Omega)$ for $1 \le p < \infty$, introduced on page 29, can be equivalently defined to be the closure of $C^k(\overline{\Omega})$ in $W^{k,p}(\Omega)$ norm.

Embedding results. We turn now on embedding and compact embedding results concerning the Sobolev spaces.

Theorem 2.15. *Let Ω be an open bounded set of $\mathbb{R}^d$ with a Lipschitz boundary. For nonnegative integers k, l such that $0 \le l \le k$, we have the continuous embeddings*

$$W^{k,p}(\Omega) \subset W^{l,p}(\Omega) \quad \text{for all} \quad 1 \le p \le \infty.$$

Moreover, for $k \ge 0$, we have

$$W^{k,r}(\Omega) \subset W^{k,p}(\Omega) \quad \text{whenever} \quad 1 \le p \le r \le \infty.$$

Since Sobolev spaces are defined through Lebesgue spaces, it follows that, roughly speaking, an element in a Sobolev space is an equivalence class of measurable functions that are equal almost everywhere. When we say that a function from a Sobolev space is continuous, we mean that it is equal almost everywhere to a continuous function, i.e., we can find a continuous function in its equivalence class.

Theorem 2.16 (Rellich–Kondrachov embedding theorem). *Let Ω be an open bounded set of $\mathbb{R}^d$ with a Lipschitz boundary, $k \in \mathbb{N}$ and $1 \le p < \infty$.*

(a) *If $k < \frac{d}{p}$, then $W^{k,p}(\Omega) \subset L^q(\Omega)$ continuously for every $q \le p^*$ and compactly for every $q < p^*$, where $p^* = \frac{dp}{d-kp}$.*
(b) *If $k = \frac{d}{p}$, then $W^{k,p}(\Omega) \subset L^r(\Omega)$ compactly for every $r < \infty$.*
(c) *If $k > \frac{d}{p}$, then $W^{k,p}(\Omega) \subset C^l(\overline{\Omega})$ compactly for every integer l such that $0 \le l < k - \frac{d}{p}$.*

The above embeddings are also valid for $W_0^{k,p}(\Omega)$ for any open bounded subset of $\mathbb{R}^d$ with no restriction on the regularity of the boundary.

For the proof of Theorem 2.16, we refer to [1, p.144]. Moreover, we have the following result which will be useful in what follows.

Corollary 2.17. *Let Ω be an open bounded set of $\mathbb{R}^d$ with a Lipschitz boundary, $k \in \mathbb{N}$ and $1 \le p < \infty$. Then the embedding $W^{k,p}(\Omega) \subset L^p(\Omega)$ is compact.*

Sobolev–Slobodeckij spaces. The Sobolev spaces of integer order have an intuitive interpretation. However, many applications require extension of the definition of Sobolev spaces to include fractional order spaces.

Definition 2.18. Let Ω be an open bounded subset of $\mathbb{R}^d$. Let k be a nonnegative real number of the form $k = [k] + r > 0$, where $[k]$ is the integer part of k, $[k] \geq 0$ and $r \in (0, 1)$. We define the Sobolev–Slobodeckij space $W^{k,p}(\Omega)$ to be the subspace of $W^{[k],p}(\Omega)$ with finite norm

$$\|u\|_{W^{k,p}(\Omega)} = \left(\|u\|^p_{W^{[k],p}(\Omega)} + \sum_{|m|=[k]} \int_\Omega \int_\Omega \frac{|D^m u(x) - D^m u(y)|^p}{\|x - y\|^{d+pr}_{\mathbb{R}^d}} \, dxdy \right)^{1/p}$$

where $1 < p < \infty$. If $p = \infty$, $W^{k,\infty}(\Omega)$ is defined to be the subspace of $W^{[k],\infty}(\Omega)$ for which the norm

$$\|u\|_{W^{k,\infty}(\Omega)} = \|u\|_{W^{[k],\infty}(\Omega)} + \max_{|m|=[k]} \operatorname*{ess\,sup}_{\substack{x,y \in \Omega \\ x \neq y}} \frac{|D^m u(x) - D^m u(y)|}{\|x - y\|^r_{\mathbb{R}^d}}$$

is finite.

For a nonnegative real number k, the Sobolev space $W^{k,2}(\Omega)$ is denoted by $H^k(\Omega)$. Also, in the case of vector-valued functions, we use the notation $H^k(\Omega; \mathbb{R}^s) = W^{k,2}(\Omega; \mathbb{R}^s)$. We recall Theorems 7.9 and 7.10 of [252], which provide compact embeddings results for fractional order Sobolev spaces.

Theorem 2.19. *Let Ω be an open bounded set of $\mathbb{R}^d$ with a Lipschitz boundary and let r_1, r_2 be real numbers such that $0 \leq r_2 < r_1 \leq 1$. Then the embedding $H^{r_1}(\Omega) \subset H^{r_2}(\Omega)$ is compact.*

We use now Corollary 2.17 and Theorem 2.19 to provide the following examples of evolution triples. The examples below can be easily generalized to vector-valued functions.

Example 2.20. (i) Let Ω be an open bounded subset of $\mathbb{R}^d$ with no restriction on the regularity of its boundary and set $V = W_0^{k,p}(\Omega)$ and $H = L^2(\Omega)$ with $2 \leq p < \infty$ and $k \in \mathbb{N}$. Then (V, H, V^*) is an evolution triple with $V^* = W^{-k,q}(\Omega)$, $1/p + 1/q = 1$ and the corresponding embeddings are compact. This example serves as a prototype of an evolution triple of spaces.

(ii) Let Ω be an open bounded connected set of $\mathbb{R}^d$ with a Lipschitz boundary, let V be any closed subspace of $H^1(\Omega)$ such that $H_0^1(\Omega) \subset V \subset H^1(\Omega)$ and let $H = L^2(\Omega)$. Since $C_0^\infty(\Omega) \subset H_0^1(\Omega) \subset V$, we know that V is always dense in H. Thus the spaces (V, H, V^*) form an evolution triple of spaces with compact embeddings $V \subset H \subset V^*$.

(iii) Let Ω be an open bounded connected set of $\mathbb{R}^d$ with a Lipschitz boundary and let $\delta \in (\frac{1}{2}, 1)$. Then we have the continuous dense and compact embeddings

$H^1(\Omega) \subset H^\delta(\Omega) \subset H^{1/2}(\Omega) \subset H$ with $H = L^2(\Omega)$. This implies that $(H^\delta(\Omega), H, (H^\delta(\Omega))^*)$ is also an evolution triple of spaces.

Trace operator. Functions from Sobolev spaces are uniquely defined only almost everywhere in Ω and the boundary of Ω has measure zero in $\mathbb{R}^d$. Nevertheless, it is possible to define the trace of a function from Sobolev space on the boundary in such a way that for a Sobolev function that is continuous up to the boundary, its trace coincides with its boundary value. More precisely, we have the following result.

Theorem 2.21. *Let Ω be an open bounded set of $\mathbb{R}^d$ with a Lipschitz boundary $\partial\Omega = \Gamma$ and $1 \le p < \infty$. Then there exists a unique linear continuous operator $\gamma\colon W^{1,p}(\Omega) \to L^p(\Gamma)$ such that*

(a) *$\gamma u = u|_\Gamma$ if $u \in C^1(\overline{\Omega})$.*
(b) *$\|\gamma u\|_{L^p(\Gamma)} \le c\,\|u\|_{W^{1,p}(\Omega)}$ with $c > 0$ depending only on p and Ω.*
(c) *If $u \in W^{1,p}(\Omega)$, then $\gamma u = 0$ in $L^p(\Gamma)$ if and only if $u \in W_0^{1,p}(\Omega)$.*
(d) *If $1 < p < \infty$, then $\gamma(W^{1,p}(\Omega)) = W^{1-\frac{1}{p},p}(\Gamma)$.*
(e) *If $1 < p < d$, then $\gamma\colon W^{1,p}(\Omega) \to L^r(\Gamma)$ is compact for any r such that $1 \le r < \frac{dp-p}{d-p}$.*
(f) *If $p \ge d$, then $\gamma\colon W^{1,p}(\Omega) \to L^r(\Gamma)$ is compact for any $r \ge 1$.*

The function γu is called the trace *of the function u on $\partial\Omega$ and the operator $\gamma\colon W^{1,p}(\Omega) \to L^p(\Gamma)$ is called the* trace operator.

The trace operator introduced in Theorem 2.21 is neither an injection nor a surjection from $W^{1,p}(\Omega)$ to $L^p(\Gamma)$. The range $\gamma(W^{1,p}(\Omega)) = W^{1-\frac{1}{p},p}(\Gamma)$ is a space smaller than $L^p(\Gamma)$. Usually we use the same symbol u for the trace of $u \in W^{1,p}(\Omega)$, i.e., we write u instead of γu.

We denote by $H^{-1/2}(\Gamma)$ the dual space of $H^{1/2}(\Gamma)$. The duality pairing between $H^{-1/2}(\Gamma)$ and $H^{1/2}(\Gamma)$ is an extension of the $L^2(\Gamma)$ inner product. More precisely, if $w \in L^2(\Gamma)$, then $w \in H^{-1/2}(\Gamma)$ and

$$\langle w, v\rangle_{H^{-1/2}(\Gamma)\times H^{1/2}(\Gamma)} = \int_\Gamma wv\, d\Gamma \quad \text{for all } v \in H^{1/2}(\Gamma).$$

The following trace theorem holds for fractional Sobolev spaces and follows from Theorem 1.5.1.2 of [92].

Theorem 2.22. *Let Ω be an open bounded set of $\mathbb{R}^d$ with $C^{k,1}$ boundary Γ. Assume that $1 \le p \le \infty$, σ is a nonnegative real number such that $\sigma - \frac{1}{p}$ is not an integer, $\sigma \le k+1$, $\sigma - \frac{1}{p} = l + r$, $r \in (0,1)$, and $l \ge 0$ is an integer. Then there exists a unique linear continuous and surjective operator*

$$\gamma\colon W^{\sigma,p}(\Omega) \to W^{\sigma-\frac{1}{p},p}(\Gamma)$$

such that $\gamma u = u|_\Gamma$ if $u \in C^{k,1}(\overline{\Omega})$.

Sobolev spaces in contact mechanics. In the study of contact problems, we frequently use Sobolev-type function spaces associated to the deformation and divergence operators. To introduce them, we start with the following notation. First, we denote by $\mathbb{R}^d$ the d-dimensional real linear space, with $d = 1, 2, 3$ in applications. The symbol $\mathbb{S}^d$ stands for the space of second-order symmetric tensors on $\mathbb{R}^d$ or, equivalently, the space of symmetric matrices of order d. The canonical inner products and the corresponding norms on $\mathbb{R}^d$ and $\mathbb{S}^d$ are

$$\boldsymbol{u} \cdot \boldsymbol{v} = u_i v_i, \quad \|\boldsymbol{v}\|_{\mathbb{R}^d} = (\boldsymbol{v} \cdot \boldsymbol{v})^{1/2} \quad \text{for all } \boldsymbol{u} = (u_i),\ \boldsymbol{v} = (v_i) \in \mathbb{R}^d,$$

$$\boldsymbol{\sigma} : \boldsymbol{\tau} = \sigma_{ij} \tau_{ij}, \quad \|\boldsymbol{\tau}\|_{\mathbb{S}^d} = (\boldsymbol{\tau} : \boldsymbol{\tau})^{1/2} \quad \text{for all } \boldsymbol{\sigma} = (\sigma_{ij}),\ \boldsymbol{\tau} = (\tau_{ij}) \in \mathbb{S}^d,$$

respectively. Here and throughout this section, the indices i and j run between 1 and d, and, unless stated otherwise, the summation convention over repeated indices is used. Moreover, we denote vectors and tensors by bold-face letters, as usual in Mechanics, and we keep this convention everywhere in Part III of the book.

Next, let Ω be an open bounded set of $\mathbb{R}^d$. In general, displacements will be sought in the space $H^1(\Omega;\mathbb{R}^d)$ or its subspaces. Given $\Gamma_D \subset \Gamma$ with $\text{meas}(\Gamma_D) > 0$, we introduce the spaces

$$H = L^2(\Omega;\mathbb{R}^d), \quad \mathcal{H} = \big\{\, \boldsymbol{\tau} = (\tau_{ij}) \mid \tau_{ij} = \tau_{ji} \in L^2(\Omega) \,\big\} = L^2(\Omega;\mathbb{S}^d),$$

$$V = \big\{\, \boldsymbol{v} \in H^1(\Omega;\mathbb{R}^d) \mid \boldsymbol{v} = \mathbf{0} \text{ on } \Gamma_D \,\big\}, \quad \mathcal{H}_1 = \{\, \boldsymbol{\tau} \in \mathcal{H} \mid \text{Div}\, \boldsymbol{\tau} \in H \,\}.$$

Recall that condition $\boldsymbol{v} = \mathbf{0}$ on Γ_D in the definition of the space V is understood in the sense of trace, i.e., $\gamma \boldsymbol{v} = \mathbf{0}$ a.e. on Γ_D. It is well known that the spaces H, $\mathcal{H}$, V, and $\mathcal{H}_1$ are Hilbert spaces equipped with the inner products

$$\langle \boldsymbol{u}, \boldsymbol{v} \rangle_H = \int_\Omega \boldsymbol{u} \cdot \boldsymbol{v}\, dx, \qquad \langle \boldsymbol{\sigma}, \boldsymbol{\tau} \rangle_{\mathcal{H}} = \int_\Omega \boldsymbol{\sigma} : \boldsymbol{\tau}\, dx,$$

$$\langle \boldsymbol{u}, \boldsymbol{v} \rangle_V = \langle \boldsymbol{\varepsilon}(\boldsymbol{u}), \boldsymbol{\varepsilon}(\boldsymbol{v}) \rangle_{\mathcal{H}}, \qquad \langle \boldsymbol{\sigma}, \boldsymbol{\tau} \rangle_{\mathcal{H}_1} = \langle \boldsymbol{\sigma}, \boldsymbol{\tau} \rangle_{\mathcal{H}} + \langle \text{Div}\, \boldsymbol{\sigma}, \text{Div}\, \boldsymbol{\tau} \rangle_H,$$

where $\boldsymbol{\varepsilon}\colon H^1(\Omega;\mathbb{R}^d) \to \mathcal{H}$ and $\text{Div}\colon \mathcal{H}_1 \to H$ denote the *deformation* and the *divergence* operator, respectively, given by

$$\boldsymbol{\varepsilon}(\boldsymbol{u}) = \big(\varepsilon_{ij}(\boldsymbol{u})\big), \quad \varepsilon_{ij}(\boldsymbol{u}) = \frac{1}{2}(u_{i,j} + u_{j,i}), \quad \text{Div}\, \boldsymbol{\sigma} = (\sigma_{ij,j}). \tag{2.5}$$

Here and below an index that follows a comma indicates a derivative with respect to the corresponding component of the variable. Therefore, since the summation convention over repeated indices is adopted, the *divergence* of the stress field is given by

$$\sigma_{ij,j} = \sum_{j=1}^{d} \frac{\partial \sigma_{ij}}{\partial x_j}.$$

The associated norms in H, $\mathcal{H}$, V, and $\mathcal{H}_1$ are denoted by $\|\cdot\|_H$, $\|\cdot\|_{\mathcal{H}}$, $\|\cdot\|_V$, and $\|\cdot\|_{\mathcal{H}_1}$, respectively. Since the trace operator is continuous, it follows that

V is a closed subspace of $H^1(\Omega;\mathbb{R}^d)$. Moreover, as it follows from the discussion in Example 2.20 we obtain that (V, H, V^*) is an evolution triple of spaces with compact embeddings where, recall, V^* denotes the dual space of V. In addition, since meas$(\Gamma_D) > 0$, the following Korn inequality holds:

$$\|\boldsymbol{v}\|_{H^1(\Omega;\mathbb{R}^d)} \le c\,\|\boldsymbol{\varepsilon}(\boldsymbol{v})\|_{\mathcal{H}} \quad \text{for all } \boldsymbol{v} \in V,$$

where $c > 0$ depends only on Ω and Γ_D. This implies that the norm $\|\cdot\|_V = \|\boldsymbol{\varepsilon}(\cdot)\|_{\mathcal{H}}$ is equivalent on V with the norm $\|\cdot\|_{H^1(\Omega;\mathbb{R}^d)}$.

Finally, we comment on the Green-type formulae which play a key role in obtaining variational formulations of contact problems. Given Ω an open bounded subset of $\mathbb{R}^d$ with a Lipschitz boundary Γ and denoting by $\boldsymbol{\nu} = (\nu_i) \in \mathbb{R}^d$ the unit outward normal vector on Γ, it is a well-known classical result that

$$\int_\Omega u_{,i}\, v\,dx = \int_\Gamma u\,v\,\nu_i\,d\Gamma - \int_\Omega u\,v_{,i}\,dx \quad \text{for all } u, v \in C^1(\overline{\Omega})$$

for all $i = 1,\ldots,d$. This is often called the *Green formula* or the *divergence theorem*. This formula can be extended to functions from certain Sobolev spaces so that the smoothness of the functions is exactly enough for integrals to be well defined in the sense of Lebesgue.

Theorem 2.23 (Multidimensional integration by parts). *Let Ω be an open bounded set of $\mathbb{R}^d$ with a Lipschitz boundary Γ, and let $1 \le p < \infty$ with the conjugate exponent q. Then for $u \in W^{1,p}(\Omega)$ and $v \in W^{1,q}(\Omega)$, we have*

$$\int_\Omega (u\,v_{,i} + u_{,i}\,v)\,dx = \int_\Gamma u\,v\,\nu_i\,d\Gamma \quad \text{for all } i = 1,\ldots,d.$$

We consider now $\boldsymbol{v} = (v_1,\ldots,v_d)$, write the formula in Theorem 2.23 for v_i instead of v, and summ it over $i = 1,\ldots,d$. Then, with the notation

$$\operatorname{div}\boldsymbol{v} = \sum_{i=1}^d \frac{\partial v_i}{\partial x_i}$$

for the *divergence* of the vector field $\boldsymbol{v}$ and

$$\nabla u = (u_{,1},\ldots,u_{,d})$$

for the *gradient* of the scalar field u we arrive at the following Green-type formula.

Theorem 2.24. *Let Ω be an open bounded set of $\mathbb{R}^d$ with a Lipschitz boundary Γ, and let $1 \le p < \infty$ with the conjugate exponent q. Then for $u \in W^{1,p}(\Omega)$ and $\boldsymbol{v} \in W^{1,q}(\Omega;\mathbb{R}^d)$, the following formula holds*

$$\int_{\Omega} (u \operatorname{div} \boldsymbol{v} + \nabla u \cdot \boldsymbol{v})\, dx = \int_{\Gamma} u\, (\boldsymbol{v} \cdot \boldsymbol{\nu})\, d\Gamma, \tag{2.6}$$

where, recall, the dot denotes the inner product in $\mathbb{R}^d$.

We conclude with a second Green-type formula that is repeatedly used in the rest of the book in order to derive variational formulations of the contact problems.

Theorem 2.25. *Let Ω be an open bounded and connected set of $\mathbb{R}^d$ with a Lipschitz boundary Γ. Then*

$$\int_{\Omega} \boldsymbol{\sigma} : \boldsymbol{\varepsilon}(\boldsymbol{v})\, dx + \int_{\Omega} \operatorname{Div} \boldsymbol{\sigma} \cdot \boldsymbol{v}\, dx = \int_{\Gamma} \boldsymbol{\sigma}\boldsymbol{\nu} \cdot \boldsymbol{v}\, d\Gamma \tag{2.7}$$

for all $\boldsymbol{v} \in H^1(\Omega; \mathbb{R}^d)$ and $\boldsymbol{\sigma} \in C^1(\overline{\Omega}; \mathbb{S}^d)$.

The proof of Theorem 2.25 is based on the integration by parts formula presented in Theorem 2.23, combined with the definition of the deformation and divergence operators in (2.5).

2.4 Bochner–Lebesgue and Bochner–Sobolev Spaces

In this section we introduce briefly the Bochner integral of Banach space-valued functions, Bochner–Lebesgue spaces, and Bochner–Sobolev spaces. These spaces play an important role in the study of evolutionary problems in Chap. 5. We omit most of the proofs and we refer to [66, 83, 263] for detailed information and proofs. Throughout this section E denotes a Banach space with a norm $\|\cdot\|_E$, E^* is its dual, and $\langle \cdot, \cdot \rangle_{E^* \times E}$ represents the duality pairing between E^* and E.

Weak and strong measurability. We begin with two types of measurability of vector-valued functions.

Definition 2.26. Let $(\mathcal{O}, \Sigma, \mu)$ be a measure space.

(i) A function $s\colon \mathcal{O} \to E$ is called a *simple* (or a *step*, or a *finitely-valued*) *function*, if there exist $\{A_i\}_{i=1}^k \subseteq \Sigma$ mutually disjoint sets and $\{c_i\}_{i=1}^k \subseteq E$ such that $s = \sum_{i=1}^k c_i \chi_{A_i}$, where χ_A denotes the characteristic function of a set A, see Definition 1.60.

(ii) A function $u\colon \mathcal{O} \to E$ is said to be *measurable* (or *strongly measurable*) if there exists a sequence $\{s_n\}$ of simple functions $s_n\colon \mathcal{O} \to E$ such that

$$\lim_{n\to\infty} \|s_n(\omega) - u(\omega)\|_E = 0 \ \text{ for } \mu\text{-a.e. } \omega \in \mathcal{O}.$$

(iii) A function $u\colon \mathcal{O} \to E$ is said to be *weakly measurable* (or E^* *measurable*) if for all $e^* \in E^*$ the real-valued function

$$\mathcal{O} \ni \omega \mapsto \langle e^*, u(\omega) \rangle_{E^* \times E} \in \mathbb{R} \ \text{ is measurable.}$$

(iv) A function $u: \mathcal{O} \to E^*$ is said to be *weakly* measurable* (or *E measurable*) if for all $e \in E$ the real-valued function

$$\mathcal{O} \ni \omega \mapsto \langle u(\omega), e \rangle_{E^* \times E} \in \mathbb{R} \text{ is measurable.}$$

Some basic properties of measurable functions are gathered in the following result.

Lemma 2.27. *Let $(\mathcal{O}, \Sigma, \mu)$ be a measure space and E_1 be a Banach space. Then, the following statements hold:*

(i) *Every continuous function $u: \mathcal{O} \to E$ is measurable.*
(ii) *If $u: \mathcal{O} \to E$ is measurable, then the real-valued function $\mathcal{O} \ni \omega \mapsto \|u(\omega)\|_E \in \mathbb{R}$ is measurable.*
(iii) *If $u: \mathcal{O} \to E$ is measurable and $g: E \to E_1$ is continuous, then the composition $g \circ u: \mathcal{O} \to E_1$ is measurable.*
(iv) *If $u: \mathcal{O} \to E$ and $g: \mathcal{O} \to \mathbb{R}$ are measurable, then the product $ug: \mathcal{O} \to E$ is measurable.*
(v) *If $u: \mathcal{O} \to E$ and $g: \mathcal{O} \to E^*$ are measurable, then the duality product $\langle g, u \rangle_{E^* \times E}: \mathcal{O} \to \mathbb{R}$ is measurable.*
(vi) *If $\{u_n\}$ is a sequence of measurable functions from $\mathcal{O}$ to E such that $\lim_{n \to \infty} \|u_n(\omega) - u(\omega)\|_E = 0$ for μ-a.e. $\omega \in \mathcal{O}$, then u is measurable.*

The proofs of the statements (i)–(v) in the lemma are straightforward. The statement (vi) represents a consequence of the corresponding property for $E = \mathbb{R}$ and the following result due to Pettis.

Theorem 2.28 (Pettis measurability theorem). *Let $(\mathcal{O}, \Sigma, \mu)$ be a finite measure space. A function $u: \mathcal{O} \to E$ is strongly measurable if and only if it is weakly measurable and there exists $A \in \Sigma$ such that $\mu(A) = 0$ and $u(\mathcal{O} \setminus A)$ is a separable set of E in the norm sense (i.e., u is μ-a.e. separably valued).*

Note that the Pettis measurability theorem shows that measurability and weak measurability coincide when E is separable. Its proof can be found, for instance, in [66].

Bochner integral. We first define the Bochner integral for a simple function, then we extend this concept to a strongly measurable function.

Definition 2.29. Let $(\mathcal{O}, \Sigma, \mu)$ be a measure space. A simple function $s: \mathcal{O} \to E$ is said to be *Bochner integrable* if it is of the form

$$s = \sum_{i=1}^{k} c_i \chi_{A_i},$$

where c_i are distinct elements of E with $i = 1, \ldots, k$ and $k \in \mathbb{N}$, the sets $\{A_i\}_{i=1}^k \subseteq \Sigma$ are mutually disjoint, and $c_i = 0$ whenever $\mu(A_i) = \infty$. For any measurable set $A \in \Sigma$ the *Bochner integral* of s over A is defined by

$$\int_A s\, d\mu = \sum_{i=1}^{k} c_i \mu(A_i \cap A),$$

where $c_i \mu(A_i \cap A)$ is set to be zero whenever $c_i = 0$ and $\mu(A_i \cap A) = \infty$.

Definition 2.30. Let $(\mathcal{O}, \Sigma, \mu)$ be a measure space. A (strongly) measurable function $u: \mathcal{O} \to E$ is *Bochner integrable* if there exists a sequence of simple functions $\{s_n\}$ such that

$$\lim_{n\to\infty} \|s_n(\omega) - u(\omega)\|_E = 0 \text{ for } \mu\text{-a.e. } \omega \in \mathcal{O}$$

and

$$\lim_{n\to\infty} \int_{\mathcal{O}} \|s_n(\omega) - u(\omega)\|_E \, d\mu = 0.$$

For any measurable set $A \in \Sigma$ the *Bochner integral* of u over A is defined by

$$\int_A u\, d\mu = \lim_n \int_A s_n \, d\mu.$$

It is a routine to verify that this is a well defined notion, i.e., the Bochner integral of u is independent of the sequence of simple functions $\{s_n\}$ used.

A very convenient characterization of Bochner integrable functions is given in the following result.

Theorem 2.31 (Bochner integrability theorem). *Let $(\mathcal{O}, \Sigma, \mu)$ be a finite measure space. A strongly measurable function $u: \mathcal{O} \to E$ is Bochner integrable if and only if $\omega \mapsto \|u(\omega)\|_E$ is integrable over $\mathcal{O}$, i.e.*

$$\int_{\mathcal{O}} \|u(\omega)\|_E \, d\mu < \infty.$$

As a consequence of the previous theorem we obtain the following estimate result.

Corollary 2.32. *Let $(\mathcal{O}, \Sigma, \mu)$ be a finite measure space. If $u: \mathcal{O} \to E$ is Bochner integrable, then*

$$\left\| \int_{\mathcal{O}} u(\omega)\, d\mu \right\|_E \leq \int_{\mathcal{O}} \|u(\omega)\|_E \, d\mu.$$

The Bochner integral represents a natural generalization of the Lebesgue integral of a scalar-valued function. It enjoys many properties known from the Lebesgue integral: the linearity, the Lebesgue-dominated convergence theorem, etc., as proved in Sect. 3.10 of [66]. Nevertheless, the result we present in what follows exhibits a property of the Bochner integral that has no analogue in the theory of Lebesgue integral. To introduce it, we recall that if X and Y are Banach spaces and

$L: D \subseteq X \to Y$ is a linear operator, then the operator L is said to be *closed* if its graph defined by

$$\mathrm{Gr}\,(L) = \{(x, y) \in X \times Y \mid x \in D,\ y = Lx\}$$

is closed in $X \times Y$.

Theorem 2.33. *Assume that $(\mathcal{O}, \Sigma, \mu)$ is a finite measure space, X, Y are Banach spaces, $L: X \to Y$ is a closed operator, $u: \mathcal{O} \to X$ is Bochner integrable and $L \circ u: \mathcal{O} \to Y$ is Bochner integrable, as well. Then*

$$L\left(\int_A u\, d\mu\right) = \int_A (L \circ u)\, d\mu \quad \text{for all } A \in \Sigma.$$

Remark 2.34. If $L \in \mathcal{L}(X, Y)$, then it is clear that if $u: \mathcal{O} \to X$ is Bochner integrable, so is $L \circ u$. Indeed, this results from the inequality $\|Lu(\omega)\|_Y \leq \|L\|_{\mathcal{L}(X,Y)} \|u(\omega)\|_X$ which is valid for all $\omega \in \mathcal{O}$.

Bochner–Lebesgue spaces. Let $(\mathcal{O}, \Sigma, \mu)$ be a measure space and $1 \leq p < \infty$. Then the Bochner–Lebesgue space $L^p(\mathcal{O}; E)$ is the space defined by

$$L^p(\mathcal{O}; E) = \left\{u: \mathcal{O} \to E \mid u \text{ is Bochner integrable and } \int_{\mathcal{O}} \|u(\omega)\|_E^p\, d\mu < \infty\right\}.$$

If $p = \infty$, then the Bochner–Lebesgue space $L^p(\mathcal{O}; E)$ is the space defined by

$$L^\infty(\mathcal{O}; E) = \{u: \mathcal{O} \to E \mid u \text{ is measurable and there is } M > 0 \text{ such that } \|u(\omega)\|_E < M \ \mu\text{-a.e. } \omega \in \mathcal{O}\}.$$

Note that, as in the case $E = \mathbb{R}$, in the definition of the Bochner–Lebesgue spaces above, we identify functions with their equivalence classes. We endow $L^p(\mathcal{O}; E)$ with the norm

$$\|u\|_{L^p(\mathcal{O};E)} = \begin{cases} \left(\int_{\mathcal{O}} \|u(\omega)\|_E^p\, d\mu\right)^{1/p} & \text{if } 1 \leq p < \infty, \\ \operatorname*{ess\,sup}_{\omega \in \mathcal{O}} \|u(\omega)\|_E & \text{if } p = \infty, \end{cases}$$

where $\operatorname*{ess\,sup}_{\omega \in \mathcal{O}} \|u(\omega)\|_E = \inf\{M > 0 \mid \|u(\omega)\|_E \leq M \ \mu\text{-a.e. } \omega \in \mathcal{O}\}$.

If $\mathcal{O} = (a, b)$ is an interval in $\mathbb{R}$ we simply write $L^p(a, b; E)$ instead of $L^p((a, b); E)$, i.e., $L^p(a, b; E) = L^p((a, b); E)$. Using the Hölder inequality, it is easy to prove that, for an open bounded set $\mathcal{O} = \Omega \subseteq \mathbb{R}^d$ and $1 \leq r \leq p \leq \infty$, we have the inclusions

$$C(\overline{\Omega}; E) \subseteq L^\infty(\Omega; E) \subseteq L^p(\Omega; E) \subseteq L^r(\Omega; E) \subseteq L^1(\Omega; E).$$

In particular, if $\Omega \subseteq \mathbb{R}^d$ is open and bounded, and u is continuous on the closure $\overline{\Omega}$, then u belongs to $L^p(\Omega; E)$ for every $1 \le p \le \infty$.

Next, we follow [66,68,69,229,263] to collect the following results on Bochner–Lebesgue spaces which will be used in the rest of the book.

Theorem 2.35. *Let $(\mathcal{O}, \Sigma, \mu)$ be a measure space. Then*

(i) *The space $L^p(\mathcal{O}; E)$ is a Banach space for $1 \le p \le \infty$.*
(ii) *The family of all integrable simple functions is dense in $L^p(\mathcal{O}; E)$ for $1 \le p < \infty$.*
(iii) *If Σ is countably generated and E is separable, then $L^p(\mathcal{O}; E)$ is separable for $1 \le p < \infty$.*
(iv) *If μ is σ-finite and E is a separable reflexive Banach space, then $L^p(\mathcal{O}; E)$ is reflexive for $1 < p < \infty$.*
(v) *If, in addition, H is a Hilbert space with inner product $\langle \cdot, \cdot \rangle_H$, then $L^2(\mathcal{O}; H)$ is a Hilbert space with the inner product*

$$\langle u, v \rangle_{L^2(\mathcal{O};H)} = \int_{\mathcal{O}} \langle u(\omega), v(\omega) \rangle_H \, d\mu \quad \text{for all } u, v \in L^2(\mathcal{O}; H).$$

The next definition and theorem concern the dual of the Bochner–Lebesgue spaces.

Definition 2.36. Let $(\mathcal{O}, \Sigma, \mu)$ be a measure space and let $1 \le p \le \infty$. Then the space $L^p_w(\mathcal{O}; E^*)$ is the space of all (equivalence classes of) weakly* measurable functions $u\colon \mathcal{O} \to E^*$ such that $\|u(\cdot)\|_{E^*} \in L^p(\mathcal{O})$. The space $L^p_w(\mathcal{O}; E^*)$ is endowed with the norm

$$\|u\|_{L^p_w(\mathcal{O};E^*)} = \begin{cases} \left(\int_{\mathcal{O}} \|u(\omega)\|^p_{E^*} \, d\mu \right)^{1/p} & \text{if } 1 \le p < \infty, \\ \operatorname*{ess\,sup}_{\omega \in \mathcal{O}} \|u(\omega)\|_{E^*} & \text{if } p = \infty. \end{cases}$$

Theorem 2.37. *Let $(\mathcal{O}, \Sigma, \mu)$ be a finite measure space, $1 \le p < \infty$ and q be the exponent conjugate to p.*

(i) *Assume that E is separable. If $L \in (L^p(\mathcal{O}; E))^*$, then there exists a unique $z \in L^q_w(\mathcal{O}; E^*)$ such that*

$$L(v) = \int_{\mathcal{O}} \langle z, v \rangle_{E^* \times E} \, d\mu \quad \text{for all } v \in L^p(\mathcal{O}; E) \tag{2.8}$$

and, moreover, $\|L\|_{(L^p(\mathcal{O};E))^*} = \|z\|_{L^q_w(\mathcal{O};E^*)}$. Conversely, every functional of the form (2.8), where $z \in L^q_w(\mathcal{O}; E^*)$, is linear and bounded on $L^p(\mathcal{O}; E)$.
(ii) *Assume that E is reflexive. If $L \in (L^p(\mathcal{O}; E))^*$, then there exists a unique $z \in L^q(\mathcal{O}; E^*)$ such that*

$$L(v) = \int_{\mathcal{O}} \langle z, v \rangle_{E^* \times E} \, d\mu \quad \text{for all } v \in L^p(\mathcal{O}; E) \tag{2.9}$$

and, moreover, $\|L\|_{(L^p(\mathcal{O};E))^*} = \|z\|_{L^q(\mathcal{O};E^*)}$. *Conversely, every functional of the form* (2.9)*, where* $z \in L^q(\mathcal{O}; E^*)$*, is linear and bounded on* $L^p(\mathcal{O}; E)$*. In this sense we write* $(L^p(\mathcal{O}; E))^* \cong L^q(\mathcal{O}; E^*)$*.*

For later referral we recall below two important theorems for Bochner–Lebesgue spaces.

Theorem 2.38 (Lebesgue-dominated convergence theorem). *Let* $(\mathcal{O}, \Sigma, \mu)$ *be a measure space and* $1 \leq p < \infty$*. If* $u_n : \mathcal{O} \to E$ *is a sequence of measurable functions such that*

$$\lim_{n\to\infty} u_n(\omega) = u(\omega) \quad \text{for } \mu\text{-a.e. } \omega \in \mathcal{O}$$

and there exists $g \in L^p(\mathcal{O})$ *such that* $\|u_n(\omega)\|_E \leq g(\omega)$ *for* μ*-a.e.* $\omega \in \mathcal{O}$*, then* $u \in L^p(\mathcal{O}; E)$ *and* $u_n \to u$ *in* $L^p(\mathcal{O}; E)$*.*

Theorem 2.39 (Converse Lebesgue-dominated convergence). *Let* $(\mathcal{O}, \Sigma, \mu)$ *be a measure space and* $1 \leq p < \infty$*. If* $u_n \to u$ *in* $L^p(\mathcal{O}; E)$*, then there exist a subsequence* $\{u_{n_k}\}$ *of* $\{u_n\}$ *and* $g \in L^p(\mathcal{O})$ *such that* $\|u_{n_k}(\omega)\|_E \leq g(\omega)$ *for* μ*-a.e.* $\omega \in \mathcal{O}$ *and*

$$\lim_k u_{n_k}(\omega) = u(\omega) \quad \text{for } \mu\text{-a.e. } \omega \in \mathcal{O}.$$

Next, we recall the well-known definitions for smooth vector-valued functions.

Definition 2.40. Let $0 < T < \infty$. A function $v : [0, T] \to E$ is said to be (strongly) *differentiable* at $t_0 \in [0, T]$ if there exists an element in E, denoted by $v'(t_0)$ and called the *derivative* of v at t_0, such that

$$\lim_{h\to 0} \left\| \frac{v(t_0 + h) - v(t_0)}{h} - v'(t_0) \right\|_E = 0,$$

where the limit is taken with respect to h with $t_0 + h \in [0, T]$. The derivative at $t_0 = 0$ is defined as a right-sided limit, and that at $t_0 = T$ as a left-sided limit. The function v is said to be *differentiable* on $[0, T]$ if it is differentiable at any $t_0 \in [0, T]$. It is *differentiable a.e.* if it is differentiable a.e. on $[0, T]$. In this case the function v' is called the (strong) *derivative* of v. Higher order derivatives of v, denoted as $v^{(i)}$, $i \geq 2$, are defined recursively by $v^{(i)} = (v^{(i-1)})'$.

We define $C([0, T]; E)$ to be the space of functions $v : [0, T] \to E$ that are continuous on $[0, T]$. Given an integer $m \geq 1$, we define

$$C^m([0, T]; E) = \{v \in C([0, T]; E) \mid v^{(i)} \in C([0, T]; E) \text{ for } i = 1, \ldots, m\}.$$

The space $C^m([0, T]; E)$ is endowed with the norm

$$\|u\|_{C^m([0,T];E)} = \sum_{i=0}^{m} \max\{\|u^{(i)}(t)\|_E \mid t \in [0, T]\}. \tag{2.10}$$

We also write $u^{(0)} = u$ and, for the sake of simplicity, in the rest of the book we use the notation $C^m(0,T;E) = C^m([0,T];E)$.

The following result collects properties of vector-valued functions needed in the sequel.

Theorem 2.41. *Let $0 < T < \infty$, $m \in \mathbb{N}_0$ and $1 \le p < \infty$. Then*

(i) *$C^m(0,T;E)$ is a Banach space with the norm* (2.10).
(ii) *$C(0,T;E)$ is dense in $L^p(0,T;E)$ and the embedding $C(0,T;E) \subseteq L^p(0,T;E)$ is continuous.*
(iii) *The set of all polynomial functions $w: [0,T] \to E$ of the form $w(t) = \sum_{i=0}^{n} a_i t^i$ with $a_i \in E$ for all $i = 1, \ldots, n$ and $n \in \mathbb{N}_0$ is dense in $C(0,T;E)$ and $L^p(0,T;E)$.*
(iv) *If, in addition, E is separable, then $L^p(0,T;E)$ is also separable for $1 \le p < \infty$.*
(v) *If, in addition, E is uniformly convex (respectively, locally uniformly convex, strictly convex), then so is $L^p(0,T;E)$ for $1 < p < \infty$.*
(vi) *If, in addition, E_1 is a Banach space and the embedding $E \subseteq E_1$ is continuous, then the embedding*

$$L^p(0,T;E) \subseteq L^r(0,T;E_1)$$

is also continuous, for $1 \le r \le p \le \infty$.

The following inequality is very useful in many applications.

Theorem 2.42 (Hölder inequality). *Let $0 < T < \infty$, $1 < p < \infty$ and $1/p + 1/q = 1$. Then the following inequality holds:*

$$\int_0^T |\langle v(t), u(t)\rangle_{E^*\times E}|\, dt \le \left(\int_0^T \|u(t)\|_E^p\, dt\right)^{1/p} \left(\int_0^T \|v(t)\|_{E^*}^q\, dt\right)^{1/q}$$

for all $u \in L^p(0,T;E)$, $v \in L^q(0,T;E^*)$.

The following result shows that the integral of the duality pairing is equal to the duality pairing of the integrals and justifies various limit passages.

Proposition 2.43. *Let $1 < p < \infty$, $1/p + 1/q = 1$ and $0 < t \le T < \infty$.*

(i) *If $u \in L^p(0,T;E)$, then*

$$\left\langle v, \int_0^t u(s)\, ds \right\rangle_{E^*\times E} = \int_0^t \langle v, u(s)\rangle_{E^*\times E}\, ds \quad \text{for all } v \in E^*.$$

(ii) *If $u \in L^p(0,T;E^*)$, then*

$$\left\langle \int_0^t u(s)\, ds, v \right\rangle_{E^*\times E} = \int_0^t \langle u(s), v\rangle_{E^*\times E}\, ds \quad \text{for all } v \in E.$$

(iii) *If* $u_n \to u$ *in* $L^p(0,T;E)$, *then*

$$\int_0^t u_n(s)\,ds \to \int_0^t u(s)\,ds \ \text{ in } E.$$

(iv) *If, in addition, E is reflexive and separable, then from*

$$u_n \to u \ \textit{ in } L^p(0,T;E),$$
$$v_n \to v \ \textit{ weakly in } L^q(0,T;E^*),$$

it follows that

$$\int_0^t \langle v_n(s), u_n(s)\rangle_{E^*\times E}\,ds \to \int_0^t \langle v(s), u(s)\rangle_{E^*\times E}\,ds,$$

and from

$$u_n \to u \ \textit{ weakly in } L^p(0,T;E),$$
$$v_n \to v \ \textit{ in } L^q(0,T;E^*),$$

it follows that

$$\int_0^t \langle v_n(s), u_n(s)\rangle_{E^*\times E}\,ds \to \int_0^t \langle v(s), u(s)\rangle_{E^*\times E}\,ds.$$

Bochner–Sobolev spaces. In what follows we provide some definitions and results for Bochner–Sobolev spaces of functions defined on intervals. These results are necessary in order to study abstract differential equations in Banach spaces. The solutions of such equations naturally live in Bochner–Sobolev spaces of Banach space-valued functions. Our presentation follows Sect. 23.2 of [263].

Unless otherwise stated, $(0,T)$ will represent an open bounded real interval. We introduce the set $L^1_{loc}(0,T;E)$ of measurable functions $u\colon (0,T)\to E$ such that $u \in L^1(K;E)$ for all compact sets $K \subseteq (0,T)$ (i.e., $u\chi_K \in L^1(0,T;E)$). Note that for all $1 \le p \le \infty$ and any compact set $K \subseteq (0,T)$, we have $L^p(K;E) \subseteq L^1(K;E)$. Hence $L^p(0,T;E) \subseteq L^1_{loc}(0,T;E)$ for all $1 \le p \le \infty$. Let $\mathcal{D}(0,T)$ or $C_0^\infty(0,T)$ denote the space of all real-valued functions defined on $(0,T)$ which are infinitely differentiable on $(0,T)$ and have compact support in $(0,T)$. The next result, often called the generalized variational lemma, represents an extension of Lemma 2.9.

Lemma 2.44 (Generalized variational lemma). *Let* $u \in L^1_{loc}(0,T;E)$ *be such that*

$$\int_0^T u(s)\varphi(s)\,ds = 0 \ \text{ for all } \ \varphi \in C_0^\infty(0,T).$$

Then $u = 0$ *in* $L^1(0,T;E)$, *i.e.,* $u(t) = 0$ *for a.e.* $t \in (0,T)$.

In order to introduce the Bochner–Sobolev spaces, we need to extend the notion of distributional derivative to Banach space-valued functions.

Definition 2.45. Given $k \geq 1$ and $u, v \in L^1(0,T;E)$, we say that v is the *kth order (distributional* or *weak) derivative* of u if

$$\int_0^T v(s)\varphi(s)\,ds = (-1)^k \int_0^T u(s)\varphi^{(k)}(s)\,ds$$

for all $\varphi \in \mathcal{D}(0,T)$, where the integrals are understood in the Bochner sense. The kth order derivative is denoted by $v = u^{(k)}$.

Note that the weak derivative is a global notion, i.e., in contrast to the strong derivative introduced in Definition 2.40, it can not be defined at a point. Note also that the weak derivative, if it exists, is uniquely defined up to a set of measure zero.

The requirement on stronger integrability of u and u' leads to the following definition of Bochner–Sobolev spaces.

Let $1 \leq p \leq \infty$. We define the *Bochner–Sobolev space* by equality

$$W^{1,p}(0,T;E) = \{u \in L^p(0,T;E) \mid u' \in L^p(0,T;E)\}.$$

It is a Banach space endowed with the norm

$$\|u\|_{W^{1,p}(0,T;E)} = \begin{cases} \|u\|_{L^p(0,T;E)} + \|u'\|_{L^p(0,T;E)} & \text{if } 1 \leq p < \infty, \\ \max\{\|u\|_{L^\infty(0,T;E)}, \|u'\|_{L^\infty(0,T;E)}\} & \text{if } p = \infty. \end{cases}$$

We denote by $C_b^0(0,T;E)$ the space of *continuous bounded functions* from $(0,T)$ to E. It is a Banach space with the norm

$$\|u\|_{C_b^0(0,T;E)} = \sup\{\|u(t)\|_E \mid t \in (0,T)\}.$$

By $C^{0,0}(0,T;E)$ we denote the (closed) subspace of $C_b^0(0,T;E)$ of *uniformly continuous and bounded functions* on $(0,T)$. Furthermore, for $\lambda \in (0,1]$, we consider the space of λ*-Hölder continuous functions* (with exponent λ) defined by

$$C^{0,\lambda}(0,T;E) = \{u \in C_b^0(0,T;E) \mid \exists c > 0 \text{ such that}$$
$$\|u(s) - u(t)\|_E \leq c\,|s-t|^\lambda \text{ for all } s,t \in (0,T)\},$$

equipped with the norm

$$\|u\|_{C^{0,\lambda}(0,T;E)} = \|u\|_{C_b^0(0,T;E)} + \sup\left\{\frac{\|u(s)-u(t)\|_E}{|s-t|^\lambda} \mid s,t \in (0,T),\ s \neq t\right\}.$$

It is well known that $C^{0,\lambda}(0,T;E)$ is a Banach space and each function belonging to this space is uniformly continuous on $(0,T)$. It follows from here that each Hölderian function is a continuous function from $[0,T]$ to E.

The following results include the basic properties of Bochner–Sobolev spaces.

Proposition 2.46. (i) *The space $W^{1,p}(0,T;E)$ is a Banach space for $1 \le p \le \infty$.*

(ii) *If E is separable and $1 \le p < \infty$, then $W^{1,p}(0,T;E)$ is separable.*

(iii) *If E is separable reflexive and $1 < p < \infty$, then $W^{1,p}(0,T;E)$ is reflexive.*

(iv) *If $(H, \langle\cdot,\cdot\rangle_H)$ is a Hilbert space, then $H^1(0,T;H) = W^{1,2}(0,T;H)$ is a Hilbert space with the inner product*

$$\langle u, v\rangle_{H^1(0,T;H)} = \int_0^T \langle u(s), v(s)\rangle_H \, ds + \int_0^T \langle u'(s), v'(s)\rangle_H \, ds$$

for $u, v \in H^1(0,T;H)$.

(v) *The embedding $W^{1,p}(0,T;E) \subseteq C^{0,1-\frac{1}{p}}(0,T;E)$ is continuous for all $1 \le p \le \infty$.*

(vi) *If $1 \le p \le \infty$ and $u \in W^{1,p}(0,T;E)$, then there exists a continuous function $\widetilde{u}\colon [0,T] \to E$ which coincides with u almost everywhere. Moreover,*

$$\widetilde{u}(t) - \widetilde{u}(s) = \int_s^t u'(\tau)\, d\tau.$$

for all $s, t \in [0,T]$.

(vii) (Sobolev embedding theorem) *If $1 \le p \le \infty$, then $W^{1,p}(0,T;E)$ is contained in $C(0,T;E)$ and there exists $c > 0$ such that*

$$\|u\|_{L^\infty(0,T;E)} \le c\, \|u\|_{W^{1,p}(0,T;E)} \quad \text{for all} \quad u \in W^{1,p}(0,T;E).$$

Proposition 2.47. *Let $1 \le p \le \infty$, $u \in W^{1,p}(0,T;E)$ and $v \in W^{1,p}(0,T)$. Then*

(i) (Product rule) *the product uv belongs to $W^{1,p}(0,T;E)$ and*

$$(uv)' = u'v + uv'.$$

(ii) (Integration by parts)

$$\int_0^T u'(s)v(s)\, ds = u(T)v(T) - u(0)v(0) - \int_0^T u(s)v'(s)\, ds.$$

Next we recall the definition of an absolutely continuous function.

Definition 2.48. (i) We say that $u\colon (0,T) \to E$ is *absolutely continuous* if for each $\varepsilon > 0$ there exists $\delta > 0$ such that $\sum_i \|u(t_i) - u(s_i)\|_E \le \varepsilon$ holds for any countable collection of pairwise disjoint intervals $(s_i, t_i) \subset (0,T)$ of total length less then δ.

(ii) Let $k \in \mathbb{N}$ and $1 \le p \le \infty$. We denote by $AC^{k,p}(0,T;E)$ the space of all *absolutely continuous functions* $u\colon (0,T) \to E$ whose strong derivatives $u^{(m)}$ exist almost everywhere for $0 \le m \le k$ and belong to $L^p(0,T;E)$.

Definition 2.49. Let $1 \leq p \leq \infty$. For every $k \geq 2$, we define inductively the *kth Bochner–Sobolev space*

$$W^{k,p}(0,T;E) = \{u \in W^{1,p}(0,T;E) \mid u' \in W^{k-1,p}(0,T;E)\}$$

which is a Banach space with the norm

$$\|u\|_{W^{k,p}(0,T;E)} = \begin{cases} \sum_{i=0}^{k} \|u^{(i)}\|_{L^p(0,T;E)} & \text{if } 1 \leq p < \infty, \\ \max_{i=0,\ldots,k} \|u^{(i)}\|_{L^\infty(0,T;E)} & \text{if } p = \infty. \end{cases}$$

We write $H^k(0,T;E) = W^{k,2}(0,T;E)$. We also define $W_0^{1,p}(0,T;E)$ as the closure of the space $C_0^1(0,T;E)$ in the norm of the space $W^{1,p}(0,T;E)$ and we put $H_0^1(0,T;E) = W_0^{1,2}(0,T;E)$.

We have the following results.

Proposition 2.50. *Let E be a reflexive Banach space and $u\colon (0,T) \to E$ be absolutely continuous. Then u is strongly differentiable almost everywhere, $u' \in L^1(0,T;E)$ and, moreover,*

$$u(t) = u(0) + \int_0^t u'(s)\,ds \quad \text{for all } t \in [0,T].$$

Proposition 2.51. *Let $(0,T)$ be an open bounded real interval, $1 \leq p \leq \infty$, $k \in \mathbb{N}$ and $u \in L^p(0,T;E)$. Then the following two statements are equivalent:*

(i) *$u \in W^{k,p}(0,T;E)$.*
(ii) *There exists $\widetilde{u} \in AC^{k,p}(0,T;E)$ such that $u(t) = \widetilde{u}(t)$ a.e. $t \in (0,T)$.*

Proposition 2.52. *Let $(0,T)$ be an open bounded real interval and $1 \leq p \leq \infty$. Then*

(i) *A function $u \in W^{1,p}(0,T;E)$ belongs to $W_0^{1,p}(0,T;E)$ if and only if $u(0) = u(T) = 0$.*
(ii) (Poincaré inequality) *if $1 \leq p < \infty$, then there exists a constant $c > 0$ such that*

$$\int_0^T \|u(s)\|_E^p\,ds \leq c \int_0^T \|u'(s)\|_E^p\,ds \quad \text{for all } u \in W_0^{1,p}(0,T;E).$$

(iii) *If H is a Hilbert space, then $H^k(0,T;H)$ is a Hilbert space with the inner product*

$$\langle u, v\rangle_{H^k(0,T;H)} = \sum_{i=0}^{k} \langle u^{(i)}, v^{(i)}\rangle_{L^2(0,T;H)} \quad \text{for all } u, v \in H^k(0,T;H).$$

Evolution triples of Bochner–Sobolev spaces. We conclude this section by recalling results on Bochner–Sobolev spaces in the context of evolution triples.

Theorem 2.53. *Let (V, H, V^*) be an evolution triple of spaces, $0 < T < \infty$, $1 \le p, q \le \infty$, $1/p + 1/q = 1$ and $u \in L^p(0, T; V)$. Then, the kth order derivative of u exists and belongs to $L^q(0, T; V^*)$ if and only if there is a function $w \in L^q(0, T; V^*)$ such that*

$$\int_0^T \langle u(s), v\rangle_H \, \varphi^{(k)}(s)\, ds = (-1)^k \int_0^T \langle w(s), v\rangle_{V^* \times V}\, \varphi(s)\, ds$$

for all $v \in V$ and all $\varphi \in \mathcal{D}(0, T)$. In this case $u^{(k)} = w$ and, moreover,

$$\frac{d^k}{dt^k} \langle u(t), v\rangle_H = \langle w(t), v\rangle_{V^* \times V} = \langle u^{(k)}(t), v\rangle_{V^* \times V}$$

for all $v \in V$ and a.e. $t \in (0, T)$. Here $\frac{d^k}{dt^k} \langle u(t), v\rangle_H$ is understood to be the kth order derivative of the real-valued function $(0, T) \ni t \mapsto \langle u(t), v\rangle_H \in \mathbb{R}$.

Let (V, H, V^*) be an evolution triple of spaces and $\mathcal{V} = L^2(0, T; V)$. Recall that from Theorem 2.37 it follows that $\mathcal{V}^* = L^2(0, T; V^*)$. We consider the linear space

$$\mathcal{W} = \{u \in \mathcal{V} \mid u' \in \mathcal{V}^*\}, \tag{2.11}$$

where the derivative of u with respect to t is understood in the distributional sense. The space $\mathcal{W}$ is a linear subspace of the space $\mathcal{V}$ and will play a central role in the study of evolution problems of Chap. 5. Moreover, the following result holds.

Proposition 2.54. *Let (V, H, V^*) be an evolution triple of spaces and $0 < T < +\infty$. Then*

(i) *The space $\mathcal{W}$ defined by (2.11) is a Banach space equipped with the norm*

$$\|u\|_{\mathcal{W}} = \|u\|_{\mathcal{V}} + \|u'\|_{\mathcal{V}^*}. \tag{2.12}$$

(ii) *The embedding $\mathcal{W} \subseteq C(0, T; H)$ is continuous. More precisely, if $u \in \mathcal{W}$, then there exists a uniquely determined continuous function $\widetilde{u}\colon [0, T] \to H$ which coincides almost everywhere on $(0, T)$ with u. Writing u instead of $\widetilde{u}$, we have*

$$\max_{t \in [0,T]} \|u(t)\|_H \le c\, \|u\|_{\mathcal{W}}$$

with a constant $c > 0$.

(iii) (Integration by parts formula) *For any u, $v \in \mathcal{W}$ and any $0 \le s \le t \le T$, the following formula holds:*

$$\langle u(t), v(t)\rangle_H - \langle u(s), v(s)\rangle_H$$
$$= \int_s^t \left(\langle u'(\tau), v(\tau)\rangle_{V^*\times V} + \langle v'(\tau), u(\tau)\rangle_{V^*\times V}\right) d\tau.$$

From Propositions 2.46(vii) and 2.54(ii) we have the following corollary which will be useful in the study of evolutionary inclusions in Chap. 5.

Lemma 2.55. *Let $(0, T)$ be an open bounded real interval.*

(i) *Let $1 \le p \le \infty$, V be a Banach space, u_n, $u \in W^{1,p}(0, T; V)$ and $u_n \to u$ weakly in $W^{1,p}(0, T; V)$. Then $u_n(t) \to u(t)$ weakly in V, for all $t \in [0, T]$.*

(ii) *Let (V, H, V^*) be an evolution triple of spaces and the space $\mathcal{W}$ be defined by (2.11). If u_n, $u \in \mathcal{W}$ and $u_n \to u$ weakly in $\mathcal{W}$, then $u_n(t) \to u(t)$ weakly in H for all $t \in [0, T]$.*

Proof. (i) Let u_n, $u \in W^{1,p}(0, T; V)$ and $u_n \to u$ weakly in $W^{1,p}(0, T; V)$. Since the embedding $W^{1,p}(0, T; V) \subset C(0, T; V)$ is continuous, we know that $u_n \to u$ weakly in $C(0, T; V)$. We fix $v^* \in V^*$ and $t \in [0, T]$. Then, for $\varphi \in (C(0, T; V))^*$ defined by $\varphi(y) = \langle v^*, y(t)\rangle_{V^*\times V}$ with $y \in C(0, T; V)$, we have

$$\langle v^*, u_n(t)\rangle_{V^*\times V} = \varphi(u_n) \to \varphi(u) = \langle v^*, u(t)\rangle_{V^*\times V}.$$

Hence $u_n(t) \to u(t)$ weakly in V for all $t \in [0, T]$ which proves (i).

(ii) Let $u_n, u \in \mathcal{W}$ and $u_n \to u$ weakly in $\mathcal{W}$. Since $\mathcal{W} \subset C(0, T; H)$ continuously, we have $u_n \to u$ weakly in $C(0, T; H)$. We fix $h \in H$ and $t \in [0, T]$. Then, we define $\psi \in (C(0, T; H))^*$ by $\psi(y) = \langle y(t), h\rangle_H$ for all $y \in C(0, T; H)$ and $t \in [0, T]$. We note that

$$\langle u_n(t), h\rangle_H = \psi(u_n) \to \psi(u) = \langle u(t), h\rangle_H.$$

This implies that $u_n(t) \to u(t)$ weakly in H for all $t \in [0, T]$ which proves the property (ii). □

Finally, we recall the following compactness embedding theorem for Bochner–Sobolev space which will be crucial in the study of evolutionary problems.

Theorem 2.56. *Let $1 < p \le \infty$ and $1 < r \le \infty$. Let X, Y, Z be Banach spaces such that X and Z are reflexive, $X \subseteq Y \subseteq Z$, the embedding $X \subseteq Y$ is compact and the embedding $Y \subseteq Z$ is continuous. Then the embedding*

$$\{u \in L^p(0, T; X) \mid u' \in L^r(0, T; Z)\} \subseteq L^p(0, T; Y)$$

is compact.

Chapter 3
Elements of Nonlinear Analysis

In this chapter we present basic material on the set-valued mappings, nonsmooth analysis, subdifferential calculus, and operators of monotone type. For set-valued mappings we concentrate on measurability and continuity issues which we need in subsequent chapters. The section on nonsmooth analysis is devoted to results on the generalized differentiation for locally Lipschitz superpotentials. Next, we provide a result on the subdifferential of the integral superpotentials which is an essential tool in Chaps. 4 and 5 of the book. Finally, we recall the results on single and multivalued operators of monotone type in Banach spaces. The surjectivity results for such operators play a crucial role in our existence results for stationary and evolutionary inclusions. Most of the results presented in this chapter are stated without proofs.

3.1 Set-Valued Mappings

In this section we briefly recall some definitions and basic results on the measurability and continuity of set-valued mappings, called also multivalued functions or multifunctions. For the proofs and a more detailed presentation we refer to monographs [14, 39, 66, 67, 109, 132, 264].

Throughout the rest of the book we denote by 2^X all subsets of a set X and by $\mathcal{P}(X)$ all nonempty subsets of a set X. For a normed space X we use the notations

$$\mathcal{P}_{f(c)}(X) = \{A \subseteq X \mid A \text{ is nonempty, closed, (convex)}\},$$
$$\mathcal{P}_{(w)k(c)}(X) = \{A \subseteq X \mid A \text{ is nonempty, (weakly) compact, (convex)}\}.$$

We start with the following definitions.

Definition 3.1. Assume that X and Y are two sets, $F\colon X \to \mathcal{P}(Y)$ is a multifunction and let $A \subset Y$. Then

S. Migórski et al., *Nonlinear Inclusions and Hemivariational Inequalities*, Advances in Mechanics and Mathematics 26, DOI 10.1007/978-1-4614-4232-5_3,

(i) The *weak inverse image* of A under F is the set

$$F^-(A) = \{x \in X \mid F(x) \cap A \neq \emptyset\}.$$

(ii) The *strong inverse image* of A under F is the set

$$F^+(A) = \{x \in X \mid F(x) \subset A\}.$$

Definition 3.2. Let Y be a normed space and $A \in \mathcal{P}(Y)$. The *support function* of the set A is defined by

$$Y^* \ni y^* \mapsto \sigma(y^*, A) = \sup\{\langle y^*, a\rangle_{Y^* \times Y} \mid a \in A\} \in \mathbb{R} \cup \{+\infty\},$$

where $\langle \cdot, \cdot \rangle_{Y^* \times Y}$ denotes the duality pairing of Y^* and Y.

Definition 3.3. Let X and Y be sets, and $F: X \to \mathcal{P}(Y)$ be a multifunction. The *graph* of the multifunction F is the set

$$\mathrm{Gr}\,(F) = \{(x, y) \in X \times Y \mid y \in F(x)\}.$$

Definition 3.4. Let $(\mathcal{O}, \Sigma)$ be a measurable space. Given a separable metric space (X, d) and a multifunction $F: \mathcal{O} \to 2^X$, we say that

(i) F is *measurable* if for every $U \subset X$ open, we have $F^-(U) \in \Sigma$.
(ii) F is *graph measurable* if $\mathrm{Gr}\,(F) \in \Sigma \times \mathcal{B}(X)$.

Moreover, if X is a separable Banach space and $F: \mathcal{O} \to \mathcal{P}(X)$, we say that F is *scalarly measurable* if for every $x^* \in X^*$ the function $\mathcal{O} \ni \omega \mapsto \sigma(x^*, F(\omega)) \in \mathbb{R} \cup \{+\infty\}$ is Σ-measurable.

The next theorem summarizes the properties of measurable multifunctions and can be found in Sect. 4.2 of [67].

Theorem 3.5. *Let $(\mathcal{O}, \Sigma)$ be a measurable space, (X, d) be a separable metric space and $F: \mathcal{O} \to 2^X$ be a multifunction with closed values. Consider the following properties:*

(1) *For every $D \in \mathcal{B}(X)$, $F^-(D) \in \Sigma$.*
(2) *For every $C \subset X$ closed, $F^-(C) \in \Sigma$.*
(3) *F is measurable.*
(4) *For every $x \in X$, the function $\omega \mapsto d(x, F(\omega))$ is Σ-measurable.*
(5) *F is graph measurable.*

Then we have the following relations:

(a) (1) $\Longrightarrow$ (2) $\Longrightarrow$ (3) $\Longleftrightarrow$ (4) $\Longrightarrow$ (5).
(b) *If X is σ-compact, then* (2) $\Longleftrightarrow$ (3).
(c) *If Σ is complete and X is complete, then conditions* (1)–(5) *are equivalent.*

Concerning the scalar measurability of multifunctions, we recall the following result.

Theorem 3.6. (i) *If $(\mathcal{O}, \Sigma)$ is a complete measurable space, X is a separable Banach space and $F: \mathcal{O} \to \mathcal{P}(X)$ is graph measurable, then F is scalarly measurable.*
(ii) *If $(\mathcal{O}, \Sigma)$ is a measurable space, X is a separable Banach space and $F: \mathcal{O} \to \mathcal{P}_{wkc}(X)$, then F is measurable if and only if F is scalarly measurable.*

It follows from Theorem 3.6 that, under the assumption in (ii), for $\mathcal{P}_{wkc}$-valued multifunctions the notions of measurability and scalar measurability are equivalent.

Now we recall the concepts of continuity of multifunctions.

Definition 3.7. Let X and Y be Hausdorff topological spaces and $F: X \to 2^Y$ be a multifunction. Then

(i) F is called *upper semicontinuous at* $x_0 \in X$, if for every open set $V \subset Y$ such that $F(x_0) \subset V$, we can find a neighborhood $\mathcal{N}(x_0)$ of x_0 such that $F(\mathcal{N}(x_0)) \subset V$. We say that F is upper semicontinuous (usc), if F is upper semicontinuous at every $x_0 \in X$.
(ii) F is called *lower semicontinuous* at $x_0 \in X$, if for every open set $V \subset Y$ such that $F(x_0) \cap V \neq \emptyset$, we can find a neighborhood $\mathcal{N}(x_0)$ of x_0 such that $F(x) \cap V \neq \emptyset$ for all $x \in \mathcal{N}(x_0)$. We say that F is lower semicontinuous (lsc), if F is lower semicontinuous at every $x_0 \in X$.
(iii) F is called *continuous* (or *Vietoris continuous*) at $x_0 \in X$, if F is both usc and lsc at x_0. We say that F is continuous (or Vietoris continuous), if it is continuous at every $x_0 \in X$.

The next two propositions give equivalent conditions for semicontinuity of multifunctions between metric spaces, which are sufficient for our purpose.

Proposition 3.8. *Let X, Y be metric spaces and $F: X \to 2^Y$. Then the following statements are equivalent:*

(1) *F is usc.*
(2) *For every $C \subset Y$ closed, $F^-(C)$ is closed in X.*
(3) *If $x \in X$, $\{x_n\} \subset X$ with $x_n \to x$ and $V \subset Y$ is an open set such that $F(x) \subset V$, then we can find $n_0 \in \mathbb{N}$, depending on V, such that $F(x_n) \subset V$ for all $n \geq n_0$.*

Proposition 3.9. *Let X, Y be metric spaces and $F: X \to 2^Y$. Then the following statements are equivalent:*

(1) *F is lsc.*
(2) *For every $C \subset Y$ closed, $F^+(C)$ is closed in X.*
(3) *If $x \in X$, $\{x_n\} \subset X$ with $x_n \to x$ in X and $y \in F(x)$, then for every $n \in \mathbb{N}$, we can find $y_n \in F(x_n)$ such that $y_n \to y$ in Y.*

From the propositions above it follows that a single-valued operator $F: X \to Y$ is upper semicontinuous or lower semicontinuous in the sense of Definition 3.7 if and only if F is continuous.

Definition 3.10. Let X and Y be metric spaces and $F: X \to 2^Y$ be a multifunction. We say that F is *closed* at $x_0 \in X$, if for every sequence $\{(x_n, y_n)\} \subset \mathrm{Gr}(F)$ such that $(x_n, y_n) \to (x_0, y_0)$ in $X \times Y$, we have $(x_0, y_0) \in \mathrm{Gr}(F)$. We say that F is closed, if it is closed at every $x_0 \in X$ (i.e., $\mathrm{Gr}(F) \subset X \times Y$ is closed).

Definition 3.11. Let X and Y be metric spaces and $F: X \to \mathcal{P}_f(Y)$ be a multifunction. We say that F is *locally compact* if for every $x \in X$, we can find a neighborhood $\mathcal{N}(x)$ such that $\overline{F(\mathcal{N}(x))} \in \mathcal{P}_k(Y)$.

The following proposition provides the relations between upper semicontinuity and closedness of multifunctions.

Proposition 3.12. *Assume that X and Y are metric spaces. Then, the following statements hold:*

(1) *If a multifunction $F: X \to \mathcal{P}_f(Y)$ is usc, then F is closed.*
(2) *If a multifunction $F: X \to \mathcal{P}_f(Y)$ is closed and locally compact, then F is usc.*
(3) *A multifunction $F: X \to \mathcal{P}_k(Y)$ is usc if and only if for every $x \in X$ and every sequence $\{(x_n, y_n)\} \in \mathrm{Gr}(F)$ with $x_n \to x$ in X, there exists a converging subsequence of $\{y_n\}$ whose limit belongs to $F(x)$.*

Note that, in general, a closed multifunction is not in usc, i.e., the converse of Proposition 3.12(1) fails. A counterexample can be found in Sect. 4.1 of [66].

The next convergence theorem will be applied in the study of evolutionary inclusions and hemivariational inequalities in Chap. 5.

Theorem 3.13. *Let $0 < T < \infty$ and F be an usc multifunction from a Hausdorff locally convex space X to the closed convex subsets of a Banach space Y endowed with the weak topology. Let $\{x_n\}$ and $\{y_n\}$ be two sequences of functions such that*

(1) *$x_n: (0, T) \to X$ and $y_n: (0, T) \to Y$ are measurable functions, for all $n \in \mathbb{N}$.*
(2) *For almost all $t \in (0, T)$ and for every neighborhood $\mathcal{N}(0)$ of 0 in $X \times Y$ there exists $n_0 \in \mathbb{N}$ such that $(x_n(t), y_n(t)) \in \mathrm{Gr}(F) + \mathcal{N}(0)$ for all $n \geq n_0$.*
(3) *$x_n(t) \to x(t)$ for a.e. $t \in (0, T)$, where $x: (0, T) \to X$.*
(4) *$y_n \in L^1(0, T; Y)$ and $y_n \to y$ weakly in $L^1(0, T; Y)$, where $y \in L^1(0, T; Y)$.*

Then $(x(t), y(t)) \in \mathrm{Gr}(F)$, i.e., $y(t) \in F(x(t))$ for a.e. $t \in (0, T)$.

In nonlinear analysis, when we deal with multifunctions depending on a parameter, we often need the following classical notion of convergence of sets.

Definition 3.14. Let (X, τ) be a Hausdorff topological space and let $\{A_n\} \subset \mathcal{P}(X)$ for $n \geq 1$. We define

$$\tau\text{-}\liminf A_n = \{x \in X \mid x = \tau\text{-}\lim x_n,\ x_n \in A_n \text{ for all } n \geq 1\}$$

and

$$\tau\text{-}\limsup A_n = \{x \in X \mid x = \tau\text{-}\lim x_{n_k},\ x_{n_k} \in A_{n_k},\ n_1 < n_2 < \cdots n_k < \cdots\}.$$

The set τ- $\liminf A_n$ is called the *τ-Kuratowski lower limit* of the sets A_n and τ- $\limsup A_n$ is called the *τ-Kuratowski upper limit* of the sets A_n. If $A = \tau\text{-}\liminf A_n = \tau\text{-}\limsup A_n$ then A is called *τ-Kuratowski limit* of the sets A_n.

In what follows we need the following definition.

Definition 3.15. Let A be an arbitrary subset of a normed space X. The *convex hull* of A, denoted by $\mathrm{conv}\,(A)$, is the smallest convex set containing A. In other words, the convex hull of an arbitrary set A is the intersection of all convex sets containing A. The *closed convex hull* of A, denoted by $\overline{\mathrm{conv}}\,(A)$, is the closure of the convex hull of A.

The following result concerns the pointwise behavior of weakly convergent sequences in $L^p(\mathcal{O}; E)$ spaces.

Proposition 3.16. *Let $(\mathcal{O}, \Sigma, \mu)$ be a σ-finite measure space, E be a Banach space, and $1 \le p < \infty$. If u_n, $u \in L^p(\mathcal{O}; E)$, $u_n \to u$ weakly in $L^p(\mathcal{O}; E)$ and $u_n(\omega) \in G(\omega)$ for μ-a.e. $\omega \in \mathcal{O}$ and all $n \in \mathbb{N}$ where $G(\omega) \in \mathcal{P}_{wk}(E)$ for μ-a.e. $\omega \in \mathcal{O}$, then*

$$u(\omega) \in \overline{\mathrm{conv}}\left(w\text{-}\limsup \{u_n(\omega)\}_{n \in \mathbb{N}}\right) \quad \text{for } \mu\text{-a.e. on } \mathcal{O}.$$

We conclude this section with the statements of two basic results on the fundamental problem of the existence of a measurable selection for a measurable multifunction. We recall that, given a multifunction $F\colon \mathcal{O} \to \mathcal{P}(X)$, a function $f\colon \mathcal{O} \to X$ is said to be a *selection* of F, if $f(\omega) \in F(\omega)$ for all $\omega \in \mathcal{O}$.

Theorem 3.17. *If $(\mathcal{O}, \Sigma)$ is a measurable space, X is a separable, complete metric space, and $F\colon \mathcal{O} \to \mathcal{P}_f(X)$ is measurable, then F admits a Σ-measurable selection.*

Theorem 3.18. *If $(\mathcal{O}, \Sigma)$ is a complete measurable space, X is locally compact, separable metric space, and $F\colon \mathcal{O} \to \mathcal{P}(X)$ is a multifunction such that* $\mathrm{Gr}\,(F) \in \Sigma \times \mathcal{B}(X)$*, then F admits a Σ-measurable selection.*

Measurable selections facilitate significantly the analysis of measurable multifunctions and are useful in a large number of applications.

3.2 Nonsmooth Analysis

The purpose of this section is to present the background of the theory of generalized differentiation for a locally Lipschitz function. We also elaborate on the classes of functions which are regular in the sense of Clarke and prove some results needed in what follows. Throughout this section X will represent a Banach space with a norm $\|\cdot\|_X$, X^* is its dual and $\langle \cdot, \cdot \rangle_{X^* \times X}$ denotes the duality pairing between X^* and X.

Generalized gradients. We begin with the definitions of generalized directional derivative and the Clarke subdifferential for a class of locally Lipschitz functions. To introduce these notions we recall the following definitions.

Definition 3.19 (Lipschitz function). Let U be a subset of X. A function $\varphi : U \to \mathbb{R}$ is said to be *Lipschitz* on U, if there exists $K > 0$ such that

$$|\varphi(y) - \varphi(z)| \le K \, \|y - z\|_X \quad \text{for all} \quad y, z \in U. \tag{3.1}$$

The constant K is called the *Lipschitz constant*. The inequality (3.1) is also referred to as a *Lipschitz condition*.

Definition 3.20 (Locally Lipschitz function). Let U be a subset of X. A function $\varphi : U \to \mathbb{R}$ is said to be *locally Lipschitz* on U, if for all $x \in U$ there exists a neighborhood $\mathcal{N}(x)$ and $K_x > 0$ such that

$$|\varphi(y) - \varphi(z)| \le K_x \, \|y - z\|_X \quad \text{for all} \quad y, z \in \mathcal{N}(x).$$

The constant K_x in the previous inequality is called the *Lipschitz constant* of φ near x. It is easy to see that a function $\varphi : U \to \mathbb{R}$ defined on a subset U of X, which is Lipschitz on bounded subsets of U, is locally Lipschitz. The converse assertion is not generally true, as it results from the example presented in Sect. 2.5 of [38]. Nevertheless, these two properties are equivalent if $\dim X < \infty$.

Definition 3.21 (Generalized directional derivative). The *generalized directional derivative* (in the sense of Clarke) of the locally Lipschitz function $\varphi : U \subseteq X \to \mathbb{R}$ at the point $x \in U$ in the direction $v \in X$, denoted $\varphi^0(x; v)$, is defined by

$$\varphi^0(x; v) = \limsup_{y \to x, \lambda \downarrow 0} \frac{\varphi(y + \lambda v) - \varphi(y)}{\lambda}.$$

We note that, in contrast to the usual directional derivative, the generalized directional derivative φ^0 is always defined.

Definition 3.22 (Generalized gradient). Let $\varphi : U \subseteq X \to \mathbb{R}$ be a locally Lipschitz function. The *(Clarke) subdifferential* or the *generalized gradient* in the sense of Clarke of φ at $x \in U$, denoted $\partial\varphi(x)$, is the subset of a dual space X^* defined by

$$\partial\varphi(x) = \{\zeta \in X^* \mid \varphi^0(x; v) \ge \langle \zeta, v \rangle_{X^* \times X} \quad \text{for all} \quad v \in X\}.$$

The next proposition provides basic properties of the generalized directional derivative and the generalized gradient.

Proposition 3.23. *If $\varphi : U \to \mathbb{R}$ is a locally Lipschitz function on a subset U of X, then*

(i) *For every* $x \in U$, *the function* $X \ni v \mapsto \varphi^0(x; v) \in \mathbb{R}$ *is positively homogeneous (i.e.,* $\varphi^0(x; \lambda v) = \lambda\varphi^0(x; v)$ *for all* $\lambda \geq 0$*), subadditive (i.e.,* $\varphi^0(x; v_1 + v_2) \leq \varphi^0(x; v_1) + \varphi^0(x; v_2)$ *for all* $v_1, v_2 \in X$*), and satisfies the inequality* $|\varphi^0(x; v)| \leq K_x\|v\|_X$ *with* $K_x > 0$ *being the Lipschitz constant of* φ *near* x*. Moreover, it is Lipschitz continuous and* $\varphi^0(x; -v) = (-\varphi)^0(x; v)$ *for all* $v \in X$.

(ii) *The function* $U \times X \ni (x, v) \mapsto \varphi^0(x; v) \in \mathbb{R}$ *is upper semicontinuous, i.e., for all* $x \in U$, $v \in X$, $\{x_n\} \subset U$, $\{v_n\} \subset X$ *such that* $x_n \to x$ *in* U *and* $v_n \to v$ *in* X, *we have* $\limsup \varphi^0(x_n; v_n) \leq \varphi^0(x; v)$.

(iii) *For every* $v \in X$ *we have* $\varphi^0(x; v) = \max\{\langle \zeta, v\rangle_{X^* \times X} \mid \zeta \in \partial\varphi(x)\}$.

(iv) *For every* $x \in U$ *the gradient* $\partial\varphi(x)$ *is a nonempty, convex, and weakly* compact subset of* X^* *which is bounded by the Lipschitz constant* $K_x > 0$ *of* φ *near* x.

(v) *The graph of the generalized gradient* $\partial\varphi$ *is closed in* $X\times$ $(w^*\text{–}X^*)$ *topology, i.e., if* $\{x_n\} \subset U$ *and* $\{\zeta_n\} \subset X^*$ *are sequences such that* $\zeta_n \in \partial\varphi(x_n)$ *and* $x_n \to x$ *in* X, $\zeta_n \to \zeta$ *weakly* in* X^*, *then* $\zeta \in \partial\varphi(x)$ *where, recall,* $(w^*\text{–}X^*)$ *denotes the space* X^* *equipped with weak* topology.*

(vi) *The multifunction* $U \ni x \mapsto \partial\varphi(x) \subseteq X^*$ *is upper semicontinuous from* U *into* $w^*\text{–}X^*$, *see Definition* 3.7.

Proof. The properties (i)–(v) can be found in Propositions 2.1.1, 2.1.2 and 2.1.5 of [48]. For the proof of (vi), we observe that from (iii), the multifunction $\partial\varphi(\cdot)$ is locally relatively compact (i.e., for every $x \in U$, there exists a neighborhood $\mathcal{N}(x)$ of x such that $\overline{\partial\varphi(\mathcal{N}(x))}$ is a weakly* compact subset of X^*). Thus, due to Proposition 3.12, since the graph of $\partial\varphi$ is closed in $X\times$ $(w^*\text{–}X^*)$ topology, we obtain the upper semicontinuity of $x \mapsto \partial\varphi(x)$. □

Relation to derivatives. In order to state the relations between the generalized directional derivative and classical notions of differentiability, we need to introduce the following definitions.

Definition 3.24 (Classical (one-sided) directional derivative). Let $\varphi\colon U \to \mathbb{R}$ be defined on a subset U of X. The *directional derivative* of φ at $x \in U$ in the direction $v \in X$ is defined by

$$\varphi'(x; v) = \lim_{\lambda \downarrow 0} \frac{\varphi(x + \lambda v) - \varphi(x)}{\lambda}, \tag{3.2}$$

whenever this limit exists.

Definition 3.25 (Regular function). A locally Lipschitz function $\varphi\colon U \to \mathbb{R}$ on a subset U of X is said to be *regular* (in the sense of Clarke) at $x \in U$, if

(i) For all $v \in X$ the directional derivative $\varphi'(x; v)$ exists.
(ii) For all $v \in X$, $\varphi'(x; v) = \varphi^0(x; v)$.

The function φ is regular (in the sense of Clarke) on U if it is regular at every point $x \in U$.

Remark 3.26. Using Definitions 3.21 and 3.24, it is easy to see that for all $x \in U$ and all $v \in X$ such that $\varphi'(x; v)$ exists, we have $\varphi'(x; v) \le \varphi^0(x; v)$.

Definition 3.27 (Gâteaux derivative). Let $\varphi: U \to \mathbb{R}$ be defined on a subset U of X. We say that φ is *Gâteaux differentiable* at $x \in U$ provided that the limit in (3.2) exists for all $v \in X$ and there exists a (necessarily unique) element $\varphi'_G(x) \in X^*$ (called the Gâteaux derivative) such that

$$\varphi'(x; v) = \langle \varphi'_G(x), v \rangle_{X^* \times X} \quad \text{for all} \quad v \in X. \tag{3.3}$$

Definition 3.28 (Fréchet derivative). Let $\varphi: U \to \mathbb{R}$ be defined on a subset U of X. We say that φ is *Fréchet differentiable* at $x \in U$ provided that (3.3) holds at the point x and, in addition, the convergence in (3.2) is uniform with respect to v in bounded subsets of X. In this case, we write $\varphi'(x)$ instead of $\varphi'_G(x)$ and we call $\varphi'(x)$ the Fréchet derivative of φ in x.

The two notions of differentiability presented above are not equivalent, even in finite-dimensional spaces, as it results from the following example.

Example 3.29. Let $\varphi: \mathbb{R}^2 \to \mathbb{R}$ be defined by

$$\varphi(x) = \begin{cases} \dfrac{x_1^3 x_2}{x_1^4 + x_2^2} & \text{if } x = (x_1, x_2) \neq (0,0), \\[2ex] 0 & \text{if } x = (0,0). \end{cases}$$

Then, it is easy to see that φ is Gâteaux differentiable at $x=(0,0)$ with $\varphi'_G(0,0) = 0$. Nevertheless, φ it is not Fréchet differentiable at $x = (0,0)$. Indeed, for $h = (h_1, h_2) \in \mathbb{R}^2$, we have

$$\frac{\varphi(h)}{\|h\|} = \frac{h_1^3 h_2}{h_1^4 + h_2^2} \frac{1}{\sqrt{h_1^2 + h_2^2}}$$

and, therefore, if we move along the curve $h_1^2 = h_2$, we have

$$\frac{\varphi(h)}{\|h\|} = \frac{1}{2\sqrt{1 + h_1^2}} \to \frac{1}{2} \quad \text{as } h_1 \to 0.$$

Note that, even if the notions of differentiability presented above are not equivalent, the following relations between Gâteaux and Fréchet derivative hold:

(i) If φ is Fréchet differentiable at $x \in U$, then φ is Gâteaux differentiable at x.
(ii) If φ is Gâteaux differentiable in a neighborhood of x and φ'_G is continuous at x, then φ is Fréchet differentiable at x and $\varphi'(x) = \varphi'_G(x)$.

Besides its local character, the notions of differentiability introduced above have also a global character. For instance, we say that a function $\varphi: U \subset X \to \mathbb{R}$ is

Fréchet differentiable in U if it is Fréchet differentiable at any point in U. Moreover, if $\varphi: U \subset X \to \mathbb{R}$ is Fréchet differentiable in U and $\varphi': U \to X^*$ is continuous, then we say that φ is *continuously differentiable* and write $\varphi \in C^1(U)$.

The following notion of strict differentiability is intermediate between Gâteaux and continuous differentiability.

Definition 3.30 (Hadamard derivative). A function $\varphi: U \to \mathbb{R}$ defined on a subset U of X is *Hadamard (strictly) differentiable* at $x \in U$, if there exists an element $D_s\varphi(x) \in X^*$ (called the Hadamard derivative) such that

$$\lim_{y \to x,\, \lambda \downarrow 0} \frac{\varphi(y + \lambda v) - \varphi(y)}{\lambda} = \langle D_s\varphi(x), v \rangle_{X^* \times X} \quad \text{for all } v \in X,$$

provided the convergence is uniform for v in compact sets.

It is well known that if φ is strictly differentiable, then the Clarke subdifferential $\partial\varphi(x)$ reduces to a singleton. Also, if $\varphi: U \to \mathbb{R}$ is locally Lipschitz on U, then Gâteaux and Hadamard differentiabilities are equivalent and when $\varphi: \mathbb{R}^d \to \mathbb{R}$, then Fréchet and Hadamard differentiabilities coincide.

The following concept of the subgradient of a convex function generalizes the classical notion of gradient.

Definition 3.31 (Convex subdifferential). Let U be an open convex subset of X and $\varphi: U \to \mathbb{R}$ be a convex function. An element $x^* \in X^*$ is called the *subgradient* of φ at $x \in U$ if the following inequality holds

$$\varphi(v) \geq \varphi(x) + \langle x^*, v - x \rangle_{X^* \times X} \quad \text{for all } v \in U. \tag{3.4}$$

The set of all $x^* \in X^*$ satisfying (3.4) is called the (convex) *subdifferential* of φ at x, and is denoted by $\partial\varphi(x)$. Sometimes we refer to $\partial\varphi$ as the subdifferential of the convex function φ or the subdifferential of φ in the sense of convex analysis.

The following two propositions collect the properties related to differentiability, subdifferentiability of locally Lipschitz and regular functions. Their proofs can be found in Chaps. 2.2 and 2.3 of [48].

Proposition 3.32. *Let $\varphi: U \to \mathbb{R}$ be defined on a subset U of X. Then*

(i) *The function φ is strictly differentiable at $x \in U$ if and only if φ is locally Lipschitz near x and $\partial\varphi(x)$ is a singleton (which is necessarily the strict derivative of φ at x). In particular, if φ is continuously differentiable at $x \in U$, then $\varphi^0(x; v) = \varphi'(x; v) = \langle \varphi'(x), v \rangle_{X^* \times X}$ for all $v \in X$ and $\partial\varphi(x) = \{\varphi'(x)\}$.*

(ii) *If φ is regular at $x \in U$ and $\varphi'(x)$ exists, then φ is strictly differentiable at x.*

(iii) *If φ is regular at $x \in U$, $\varphi'(x)$ exists and $\psi: U \to \mathbb{R}$ is locally Lipschitz near x, then $\partial(\varphi + \psi)(x) = \{\varphi'(x)\} + \partial\psi(x)$.*

(iv) *If φ is Gâteaux differentiable at $x \in U$, then $\varphi'_G(x) \in \partial\varphi(x)$.*

(v) *If U is an open convex set and φ is convex, then the Clarke subdifferential $\partial\varphi(x)$ at any $x \in U$ coincides with the subdifferential of φ at x in the sense of convex analysis.*

(vi) *If U is an open convex set and φ is convex, then the Clarke subdifferential $\partial\varphi: U \to 2^{X^*}$ is a monotone operator (see Definition 3.54* (a) *on page 81).*

Proposition 3.33. *Let $\varphi: U \to \mathbb{R}$ be defined on a subset U of X. Then, the following statements hold:*

(i) *If φ is strictly differentiable at $x \in U$, then φ is regular at x.*

(ii) *If U is an open convex set and φ is a convex function, then φ is locally Lipschitz and regular on U.*

(iii) *If φ is regular at $x \in U$ and there exists the Gâteaux derivative $\varphi'_G(x)$ of φ at x, then $\partial\varphi(x) = \{\varphi'_G(x)\}$.*

Recall that, using a corollary of the celebrated theorem of Rademacher (see, e.g., Theorem 5.6.16 in [66] or Corollary 4.19 in [49]), it can be proved that if a function $\varphi: U \to \mathbb{R}$ is locally Lipschitz on an open set $U \subset \mathbb{R}^d$, then φ is Fréchet differentiable almost everywhere on U. Using this result we have the following useful characterization of the Clarke subdifferential, which says that, in the case when X is a finite-dimensional space, the generalized gradient is "blind to sets of measure zero."

Proposition 3.34 (Generalized gradient formula). *Let $\varphi: U \subset \mathbb{R}^d \to \mathbb{R}$ be a locally Lipschitz function near $x \in U$, N be any set of Lebesgue measure zero in $\mathbb{R}^d$ and N_φ be the set of Lebesgue measure zero outside of which φ is Fréchet differentiable. Then*

$$\partial\varphi(x) = \mathrm{conv}\,\{\lim \varphi'(x_i) \mid x_i \to x,\ x_i \notin (N \cup N_\varphi)\},$$

where φ' denotes the Fréchet derivative of φ and conv *is the convex hull.*

Basic calculus. In the remaining part of this section, following Sect. 2.3 of [48], we recall the basic calculus rules for the generalized directional derivative and the Clarke subdifferential which are needed in the sequel.

Proposition 3.35. *Let $\varphi, \varphi_1, \varphi_2: U \to \mathbb{R}$ be locally Lipschitz functions on a subset U of X. Then*

(i) (Scalar multiples) *the equality $\partial(\lambda\varphi)(x) = \lambda\partial\varphi(x)$ holds, for all $\lambda \in \mathbb{R}$ and all $x \in U$.*

(ii) (Sum rules) *the inclusion*

$$\partial(\varphi_1 + \varphi_2)(x) \subseteq \partial\varphi_1(x) + \partial\varphi_2(x) \tag{3.5}$$

holds for all $x \in U$ *or, equivalently,*

$$(\varphi_1 + \varphi_2)^0(x; v) \le \varphi_1^0(x; v) + \varphi_2^0(x; v) \tag{3.6}$$

for all $x \in U$, $v \in X$.

(iii) *If one of* φ_1, φ_2 *is strictly differentiable at* $x \in U$, *then* (3.5) *and* (3.6) *hold with equalities.*

(iv) *If* φ_1, φ_2 *are regular at* $x \in U$, *then* $\varphi_1 + \varphi_2$ *is regular at* $x \in U$ *and we also have equalities in* (3.5) *and* (3.6).

Note that the extension of (3.5) and (3.6) to finite nonnegative linear combinations is immediate.

We turn now to a mean value property of locally Lipschitz functions. To this end, for the convenience of the reader, we recall that the line segment $[x, y]$ used below is defined by $[x, y] = \{z = \lambda x + (1 - \lambda)y \mid \lambda \in [0, 1]\}$.

Proposition 3.36 (Lebourg mean value theorem). *Let* $x, y \in X$ *and* $\varphi: U \subseteq X \to \mathbb{R}$ *be a locally Lipschitz function defined on an open subset* U *containing the line segment* $[x, y]$. *Then there exist* $z \in [x, y]$ *and* $\zeta \in \partial\varphi(z)$ *such that* $\varphi(y) - \varphi(x) = \langle \zeta, y - x \rangle_{X^* \times X}$.

The proof of the following result can be found in Lemma 4.2 of [184] and follows from the chain rule for the generalized gradient.

Proposition 3.37. *Let* X *and* Y *be Banach spaces,* $\psi: Y \to \mathbb{R}$ *be locally Lipschitz and* $T: X \to Y$ *be given by* $Tx = Ax + y$ *for* $x \in X$, *where* $A \in \mathcal{L}(X, Y)$ *and* $y \in Y$ *is fixed. Then the function* $\varphi: X \to \mathbb{R}$ *defined by* $\varphi(x) = \psi(Tx)$ *is locally Lipschitz and*

(i) $\varphi^0(x; v) \le \psi^0(Tx; Av)$ for all $x, v \in X$.

(ii) $\partial\varphi(x) \subseteq A^* \partial\psi(Tx)$ for all $x \in X$.

where $A^* \in \mathcal{L}(Y^*, X^*)$ *denotes the adjoint operator to* A. *Moreover, if* ψ *(or* $-\psi$*) is regular, then* φ *(or* $-\varphi$*) is regular and in* (i) *and* (ii) *we have equalities. These equalities are also true if instead of the regularity condition, we assume that* A *is surjective.*

The following result concerns the partial generalized gradients under the regularity hypothesis.

Proposition 3.38. *Let* X_1 *and* X_2 *be Banach spaces. If* $\varphi: X_1 \times X_2 \to \mathbb{R}$ *is locally Lipschitz and either* φ *or* $-\varphi$ *is regular at* $(x_1, x_2) \in X_1 \times X_2$, *then*

$$\partial\varphi(x_1, x_2) \subseteq \partial_1\varphi(x_1, x_2) \times \partial_2\varphi(x_1, x_2), \tag{3.7}$$

where $\partial_1\varphi(x_1, x_2)$ *(respectively,* $\partial_2\varphi(x_1, x_2)$*) represents the partial generalized subdifferential of* $\varphi(\cdot, x_2)$ *(respectively,* $\varphi(x_1, \cdot)$*). Equivalently,*

$$\varphi^0(x_1, x_2; v_1, v_2) \leq \varphi_1^0(x_1, x_2; v_1) + \varphi_2^0(x_1, x_2; v_2) \text{ for all } (v_1, v_2) \in X_1 \times X_2,$$

where $\varphi_1^0(x_1, x_2; v_1)$ *(respectively,* $\varphi_2^0(x_1, x_2; v_2)$*) denotes the partial generalized directional derivative of* $\varphi(\cdot, x_2)$ *(respectively,* $\varphi(x_1, \cdot)$*) at the point* x_1 *(respectively,* x_2*) in the direction* v_1 *(respectively,* v_2*).*

A proof of Proposition 3.38 can be found in [48] or [66]. Note that without the regularity hypothesis, in general there is no relation between the two sets in (3.7), as it results from the Example 2.5.2 in [48].

Next, we state and prove an additional result concerning the subgradient of functions defined on the product of two Banach spaces.

Lemma 3.39. *Let* X_1 *and* X_2 *be Banach spaces,* $(x_1, x_2) \in X_1 \times X_2$, $g: X_1 \to \mathbb{R}$ *be locally Lipschitz near* x_1 *and* $h: X_2 \to \mathbb{R}$ *be locally Lipschitz near* x_2*. Then*

(1) $\varphi: X_1 \times X_2 \to \mathbb{R}$ *given by* $\varphi(y_1, y_2) = g(y_1)$ *for all* $(y_1, y_2) \in X_1 \times X_2$ *is a locally Lipschitz function near* (x_1, x_2) *and*

   (i) $\varphi^0(x_1, x_2; v_1, v_2) = g^0(x_1; v_1)$ for all $(v_1, v_2) \in X_1 \times X_2$.
   (ii) $\partial\varphi(x_1, x_2) = \partial g(x_1) \times \{0\}$.

   Moreover, if g *(respectively,* $-g$*) is regular at* x_1*, then* φ *(respectively,* $-\varphi$*) is regular at* (x_1, x_2)*.*

(2) $\psi: X_1 \times X_2 \to \mathbb{R}$ *given by* $\psi(y_1, y_2) = h(y_2)$ *for all* $(y_1, y_2) \in X_1 \times X_2$ *is a locally Lipschitz function near* (x_1, x_2) *and*

   (i) $\psi^0(x_1, x_2; v_1, v_2) = h^0(x_2; v_2)$ for all $(v_1, v_2) \in X_1 \times X_2$.
   (ii) $\partial\psi(x_1, x_2) = \{0\} \times \partial h(x_2)$.

   Moreover, if h *(respectively,* $-h$*) is regular at* x_2*, then* ψ *(respectively,* $-\psi$*) is regular at* (x_1, x_2)*.*

(3) *Let* $\theta: X_1 \times X_2 \to \mathbb{R}$ *be given by* $\theta(y_1, y_2) = g(y_1) + h(y_2)$ *for all* $(y_1, y_2) \in X_1 \times X_2$*. Then* θ *is locally Lipschitz near* (x_1, x_2)*. Moreover, if* g *(respectively,* $-g$*) is regular at* x_1 *and* h *(respectively,* $-h$*) is regular at* x_2 *then* θ *(respectively,* $-\theta$*) is regular at* (x_1, x_2) *and*

   (i) $\theta^0(x_1, x_2; v_1, v_2) = g^0(x_1; v_1) + h^0(x_2; v_2)$ for all $(v_1, v_2) \in X_1 \times X_2$.
   (ii) $\partial\theta(x_1, x_2) = \partial g(x_1) \times \partial h(x_2)$.

Proof. We prove (1) since the proof of (2) is analogous. It is clear that φ is locally Lipschitz near (x_1, x_2). The relation (i) follows from the direct calculation

$$\begin{aligned}\varphi^0(x_1, x_2; v_1, v_2) &= \limsup_{(y_1,y_2)\to(x_1,x_2),\lambda\downarrow 0} \frac{\varphi((y_1, y_2) + \lambda(v_1, v_2)) - \varphi(y_1, y_2)}{\lambda} \\ &= \limsup_{(y_1,y_2)\to(x_1,x_2),\lambda\downarrow 0} \frac{g(y_1 + \lambda v_1) - g(y_1)}{\lambda}\end{aligned}$$

$$= \limsup_{y_1 \to x_1, \lambda \downarrow 0} \frac{g(y_1 + \lambda v_1) - g(y_1)}{\lambda}$$

$$= g^0(x_1; v_1)$$

for all $(v_1, v_2) \in X_1 \times X_2$, which shows the condition (i).

For the proof of (ii), let $(x_1^*, x_2^*) \in \partial\varphi(x_1, x_2)$. By the definition of the subdifferential, we have

$$\langle x_1^*, v_1 \rangle_{X_1^* \times X_1} + \langle x_2^*, v_2 \rangle_{X_2^* \times X_2} \le \varphi^0(x_1, x_2; v_1, v_2)$$

for every $(v_1, v_2) \in X_1 \times X_2$. Choosing $(v_1, v_2) = (v_1, 0)$, we obtain

$$\langle x_1^*, v_1 \rangle_{X_1^* \times X_1} \le \varphi^0(x_1, x_2; v_1, 0) = g^0(x_1; v_1)$$

for every $v_1 \in X_1$ which means that $x_1^* \in \partial g(x_1)$. Taking $(v_1, v_2) = (0, v_2)$, we get $\langle x_2^*, v_2 \rangle_{X_2^* \times X_2} \le g^0(x_1; 0) = 0$ for $v_2 \in X_2$. Since $v_2 \in X_2$ is arbitrary, we have $\langle x_2^*, v_2 \rangle_{X_2^* \times X_2} = 0$ and then $x_2^* = 0$.

Conversely, let $(x_1^*, x_2^*) \in \partial g(x_1) \times \{0\}$. For all $(v_1, v_2) \in X_1 \times X_2$, we have

$$\begin{aligned} \langle x_1^*, v_1 \rangle_{X_1^* \times X_1} + \langle x_2^*, v_2 \rangle_{X_2^* \times X_2} &= \langle x_1^*, v_1 \rangle_{X_1^* \times X_1} \\ &\le g^0(x_1; v_1) \\ &= \varphi^0(x_1, x_2; v_1, v_2) \end{aligned}$$

which implies that $(x_1^*, x_2^*) \in \partial\varphi(x_1, x_2)$.

Now, in addition, we suppose that g is regular at x_1. Then from (i), we have

$$\begin{aligned} \varphi^0(x_1, x_2; v_1, v_2) = g^0(x_1; v_1) = g'(x_1; v_1) &= \lim_{\lambda \downarrow 0} \frac{g(x_1 + \lambda v_1) - g(x_1)}{\lambda} \\ &= \lim_{\lambda \downarrow 0} \frac{\varphi((x_1, x_2) + \lambda(v_1, v_2)) - \varphi(x_1, x_2)}{\lambda} \\ &= \varphi'(x_1, x_2; v_1, v_2) \end{aligned}$$

for all $(v_1, v_2) \in X_1 \times X_2$, which shows that φ is regular at (x_1, x_2).

We turn to the proof of (3). It is obvious that θ is locally Lipschitz near (x_1, x_2) and $\theta(y_1, y_2) = \varphi(y_1, y_2) + \psi(y_1, y_2)$ for all $(y_1, y_2) \in X_1 \times X_2$, where φ and ψ are defined in (1) and (2), respectively. In view of (1) and (2), we know that φ and ψ (respectively, $-\varphi$ and $-\psi$) are regular at $(x_1, x_2) \in X_1 \times X_2$. Hence θ (respectively, $-\theta$) is regular at (x_1, x_2) as the sum of regular functions (see Proposition 3.35(iv)). Furthermore, for all $(v_1, v_2) \in X_1 \times X_2$, we have

$$
\begin{aligned}
\theta^0(x_1, x_2; v_1, v_2) = \theta'(x_1, x_2; v_1, v_2) &= \lim_{\lambda \downarrow 0} \frac{\theta((x_1, x_2) + \lambda(v_1, v_2)) - \theta(x_1, x_2)}{\lambda} \\
&= \lim_{\lambda \downarrow 0} \frac{g(x_1 + \lambda v_1) - g(x_1)}{\lambda} + \lim_{\lambda \downarrow 0} \frac{h(x_2 + \lambda v_2) - h(x_2)}{\lambda} \\
&= g'(x_1; v_1) + h'(x_2; v_2) \\
&= g^0(x_1; v_1) + h^0(x_2; v_2)
\end{aligned}
$$

which implies (3)(i). To show (3)(ii), we use the property (3)(i) and proceed as in the proof of (1)(ii). The proof of the lemma is complete. □

In what follows we turn to some additional properties of locally Lipschitz functions which are regular in the sense of Clarke. We consider the classes of nonconvex functions which are the pointwise maxima, minima, or differences of convex functions. The next result corresponds to Proposition 2.3.12 of [48] and Proposition 5.6.29 of [66].

Proposition 3.40. *Let $\varphi_1, \varphi_2 \colon U \to \mathbb{R}$ be locally Lipschitz functions on U, U be a subset of X and $\varphi = \max\{\varphi_1, \varphi_2\}$. Then φ is locally Lipschitz on U and*

$$\partial\varphi(x) \subseteq \operatorname{conv}\{\partial\varphi_k(x) \mid k \in I(x)\} \text{ for all } x \in U, \tag{3.8}$$

where $I(x) = \{k \in \{1, 2\} \mid \varphi(x) = \varphi_k(x)\}$ is the active index set at x. If in addition, φ_1 and φ_2 are regular at x, then φ is regular at x and (3.8) *holds with equality.*

Corollary 3.41. *Let $\varphi_1, \varphi_2 \colon U \to \mathbb{R}$ be strictly differentiable functions at $x \in U$, U be a subset of X and $\varphi = \min\{\varphi_1, \varphi_2\}$. Then $-\varphi$ is locally Lipschitz on U, regular at x, and $\partial\varphi(x) = \operatorname{conv}\{\partial\varphi_k(x) \mid k \in I(x)\}$, where $I(x)$ is the active index set at x defined in Proposition* 3.40.

Proof. Since φ_1 and φ_2 are strictly differentiable at $x \in U$, the functions $-\varphi_1$ and $-\varphi_2$ also have the same property. From Propositions 3.32(i) and 3.33(i), it follows that $-\varphi_1$ and $-\varphi_2$ are locally Lipschitz near x and regular at x. Let $f_1 = -\varphi_1$, $f_2 = -\varphi_2$ and $f = \max\{f_1, f_2\}$. It follows from Proposition 3.40 that f is locally Lipschitz near x, regular at x, and $\partial f(x) = \operatorname{conv}\{\partial f_k(x) \mid k \in I(x)\}$. On the other hand, we have

$$f = \max\{f_1, f_2\} = \max\{-\varphi_1, -\varphi_2\} = -\min\{\varphi_1, \varphi_2\} = -\varphi$$

and

$$
\begin{aligned}
-\partial\varphi(x) = \partial(-\varphi)(x) = \partial f(x) &= \operatorname{conv}\{\partial(-\varphi_k)(x) \mid k \in I(x)\} \\
&= \operatorname{conv}\{-\partial\varphi_k(x) \mid k \in I(x)\} = -\operatorname{conv}\{\partial\varphi_k(x) \mid k \in I(x)\},
\end{aligned}
$$

which concludes the proof. □

The next proposition represents a generalization of Lemma 14 of [178].

Proposition 3.42. *Let* $\varphi_1, \varphi_2 \colon U \to \mathbb{R}$ *be convex functions,* U *be an open convex subset of* X, $\varphi = \varphi_1 - \varphi_2$ *and* $x \in U$. *Assume that* $\partial\varphi_1(x)$ *is a singleton (or* $\partial\varphi_2(x)$ *is a singleton). Then* $-\varphi$ *is regular at* x *(or* φ *is regular at* x*) and*

$$\partial\varphi(x) = \partial\varphi_1(x) - \partial\varphi_2(x), \tag{3.9}$$

where $\partial\varphi_k$ *is the subdifferential in the sense of convex analysis of the function* φ_k, $k = 1, 2$.

Proof. From Proposition 3.33(ii) we know that φ_k, $k = 1, 2$ are locally Lipschitz and regular on U. Suppose that $\partial\varphi_1(x)$ is a singleton. By Proposition 3.32(i), the function φ_1 is strictly differentiable at x. Thus $-\varphi_1$ is also strictly differentiable at x and, by Proposition 3.33(i), it follows that $-\varphi_1$ is regular at x. Hence $-\varphi = -\varphi_1 + \varphi_2$ is regular at x as the sum of two regular functions. Moreover, from Proposition 3.35(i) and (iii), we have

$$\begin{aligned} -\partial\varphi(x) = \partial(-\varphi)(x) &= \partial(-\varphi_1 + \varphi_2)(x) \\ &= \partial(-\varphi_1)(x) + \partial\varphi_2(x) \\ &= -\partial\varphi_1(x) + \partial\varphi_2(x) \end{aligned}$$

which implies (3.9).

If $\partial\varphi_2(x)$ is a singleton then, as above, by using Proposition 3.33(i), we deduce φ_2 is strictly differentiable at x which in turn implies that $-\varphi_2$ is strictly differentiable and regular at x. So $\varphi = \varphi_1 + (-\varphi_2)$ is regular at x, being the sum of two regular functions. Moreover, by Proposition 3.35(i) and (iii), we obtain

$$\partial\varphi(x) = \partial(\varphi_1 + (-\varphi_2))(x) = \partial\varphi_1(x) + \partial(-\varphi_2)(x) = \partial\varphi_1(x) - \partial\varphi_2(x)$$

which gives the equality (3.9). In view of convexity of φ_k, $k = 1, 2$, note that the Clarke subdifferentials of these functions coincide with the subdifferentials in the sense of convex analysis, which completes the proof. □

The next result provides a continuity criterium for a function defined on the product of two Banach spaces.

Lemma 3.43. *Let* X *and* Y *be Banach spaces and* $\varphi \colon X \times Y \to \mathbb{R}$ *be such that*

(i) $\varphi(\cdot, y)$ *is continuous on* X *for all* $y \in Y$.
(ii) $\varphi(x, \cdot)$ *is locally Lipschitz on* Y *for all* $x \in X$.
(iii) *There is a constant* $c_0 > 0$ *such that for all* $\zeta \in \partial\varphi(x, y)$ *we have*

$$\|\zeta\|_{Y^*} \le c_0 \,(1 + \|x\|_X + \|y\|_Y) \quad \text{for all } x \in X,\ y \in Y,$$

where $\partial\varphi$ *denotes the generalized gradient of* $\varphi(x, \cdot)$.

Then φ is continuous on $X \times Y$.

Proof. Let $x \in X$ and $y_1, y_2 \in Y$. By the Lebourg mean value theorem (see Proposition 3.36), we can find y^* in the interval $[y_1, y_2]$ and $\zeta \in \partial\varphi(x, y^*)$ such that $\varphi(x, y_1) - \varphi(x, y_2) = \langle \zeta, y_1 - y_2 \rangle_{Y^* \times Y}$. Hence

$$\begin{aligned} |\varphi(x, y_1) - \varphi(x, y_2)| &\le \|\zeta\|_{Y^*} \|y_1 - y_2\|_Y \\ &\le c_0 \left(1 + \|x\|_X + \|y^*\|_Y\right) \|y_1 - y_2\|_Y \\ &\le c_1 \left(1 + \|x\|_X + \|y_1\|_Y + \|y_2\|_Y\right) \|y_1 - y_2\|_Y \end{aligned}$$

for some $c_1 > 0$. Let $(x_0, y_0) \in X \times Y$, $\{x_n\} \subset X$ and $\{y_n\} \subset Y$ be such that $x_n \to x_0$ in X and $y_n \to y_0$ in Y. We have

$$\begin{aligned} |\varphi(x_n, y_n) - \varphi(x_0, y_0)| &\le |\varphi(x_n, y_n) - \varphi(x_n, y_0)| + |\varphi(x_n, y_0) - \varphi(x_0, y_0)| \\ &\le c_1 \left(1 + \|x_n\|_X + \|y_n\|_Y + \|y_0\|_Y\right) \|y_n - y_0\|_Y \\ &\quad + |\varphi(x_n, y_0) - \varphi(x_0, y_0)|. \end{aligned}$$

Since $\|x_n\|_X, \|y_n\|_Y \le c_2$ with a constant $c_2 > 0$ and $\varphi(\cdot, y_0)$ is continuous, we deduce that $\varphi(x_n, y_n) \to \varphi(x_0, y_0)$, which completes the proof. □

We conclude this section with a result on measurability of the multifunction of the subdifferential type.

Proposition 3.44. *Let X be a separable reflexive Banach space, $0 < T < \infty$ and $\varphi\colon (0, T) \times X \to \mathbb{R}$ be a function such that $\varphi(\cdot, x)$ is measurable on $(0, T)$ for all $x \in X$ and $\varphi(t, \cdot)$ is locally Lipschitz on X for all $t \in (0, T)$. Then the multifunction $(0, T) \times X \ni (t, x) \mapsto \partial\varphi(t, x) \subset X^*$ is measurable, where $\partial\varphi$ denotes the Clarke generalized gradient of $\varphi(t, \cdot)$.*

Proof. Let $(t, x) \in (0, T) \times X$. First, note that since X is separable, by Definition 3.21 we may express the generalized directional derivative of $\varphi(t, \cdot)$ as the upper limit of the quotient $\frac{1}{\lambda}(\varphi(t, y + \lambda v) - \varphi(t, y))$, where $\lambda \downarrow 0$ taking rational values and $y \to x$ taking values in a countable dense subset of X. Therefore,

$$\begin{aligned} \varphi^0(t, x; v) &= \limsup_{y \to x, \lambda \downarrow 0} \frac{\varphi(t, y + \lambda v) - \varphi(t, y)}{\lambda} \\ &= \inf_{r > 0} \sup_{\substack{\|y - x\|_X \le r \\ 0 < \lambda < r}} \frac{\varphi(t, y + \lambda v) - \varphi(t, y)}{\lambda} \\ &= \inf_{r > 0} \sup_{\substack{\|y - x\|_X \le r,\, 0 < \lambda < r \\ y \in D,\, \lambda \in \mathbb{Q}}} \frac{\varphi(t, y + \lambda v) - \varphi(t, y)}{\lambda} \end{aligned}$$

for all $v \in X$, where $D \subset X$ is a countable dense set. From this it follows that, for all $v \in X$, the function $(t, x) \mapsto \varphi^0(t, x; v)$ is Borel measurable as "the countable" lim sup of measurable functions of (t, x) (note that by hypotheses, the function $(t, x) \mapsto \varphi(t, x)$, being Carathéodory, is jointly measurable).

Next, let $F: (0, T) \times X \to 2^{X^*}$ be defined by $F(t, x) = \partial\varphi(t, x)$ for all $(t, x) \in (0, T) \times X$. We know from Proposition 3.23(iv) that for every $(t, x) \in (0, T) \times X$, the set $\partial\varphi(t, x)$ is nonempty, convex, and weakly* compact in X^*. From Corollary 3.6.16 of [66], it follows that if X is a reflexive Banach space, then X is separable if and only if X^* is separable. Hence X^* is a separable Banach space. Since the weak and weak* topologies on the dual space of a reflexive Banach space coincide (see e.g. page 16), the multifunction F is $\mathcal{P}_{wkc}(X^*)$-valued. Using the definition of the support function (see Definition 3.2), from Proposition 3.23(iii), we have

$$\begin{aligned}\sigma(v, F(t, x)) &= \sup\{\langle \zeta, v\rangle_{X^* \times X} \mid \zeta \in F(t, x)\}\\ &= \max\{\langle \zeta, v\rangle_{X^* \times X} \mid \zeta \in F(t, x)\}\\ &= \varphi^0(t, x; v)\end{aligned}$$

for all $v \in X$. Since the function $(t, x) \mapsto \varphi^0(t, x; v)$ is measurable for all $v \in X$, we get that the function $(t, x) \mapsto \sigma(v, F(t, x))$ is measurable for all $v \in X$, i.e., F is scalarly measurable. We use now Theorem 3.6 to see that F is measurable, which completes the proof. □

3.3 Subdifferential of Superpotentials

In this section our goal is to state and prove a result on the subdifferentiability of integral functionals defined on Bochner–Lebesgue spaces. Everywhere below we suppose that $(\mathcal{O}, \Sigma, \mu)$ is a complete finite measure space and X is a separable Banach space. Given a multifunction $F: \mathcal{O} \to \mathcal{P}(X)$ and $1 \le p \le \infty$, the set of selections of F that belong to $L^p(\mathcal{O}; X)$ is defined by

$$S_F^p = \{f \in L^p(\mathcal{O}; X) \mid f(\omega) \in F(\omega)\ \mu\text{-a.e.}\}.$$

We also recall the following definition.

Definition 3.45. A function $\psi: \mathcal{O} \times X \to \overline{\mathbb{R}} = \mathbb{R} \cup \{+\infty\}$ is said to be

(i) a *normal integrand* if it is $\Sigma \times \mathcal{B}(X)$-measurable and for μ-almost all $\omega \in \mathcal{O}$ $x \mapsto \psi(\omega, x)$ is lower semicontinuous.
(ii) a *convex normal integrand* if it is a normal integrand and for μ-almost all $\omega \in \mathcal{O}$, $x \mapsto \psi(\omega, x)$ is convex.
(iii) a *Carathéodory integrand* if it is $\Sigma \times \mathcal{B}(X)$-measurable and for μ-almost all $\omega \in \mathcal{O}$, $x \mapsto \psi(\omega, x)$ is finite and continuous.

(iv) a *proper convex normal integrand* if it is a convex normal integrand and for μ-almost all $\omega \in \Omega$, $x \mapsto \psi(\omega, x)$ is proper (i.e. $\psi(\omega, x) > -\infty$ for all $x \in X$ and there is $x_0 \in X$ such that $\psi(\omega, x_0) < +\infty$).

We consider the integral functional $\Psi: L^p(\mathcal{O}; X) \to \mathbb{R}$ given by

$$\Psi(v) = \int_{\mathcal{O}} \psi(\omega, v(\omega))\, d\mu(\omega) \text{ for all } v \in L^p(\mathcal{O}; X),$$

where $\psi: \mathcal{O} \times X \to \mathbb{R}$ is a given integrand. Under the above notation, exploiting the lines of the proof of Proposition 5.5.21 in [66], we can show the following result on the subdifferential of the convex functionals.

Proposition 3.46. *Assume that $\psi: \mathcal{O} \times X \to \mathbb{R} \cup \{+\infty\}$ is a proper convex normal integrand and there exists at least one element $v \in L^p(\mathcal{O}; X)$ with $1 \le p \le \infty$, such that $\Psi(v) < +\infty$. Then $\partial\Psi(v) = S^q_{\partial\psi(\cdot, v(\cdot))}$ with $1/p + 1/q = 1$.*

We now turn to the subdifferentiation of the integral functionals defined on the Bochner–Lebesgue spaces. To this end, we assume in what follows that E_1 and E_2 are reflexive Banach spaces and we consider a function j which satisfies the following hypothesis.

$$\left.\begin{array}{l}
j: \mathcal{O} \times (0, T) \times E_1 \times E_2 \to \mathbb{R} \text{ is such that} \\[4pt]
\text{(a) } j(\cdot, \cdot, \eta, \xi) \text{ is measurable on } \mathcal{O} \times (0, T) \text{ for all } \eta \in E_1, \\
\quad \xi \in E_2 \text{ and there exists } e \in L^2(\mathcal{O}; E_2) \text{ such that for all} \\
\quad w \in L^2(\mathcal{O}; E_1), \text{ we have } j(\cdot, \cdot, w(\cdot), e(\cdot)) \in L^1(\mathcal{O} \times (0, T)). \\[4pt]
\text{(b) } j(x, t, \cdot, \xi) \text{ is continuous on } E_1 \text{ for all } \xi \in E_2, \text{ a.e.} \\
\quad (x, t) \in \mathcal{O} \times (0, T) \text{ and } j(x, t, \eta, \cdot) \text{ is locally Lipschitz} \\
\quad \text{on } E_2 \text{ for all } \eta \in E_1, \text{ a.e. } (x, t) \in \mathcal{O} \times (0, T). \\[4pt]
\text{(c) } \|\partial j(x, t, \eta, \xi)\|_{E_2^*} \le \overline{c}_0(t) + \overline{c}_1 \|\xi\|_{E_2} + \overline{c}_2 \|\eta\|_{E_1} \text{ for all} \\
\quad \eta \in E_1, \xi \in E_2, \text{ a.e. } (x, t) \in \mathcal{O} \times (0, T) \text{ with } \overline{c}_0, \overline{c}_1, \overline{c}_2 > 0, \\
\quad \overline{c}_0 \in L^\infty(0, T). \\[4pt]
\text{(d) Either } j(x, t, \eta, \cdot) \text{ or } -j(x, t, \eta, \cdot) \text{ is regular on } E_2 \text{ for all} \\
\quad \eta \in E_1, \text{ a.e. } (x, t) \in \mathcal{O} \times (0, T). \\[4pt]
\text{(e) } j^0(x, t, \cdot, \cdot; \rho) \text{ is upper semicontinuous on } E_1 \times E_2 \text{ for all} \\
\quad \rho \in E_2, \text{ a.e. } (x, t) \in \mathcal{O} \times (0, T). \\[4pt]
\text{(f) } j^0(x, t, \eta, \xi; -\xi) \le \overline{d}_0 (1 + \|\eta\|_{E_1} + \|\xi\|_{E_2}) \text{ for all } \eta \in E_1, \\
\quad \xi \in E_2, \text{ a.e. } (x, t) \in \mathcal{O} \times (0, T) \text{ with } \overline{d}_0 \ge 0.
\end{array}\right\} \quad (3.10)$$

Next, we define the superpotential $J: (0, T) \times L^2(\mathcal{O}; E_1) \times L^2(\mathcal{O}; E_2) \to \mathbb{R}$ by equality

$$J(t, w, u) = \int_{\mathcal{O}} j(x, t, w(x), u(x))\, d\mu \tag{3.11}$$

for all $w \in L^2(\mathcal{O}; E_1)$, $u \in L^2(\mathcal{O}; E_2)$ and a.e. $t \in (0, T)$. Our main results in this section is the following.

Theorem 3.47. *Let E_1 and E_2 be separable reflexive Banach spaces, and assume that the hypotheses* (3.10)(a)–(c) *are satisfied. Then the functional J defined by* (3.11) *satisfies*

(i) *$J(t, \cdot, \cdot)$ is well defined and finite on $L^2(\mathcal{O}; E_1) \times L^2(\mathcal{O}; E_2)$ for a.e. $t \in (0, T)$.*
(ii) *$J(\cdot, w, u)$ is measurable on $(0, T)$ for all $w \in L^2(\mathcal{O}; E_1)$, $u \in L^2(\mathcal{O}; E_2)$.*
(iii) *$J(t, w, \cdot)$ is Lipschitz on bounded subsets of $L^2(\mathcal{O}; E_2)$ for all $w \in L^2(\mathcal{O}; E_1)$, a.e. $t \in (0, T)$.*
(iv) *For all $w \in L^2(\mathcal{O}; E_1)$, $u, v \in L^2(\mathcal{O}; E_2)$, a.e. $t \in (0, T)$, we have*

$$J^0(t, w, u; v) \le \int_{\mathcal{O}} j^0(x, t, w(x), u(x); v(x))\, d\mu.$$

(v) *For all $w \in L^2(\mathcal{O}; E_1)$, $u \in L^2(\mathcal{O}; E_2)$, a.e. $t \in (0, T)$, we have*

$$\partial J(t, w, u) \subseteq \int_{\mathcal{O}} \partial j(x, t, w(x), u(x))\, d\mu.$$

This inclusion is understood in the sense that for each $u^* \in \partial J(t, w, u) \subset L^2(\mathcal{O}; E_2^*)$, there exists a mapping $\mathcal{O} \ni x \mapsto \zeta(x) \in E_2^*$ such that $\zeta(x) \in \partial j(x, t, w(x), u(x))$ for a.e. $x \in \mathcal{O}$, $\langle \zeta(\cdot), z \rangle_{E_2^* \times E_2} \in L^1(\mathcal{O})$ for all $z \in E_2$, and

$$\langle u^*, v \rangle_{L^2(\mathcal{O}; E_2^*) \times L^2(\mathcal{O}; E_2)} = \int_{\mathcal{O}} \langle \zeta(x), v(x) \rangle_{E_2^* \times E_2}\, d\mu$$

for all $v \in L^2(\mathcal{O}; E_2)$.
(vi) *For all $w \in L^2(\mathcal{O}; E_1)$, $u \in L^2(\mathcal{O}; E_2)$, a.e. $t \in (0, T)$, we have*

$$\|\partial J(t, w, u)\|_{L^2(\mathcal{O}; E_2^*)} \le c_0(t) + c_1 \|u\|_{L^2(\mathcal{O}; E_2)} + c_2 \|w\|_{L^2(\mathcal{O}; E_1)}$$

with $c_0(t) = \sqrt{3\, \mathrm{meas}(\mathcal{O})}\, \overline{c}_0(t)$, $c_1 = \sqrt{3}\, \overline{c}_1$ and $c_2 = \sqrt{3}\, \overline{c}_2$.
(vii) *If, in addition,* (3.10)(d) *is satisfied, then* (iv) *and* (v) *hold with equalities.*
(viii) *If, in addition,* (3.10)(d) *is satisfied, then $J(t, w, \cdot)$ or $-J(t, w, \cdot)$ is regular for all $w \in L^2(\mathcal{O}; E_1)$, a.e. $t \in (0, T)$, respectively.*
(ix) *If, in addition,* (3.10)(d) *and* (e) *are satisfied, then the multifunction $\partial J(t, \cdot, \cdot)$: $L^2(\mathcal{O}; E_1) \times L^2(\mathcal{O}; E_2) \to 2^{L^2(\mathcal{O}; E_2^*)}$ has a closed graph in $L^2(\mathcal{O}; E_1) \times L^2(\mathcal{O}; E_2) \times (w\text{-}L^2(\mathcal{O}; E_2^*))$ topology for a.e. $t \in (0, T)$.*

(x) *If, in addition,* (3.10)(f) *holds, then*

$$J^0(t, w, u; -u) \le d_0 \left(1 + \|w\|_{L^2(\mathcal{O};E_1)} + \|u\|_{L^2(\mathcal{O};E_2)}\right)$$

for all $w \in L^2(\mathcal{O}; E_1)$, $u \in L^2(\mathcal{O}; E_2)$, *a.e.* $t \in (0, T)$ *with* $d_0 \ge 0$.

Proof. For the proof of (i), we first observe that from (3.10)(b) and (c) and Lemma 3.43, $j(x, t, \cdot, \cdot)$ is continuous on $E_1 \times E_2$ for a.e. $(x, t) \in \mathcal{O} \times (0, T)$. This, together with (3.10)(a), implies that j is a Carathéodory function. By Lemma 1.68, we infer that $j(\cdot, \cdot, \widehat{w}(\cdot, \cdot), \widehat{u}(\cdot, \cdot))$ is measurable on $\mathcal{O} \times (0, T)$ for all $\widehat{w} \in L^2(\mathcal{O} \times (0, T); E_1)$, $\widehat{u} \in L^2(\mathcal{O} \times (0, T); E_2)$. Thus, by Lemma 1.66, the function $j(\cdot, t, w(\cdot), u(\cdot))$ is measurable on $\mathcal{O}$ for all $w \in L^2(\mathcal{O}; E_1)$, $u \in L^2(\mathcal{O}; E_2)$, a.e. $t \in (0, T)$. Hence $J(t, \cdot, \cdot)$ is well defined on $L^2(\mathcal{O}; E_1) \times L^2(\mathcal{O}; E_2)$, for a.e. $t \in (0, T)$.

Next, let $w \in L^2(\mathcal{O}; E_1)$ and $t \in (0, T) \setminus N$ with $\text{meas}(N) = 0$ be fixed. Let $e \in L^2(\mathcal{O}; E_2)$ be such as in (3.10)(a). From Fubini's theorem (see Theorem 1.69 on page 22), we know that $j(\cdot, t, w(\cdot), e(\cdot)) \in L^1(\mathcal{O})$. Hence $J(t, w, e) < +\infty$. Let $u \in L^2(\mathcal{O}; E_2)$. By Proposition 3.36, for a.e. $x \in \mathcal{O}$, we can find $z(x) \in [u(x), e(x)]$ and $\zeta_t \in \partial j(x, t, w(x), z(x))$ such that

$$j(x, t, w(x), u(x)) - j(x, t, w(x), e(x)) = \langle \zeta_t(x), u(x) - e(x)\rangle_{E_2^* \times E_2}. \tag{3.12}$$

Hence,

$$\begin{aligned}
&|j(x, t, w(x), u(x)) - j(x, t, w(x), e(x))| \\
&\quad \le \|\zeta_t(x)\|_{E_2^*} \|u(x) - e(x)\|_{E_2} \\
&\quad \le (\overline{c}_0 + \overline{c}_2\|w(x)\|_{E_1} + \overline{c}_1\|z(x)\|_{E_2})\, \|u(x) - e(x)\|_{E_2} \\
&\quad \le (\overline{c}_0 + \overline{c}_2\|w(x)\|_{E_1} + \overline{c}_1\|u(x)\|_{E_2} + \overline{c}_1\|e(x)\|_{E_2})\, \|u(x) - e(x)\|_{E_2}.
\end{aligned}$$

Then, from the Hölder inequality, we have

$$\begin{aligned}
&\int_{\mathcal{O}} |j(x, t, w(x), u(x)) - j(x, t, w(x), e(x))|\, d\mu \\
&\le \int_{\mathcal{O}} (\overline{c}_0 + \overline{c}_2\|w(x)\|_{E_1} + \overline{c}_1\|z(x)\|_{E_2})\, \|u(x) - e(x)\|_{E_2}\, d\mu \\
&\le \overline{c} \left(\sqrt{\text{meas}(\mathcal{O})} + \|w\|_{L^2(\mathcal{O};E_1)} + \|u\|_{L^2(\mathcal{O};E_2)} + \|e\|_{L^2(\mathcal{O};E_2)}\right) \|u - e\|_{L^2(\mathcal{O};E_2)}
\end{aligned} \tag{3.13}$$

for some $\overline{c} > 0$. Therefore,

$$\begin{aligned} J(t,w,u) &\le |J(t,w,u) - J(t,w,e)| + |J(x,t,w,e)| \\ &\le \int_{\mathcal{O}} |j(x,t,w(x),u(x)) - j(x,t,w(x),e(x))|\, d\mu + |J(t,w,e)| \\ &\le \overline{c} \left(\sqrt{\text{meas}(\mathcal{O})} + \|w\|_{L^2(\mathcal{O};E_1)} + \|u\|_{L^2(\mathcal{O};E_2)} + \|e\|_{L^2(\mathcal{O};E_2)}\right) \\ &\quad \times \left(\|u\|_{L^2(\mathcal{O};E_2)} + \|e\|_{L^2(\mathcal{O};E_2)}\right) + |J(t,w,e)| \end{aligned}$$

which gives $J(t,w,u) < +\infty$ for all $w \in L^2(\mathcal{O}; E_1)$, $u \in L^2(\mathcal{O}; E_2)$ and a.e. $t \in (0,T)$. We conclude from above that (i) follows.

Analogously as in the proof of (i), from (3.12), we have

$$\begin{aligned} j(x,t,w(x),u(x)) &= j(x,t,w(x),e(x)) + \langle \zeta_t(x), u(x) - e(x)\rangle_{E_2^* \times E_2} \\ &\le |j(x,t,w(x),e(x))| + (\overline{c}_0 + \overline{c}_2\|w(x)\|_{E_1} + \overline{c}_1\|u(x)\|_{E_2} + \overline{c}_1\|e(x)\|_{E_2}) \\ &\quad \times (\|u(x)\|_{E_2} + \|e(x)\|_{E_2}) \end{aligned}$$

for all $w \in L^2(\mathcal{O}; E_1)$, $u \in L^2(\mathcal{O}; E_2)$ and a.e. $(x,t) \in \mathcal{O} \times (0,T)$. From (3.10)(a), it is easy to see that the function $(x,t) \mapsto j(x,t,w(x),u(x))$ is integrable on $\mathcal{O} \times (0,T)$. On the other hand, from the Fubini theorem, we infer that $J(\cdot,w,u)$ is measurable on $(0,T)$ for all $w \in L^2(\mathcal{O}; E_1)$, $u \in L^2(\mathcal{O}; E_2)$ and, therefore, (ii) is satisfied.

Next, we establish the Lipschitzness of $J(t,w,\cdot)$ on bounded sets, for all $w \in L^2(\mathcal{O}; E_1)$ and a.e. $t \in (0,T)$. Let $u_1, u_2 \in L^2(\mathcal{O}; E_2)$ be such that $\|u_1\|_{L^2(\mathcal{O};E_2)}$, $\|u_2\|_{L^2(\mathcal{O};E_2)} \le r$, where $r > 0$. Since the argument in (3.13) is still valid if we replace u by u_1 and e by u_2, we have

$$\begin{aligned} |J(t,w,u_1) - J(t,w,u_2)| \le \overline{c} \Big(\sqrt{\text{meas}(\mathcal{O})} + \|w\|_{L^2(\mathcal{O};E_1)} + \\ + \|u_1\|_{L^2(\mathcal{O};E_2)} + \|u_2\|_{L^2(\mathcal{O};E_2)}\Big)\|u_1 - u_2\|_{L^2(\mathcal{O};E_2)} \end{aligned}$$

for all $w \in L^2(\mathcal{O}; E_1)$ and a.e. $t \in (0,T)$. Therefore, $J(t,w,\cdot)$ is Lipschitz on bounded subsets of $L^2(\mathcal{O}; E_2)$ for all $w \in L^2(\mathcal{O}; E_1)$ and a.e. $t \in (0,T)$, and (iii) follows.

For the proof of the inequality in (iv), let $w \in L^2(\mathcal{O}; E_1)$, $u, v \in L^2(\mathcal{O}; E_2)$ and $t \in (0,T) \setminus N$, where $\text{meas}(N) = 0$. Since E_2 is separable, we may express the generalized directional derivative of $j(x,t,w(x),\cdot)$ as the upper limit of

$$\frac{j(x,t,w(x),y + \lambda v(x)) - j(x,t,w(x),y)}{\lambda},$$

where $\lambda \downarrow 0$ taking rational values and $y \to u(x)$ taking values in a countable dense subset of E_2, i.e.,

$$
\begin{aligned}
& j^0(x,t,w(x),u(x);v(x)) \\
&= \limsup_{y\to u(x),\lambda\downarrow 0} \frac{j(x,t,w(x),y+\lambda v(x)) - j(x,t,w(x),y)}{\lambda} \\
&= \inf_{r>0} \sup_{\substack{\|y-u(x)\|_{E_2}\le r \\ 0<\lambda<r}} \frac{j(x,t,w(x),y+\lambda v(x)) - j(x,t,w(x),y)}{\lambda} \\
&= \inf_{r>0} \sup_{\substack{\|y-u(x)\|_{E_2}\le r,\, 0<\lambda<r \\ y\in D,\, \lambda\in\mathbb{Q}}} \frac{j(x,t,w(x),y+\lambda v(x)) - j(x,t,w(x),y)}{\lambda},
\end{aligned}
$$

where $D \subset E_2$ is a countable dense set. From this it follows that the function $(t,x) \mapsto j^0(x,t,w(x),u(x);v(x))$ is measurable as "the countable" lim sup of measurable functions of (t,x). Thus, the integrand on the right hand of the inequality in (iv) is a measurable function of $x \in \mathcal{O}$.

Subsequently, we apply the Fatou lemma (Theorem 1.64 on page 21) to establish the inequality (iv). First, we note that from the Lebourg mean value theorem (Proposition 3.36 on page 61), for a.e. $(x,t) \in \mathcal{O} \times (0,T)$, we obtain

$$
\frac{j(x,t,w(x),u(x)+\lambda v(x)) - j(x,t,w(x),u(x))}{\lambda} = \langle \zeta(x,t), v(x)\rangle_{E_2^*\times E_2} \quad (3.14)
$$

with some $\zeta(x,t) \in \partial j(x,t,w(x),\overline{u}(x))$ and $\overline{u}(x) \in [u(x), u(x)+\lambda v(x)]$, $\lambda \in (0,1]$. Next, since

$$
\|\zeta(x,t)\|_{E_2^*} \le \overline{c}_0 + \overline{c}_2\|w(x)\|_{E_1} + \overline{c}_1\|\overline{u}(x)\|_{E_2}
$$

and

$$
\|\overline{u}(x)\|_{E_2} \le \|u(x)\|_{E_2} + \|u(x)\|_{E_2} + \lambda\|v(x)\|_{E_2} \le 2\|u(x)\|_{E_2} + \|v(x)\|_{E_2},
$$

we have $\|\zeta(x,t)\|_{E_2^*} \le z(x,t)$ with $z \in L^\infty(0,T;L^2(\mathcal{O}))$ and, therefore, $\langle \zeta(\cdot,t), v(\cdot)\rangle_{E_2^*\times E_2}$ is bounded from above by an $L^1(\mathcal{O})$ function for a.e. $t \in (0,T)$. This shows that it is permitted to apply the Fatou lemma after taking upper limit in (3.14) as $\lambda \downarrow 0$. We have

$$
\begin{aligned}
& J^0(t,w,u;v) \\
&= \limsup_{y\to u,\lambda\downarrow 0} \frac{J(t,w,y+\lambda v) - J(t,w,y)}{\lambda}
\end{aligned}
$$

$$= \limsup_{y \to u, \lambda \downarrow 0} \int_{\mathcal{O}} \frac{j(x,t,w(x),y(x)+\lambda v(x)) - j(x,t,w(x),y(x))}{\lambda}\, d\mu$$

$$\leq \int_{\mathcal{O}} \limsup_{y \to u, \lambda \downarrow 0} \frac{j(x,t,w(x),y(x)+\lambda v(x)) - j(x,t,w(x),y(x))}{\lambda}\, d\mu$$

$$= \int_{\mathcal{O}} j^0(x,t,w(x),u(x);v(x))\, d\mu$$

which proves (iv).

The next step is to show (v). To this end we use the inequality in (iv) and, similarly as in Sect. 2.7 of [48], we reduce the problem to the convex case, and then we apply a result for subdifferentiation of convex integral functionals. Let us fix $w \in L^2(\mathcal{O}; E_1)$, $u \in L^2(\mathcal{O}; E_2)$, and $t \in (0,T) \setminus N$, where $\text{meas}(N) = 0$. Let $u^* \in \partial J(t,w,u)$. By the definition of the generalized directional derivative and (iv), we have

$$\begin{aligned} \langle u^*, v\rangle_{L^2(\mathcal{O};E_2^*)\times L^2(\mathcal{O};E_2)} &\leq J^0(t,w,u;v) \\ &\leq \int_{\mathcal{O}} j^0(x,t,w(x),u(x);v(x))\, d\mu \qquad (3.15) \end{aligned}$$

for all $v \in L^2(\mathcal{O}; E_2)$. We introduce the functional $\Psi\colon (0,T) \times L^2(\mathcal{O}; E_2) \to \mathbb{R}$ defined by

$$\Psi(t,v) = \int_{\mathcal{O}} \psi(x,t,w(x),u(x),v(x))\, d\mu,$$

where $\psi(x,t,\eta,\xi,\rho) = j^0(x,t,\eta,\xi;\rho)$ for all $\eta \in E_1$, $\xi, \rho \in E_2$ and a.e. $x \in \mathcal{O}$. Then (3.15) implies

$$\langle u^*, v\rangle_{L^2(\mathcal{O};E_2^*)\times L^2(\mathcal{O};E_2)} \leq \Psi(t,v) - \Psi(t,0) \text{ for all } v \in L^2(\mathcal{O}; E_2)$$

which means that $u^* \in \partial\Psi(t,0)$, where $\partial\Psi$ denotes the subdifferential in the sense of convex analysis of the integral functional $\Psi(t,\cdot)$. From the properties of j^0 we know that ψ is a convex Carathéodory integrand and $\Psi(t,0) = 0$. Also, from Proposition 3.46 applied with $v = 0 \in L^2(\mathcal{O}; E_2)$, we know that

$$\partial\Psi(t,0) = S^2_{\partial\psi(\cdot,t,w(\cdot),u(\cdot),0)}.$$

Moreover, we note that

$$\begin{aligned} &\partial\psi(x,t,\eta,\xi,0) \\ &\quad = \{\zeta \in E_2^* \mid \langle \zeta, \rho\rangle_{E_2^*\times E_2} \leq \psi(x,t,\eta,\xi,\rho) - \psi(x,t,\eta,\xi,0) \text{ for all } \rho \in E_2\} \\ &\quad = \{\zeta \in E_2^* \mid \langle \zeta, \rho\rangle_{E_2^*\times E_2} \leq j^0(x,t,\eta,\xi;\rho) \text{ for all } \rho \in E_2\} = \partial j(x,t,\eta,\xi) \end{aligned}$$

for all $\eta \in E_1$, $\xi \in E_2$, and a.e. $x \in \mathcal{O}$, where the subdifferentials of ψ and j are taken with respect to their last variables. Therefore, we conclude that $u^* \in \partial\Psi(t,0) = S^2_{\partial j(\cdot,t,w(\cdot),u(\cdot))}$ which proves the inclusion in (v).

We turn now to the proof of the estimate in (vi). Let $w \in L^2(\mathcal{O}; E_1)$, $u \in L^2(\mathcal{O}; E_2)$ and $t \in (0,T) \setminus N$, where $\text{meas}(N) = 0$. Let also $u^* \in \partial J(t,w,u)$. From the inclusion in (v), there exists $\zeta \in L^2(\mathcal{O}; E_2^*)$ such that $\zeta(x) \in \partial j(x,t,w(x),u(x))$ for a.e. $x \in \mathcal{O}$ and

$$\langle u^*, v\rangle_{L^2(\mathcal{O};E_2^*)\times L^2(\mathcal{O};E_2)} = \int_{\mathcal{O}} \langle \zeta(x), v(x)\rangle_{E_2^*\times E_2}\, d\mu \tag{3.16}$$

for all $v \in L^2(\mathcal{O}; E_2)$. Also, by the hypothesis (3.10)(c), we have

$$\|\zeta(x)\|_{E_2^*} \le \overline{c}_0(t) + \overline{c}_1\, \|u(x)\|_{E_2} + \overline{c}_2\, \|w(x)\|_{E_1} \text{ for a.e. } x \in \mathcal{O}.$$

Taking the squares of both sides, integrating over $\mathcal{O}$ and applying the elementary inequality $(a+b+c)^2 \le 3\,(a^2+b^2+c^2)$ for $a, b, c \in \mathbb{R}$, we obtain

$$\|\zeta\|^2_{L^2(\mathcal{O};E_2^*)} \le 3\left(\overline{c}_0^{\,2}(t)\,\text{meas}(\mathcal{O}) + \overline{c}_1^{\,2}\, \|u\|^2_{L^2(\mathcal{O};E_2)} + \overline{c}_2^{\,2}\, \|w\|^2_{L^2(\mathcal{O};E_1)}\right). \tag{3.17}$$

The inequality in (vi) is now a consequence of (3.16) and (3.17).

Next, we turn to the proof of (vii). To this end, we prove that under (3.10)(a)–(d), we have equalities in (iv) and (v). First, we consider the case when $j(x,t,\eta,\cdot)$ is regular for all $\eta \in E_1$ and a.e. $(x,t) \in \mathcal{O}\times(0,T)$, i.e., the one-sided directional derivative $j'(x,t,\eta,\xi;\rho)$ of $j(x,t,\eta,\cdot)$ exists and $j^0(x,t,\eta,\xi;\rho) = j'(x,t,\eta,\xi;\rho)$ for a.e. $(x,t) \in \mathcal{O}\times(0,T)$ and for all $\eta \in E_1$, $\xi, \rho \in E_2$. Let $w \in L^2(\mathcal{O}; E_1)$, u, $v \in L^2(\mathcal{O}; E_2)$. Exploiting (iv) and using again the Fatou lemma, we get

$$\begin{aligned}
&J^0(t,w,u;v)\\
&= \limsup_{y\to u,\lambda\downarrow 0} \frac{J(t,w,y+\lambda v) - J(t,w,y)}{\lambda}\\
&\ge \liminf_{\lambda\downarrow 0} \frac{J(t,w,u+\lambda v) - J(t,w,u)}{\lambda}\\
&= \liminf_{\lambda\downarrow 0} \int_{\mathcal{O}} \frac{j(x,t,w(x),u(x)+\lambda v(x)) - j(x,t,w(x),u(x))}{\lambda}\, d\mu\\
&\ge \int_{\mathcal{O}} \liminf_{\lambda\downarrow 0} \frac{j(x,t,w(x),u(x)+\lambda v(x)) - j(x,t,w(x),u(x))}{\lambda}\, d\mu\\
&= \int_{\mathcal{O}} j'(x,t,w(x),u(x);v(x))\, d\mu
\end{aligned}$$

$$= \int_{\mathcal{O}} j^0(x,t,w(x),u(x);v(x))\,d\mu$$
$$\geq J^0(t,w,u;v) \tag{3.18}$$

for a.e. $t \in (0,T)$. Furthermore, using the Lebourg mean value theorem (Proposition 3.36 on page 61) and the Lebesgue-dominated convergence theorem (Theorem 1.65 on page 21), we obtain

$$\begin{aligned} &J'(t,w,u;v) \\ &= \lim_{\lambda \downarrow 0} \frac{J(t,w,u+\lambda v) - J(t,w,u)}{\lambda} \\ &= \lim_{\lambda \downarrow 0} \int_{\mathcal{O}} \frac{j(x,t,w(x),u(x)+\lambda v(x)) - j(x,t,w(x),u(x))}{\lambda}\,d\mu \\ &= \int_{\mathcal{O}} \lim_{\lambda \downarrow 0} \frac{j(x,t,w(x),u(x)+\lambda v(x)) - j(x,t,w(x),u(x))}{\lambda}\,d\mu \\ &= \int_{\mathcal{O}} j'(x,t,w(x),u(x);v(x))\,d\mu \end{aligned} \tag{3.19}$$

which implies that $J'(t,w,u;v)$ exists for a.e. $t \in (0,T)$. From (3.18) and (3.19), we deduce

$$\begin{aligned} J^0(t,w,u;v) &= \int_{\mathcal{O}} j^0(x,t,w(x),u(x);v(x))\,d\mu \\ &= \int_{\mathcal{O}} j'(x,t,w(x),u(x);v(x))\,d\mu \\ &= J'(t,w,u;v) \end{aligned}$$

for a.e. $t \in (0,T)$. We conclude from the above $J^0(t,w,u;v) = J'(t,w,u;v)$ for a.e. $t \in (0,T)$. This means that $J(t,w,\cdot)$ is regular and (iv) holds with equality.

Next, we treat the case when $-j(x,t,\eta,\cdot)$ is regular for all $\eta \in E_1$ and a.e. $(x,t) \in \mathcal{O} \times (0,T)$. We proceed similarly as above and use the property (i) of Proposition 3.23 to obtain

$$\begin{aligned} J^0(t,w,u;-v) &= (-J)^0(t,w,u;v) \\ &= \int_{\mathcal{O}} (-j)^0(x,t,w(x),u(x);v(x))\,d\mu \\ &= \int_{\mathcal{O}} j^0(x,t,w(x),u(x);-v(x))\,d\mu \end{aligned}$$

for a.e. $t \in (0, T)$. Since the latter holds for all $v \in L^2(\mathcal{O}; E_2)$, we deduce that (iv) holds with equality. Also, the repetition of the reasoning in (3.18) yields the conclusion that $-J(t, w, \cdot)$ is regular. A careful examination of our previous considerations shows that under the hypothesis (3.10)(e) we have equality in the inclusion (v). This completes the proof of (vii).

The proof of (viii) is a consequence of (vii) since we already know that $J(t, w, \cdot)$ or $-J(t, w, \cdot)$ is regular for all $w \in L^2(\mathcal{O}; E_1)$ and a.e. $t \in (0, T)$.

Next, we turn to the proof of (ix) and, therefore, we suppose that (3.10)(a)–(e) are satisfied. We prove that the multifunction

$$\partial J(t, \cdot, \cdot) \colon L^2(\mathcal{O}; E_1) \times L^2(\mathcal{O}; E_2) \to 2^{L^2(\mathcal{O}; E_2^*)}$$

has a closed graph in $L^2(\mathcal{O}; E_1) \times L^2(\mathcal{O}; E_2) \times (w\text{–}L^2(\mathcal{O}; E_2^*))$ topology for a.e. $t \in (0, T)$. Let $\{w_n\} \subset L^2(\mathcal{O}; E_1)$, $\{u_n\} \subset L^2(\mathcal{O}; E_2)$, $w_n \to w$ in $L^2(\mathcal{O}; E_1)$, $u_n \to u$ in $L^2(\mathcal{O}; E_2)$, $\{u_n^*\} \subset L^2(\mathcal{O}; E_2^*)$, $u_n^* \to u^*$ weakly in $L^2(\mathcal{O}; E_2^*)$ and $u_n^* \in \partial J(t, w_n, u_n)$. The latter is equivalent to

$$\langle u_n^*, v \rangle_{L^2(\mathcal{O}; E_2^*) \times L^2(\mathcal{O}; E_2)} \le J^0(t, w_n, u_n; v)$$

for all $v \in L^2(\mathcal{O}; E_2)$ and a.e. $t \in (0, T)$. By Theorem 2.39 we may assume, by passing to subsequences, if necessary, that $w_n(x) \to w(x)$ in E_1, $u_n(x) \to u(x)$ in E_2 for a.e. $x \in \mathcal{O}$, $\|w_n(x)\|_{E_1} \le w_0(x)$ and $\|u_n(x)\|_{E_2} \le u_0(x)$ with $w_0, u_0 \in L^2(\mathcal{O})$. Let $v \in L^2(\mathcal{O}; E_2)$. By the Fatou lemma, (3.10)(e) and the equality in (iv), we have

$$\begin{aligned} \limsup J^0(t, w_n, u_n; v) &\le \limsup \int_{\mathcal{O}} j^0(x, t, w_n(x), u_n(x); v(x))\, d\mu \\ &\le \int_{\mathcal{O}} \limsup j^0(x, t, w_n(x), u_n(x); v(x))\, d\mu \\ &\le \int_{\mathcal{O}} j^0(x, t, w(x), u(x); v(x))\, d\mu \\ &= J^0(t, w, u; v) \end{aligned}$$

for a.e. $t \in (0, T)$. Hence we deduce

$$\langle u^*, v \rangle_{L^2(\mathcal{O}; E_2^*) \times L^2(\mathcal{O}; E_2)} \le \limsup J^0(t, w_n, u_n; v) \le J^0(t, w, u; v)$$

for all $v \in L^2(\mathcal{O}; E_2)$ and a.e. $t \in (0, T)$, which proves that $u^* \in \partial J(t, w, u)$. It follows from above that the graph of $\partial J(t, \cdot, \cdot)$ is closed in the aforementioned topology for a.e. $t \in (0, T)$, which completes the proof of (ix).

Finally, assume (3.10)(a)–(c) and (f). Exploiting the inequality (iv) and the Hölder inequality, we have

$$
\begin{aligned}
J^0(t,w,u;-u) &\le \int_{\mathcal{O}} j^0(x,t,w(x),u(x);-u(x))\,d\mu \\
&\le \overline{d}_0 \int_{\mathcal{O}} (1+\|w(x)\|_{E_1}+\|u(x)\|_{E_2})\,d\mu \\
&\le d_0\left(1+\|w\|_{L^2(\mathcal{O};E_1)}+\|u\|_{L^2(\mathcal{O};E_2)}\right)
\end{aligned}
$$

for all $w \in L^2(\mathcal{O}; E_1)$, $u \in L^2(\mathcal{O}; E_2)$ and a.e. $t \in (0,T)$ with $d_0 \ge 0$. This proves the property (x) and completes the proof of the theorem. □

From Theorem 3.47(i) we deduce the following.

Corollary 3.48. *Let E be a Banach space and $j\colon \mathcal{O}\times(0,T)\times E \to \mathbb{R}$ be such that*

(a) *$j(\cdot,\cdot,\xi)$ is measurable on $\mathcal{O}\times(0,T)$ for all $\xi \in E$ and there exists $e \in L^2(\mathcal{O}; E)$ such that $j(\cdot,\cdot,e(\cdot)) \in L^1(\mathcal{O}\times(0,T))$.*
(b) *$j(x,t,\cdot)$ is locally Lipschitz on E for a.e. $(x,t) \in \mathcal{O}\times(0,T)$.*
(c) *$\|\partial j(x,t,\xi)\|_{E^*} \le \overline{c}_0(t)+\overline{c}_1\|\xi\|_E$ for all $\xi \in E$, a.e. $(x,t) \in \mathcal{O}\times(0,T)$ with $\overline{c}_0,\overline{c}_1 > 0$, $\overline{c}_0 \in L^2(0,T)$.*

Then $j(\cdot,\cdot,u(\cdot)) \in L^1(\mathcal{O}\times(0,T))$ for all $u \in L^2(\mathcal{O}; E)$.

Proof. Using the Lebourg mean value theorem (Proposition 3.36 on page 61), analogously as in (3.12), for all $e \in L^2(\mathcal{O}; E)$, we have

$$
\begin{aligned}
&\int_{\mathcal{O}\times(0,T)} |j(x,t,u(x))|\,d\mu dt \\
&\le \int_{\mathcal{O}\times(0,T)} |j(x,t,u(x))-j(x,t,e(x))|\,d\mu dt + \int_{\mathcal{O}\times(0,T)} |j(x,t,e(x))|\,d\mu dt \\
&\le \int_{\mathcal{O}\times(0,T)} |\langle \zeta(x,t), u(x)-e(x)\rangle_{E^*\times E}|\,d\mu dt + \int_{\mathcal{O}\times(0,T)} |j(x,t,e(x))|\,d\mu dt \\
&\le \int_{\mathcal{O}\times(0,T)} (\overline{c}_0(t)+\overline{c}_1\|\overline{e}(x)\|_E)(\|u(x)\|_E+\|e(x)\|_E)\,d\mu dt \\
&\quad + \int_{\mathcal{O}\times(0,T)} |j(x,t,e(x))|\,d\mu dt \\
&\le \int_{\mathcal{O}\times(0,T)} (\overline{c}_0(t)+\overline{c}_1\|u(x)\|_E+\overline{c}_1\|e(x)\|_E)(\|u(x)\|_E+\|e(x)\|_E)\,d\mu dt \\
&\quad + \int_{\mathcal{O}\times(0,T)} |j(x,t,e(x))|\,d\mu dt
\end{aligned}
$$

for some $\zeta(x,t) \in \partial j(x,t,\overline{e}(x))$ and $\overline{e}(x) \in [u(x), e(x)]$ for a.e. $(x,t) \in \mathcal{O}\times(0,T)$. Hence, by the hypotheses, it is easy to deduce that $j(\cdot,\cdot,u(\cdot)) \in L^1(\mathcal{O}\times(0,T))$ for all $u \in L^2(\mathcal{O};E)$. □

Note that Theorem 3.47 represents a generalization of a result obtained in Theorem 2.7.5 in [48]. We shall use Theorem 3.47 and Corollary 3.48 in Chap. 4 in the case when $\mathcal{O} = \partial\Omega$ and Ω is an open bounded subset of $\mathbb{R}^d$, with the corresponding $(d-1)$-dimensional Lebesgue measure.

We end this section with some additional results on the Clarke subdifferential of the indefinite integral, which will be very useful in the applications we present in Chaps. 7 and 8 of the book. First, we recall the following definition.

Definition 3.49. We say that a locally Lipschitz function $\varphi\colon X \to \mathbb{R}$ defined on a Banach space X satisfies the *relaxed monotonicity condition*, if there exists $m \geq 0$ such that

$$\langle \eta_1 - \eta_2, \xi_1 - \xi_2 \rangle_{X^*\times X} \geq -m\, \|\xi_1 - \xi_2\|_X^2 \tag{3.20}$$

for all $\xi_i \in X$, $\eta_i \in \partial\varphi(\xi_i)$, $i = 1, 2$.

The relaxed monotonicity condition will be frequently used in the rest of the book. Note that this condition holds with $m = 0$, if $\varphi\colon X \to \mathbb{R}$ is a convex function. Next, we consider the function $g\colon \mathbb{R} \to \mathbb{R}$ given by

$$g(r) = \int_0^r p(s)\,ds \quad \text{for all } r \in \mathbb{R} \tag{3.21}$$

with $p \in L^\infty_{loc}(\mathbb{R})$. For $r \in \mathbb{R}$, we set

$$\underline{p}(r) = \lim_{\delta\to 0} \operatorname*{ess\,inf}_{|\tau - r|<\delta} p(\tau), \quad \overline{p}(r) = \lim_{\delta\to 0} \operatorname*{ess\,sup}_{|\tau - r|<\delta} p(\tau).$$

The properties of the subdifferential of g are collected below.

Lemma 3.50. (i) *Let $p \in L^\infty_{loc}(\mathbb{R})$. Then g is a locally Lipschitz function and $\partial g(r) \subset [\underline{p}(r), \overline{p}(r)]$, where $[\cdot,\cdot]$ denotes the line segment.*

(ii) *Let $p \in L^\infty_{loc}(\mathbb{R})$ be such that for all $r \in \mathbb{R}$*

$$\textit{there exist } \lim_{\xi\to r^-} p(\xi) = p(r-) \in \mathbb{R} \textit{ and } \lim_{\xi\to r^+} p(\xi) = p(r+) \in \mathbb{R}.$$

Then $\partial g(r) = \widehat{p}(r)$ for all $r \in \mathbb{R}$ where $\widehat{p}\colon \mathbb{R} \to 2^{\mathbb{R}}$ is a multifunction which results from p by "filling in a gap procedure," *i.e.,*

$$\widehat{p}(r) = \begin{cases} [p(r-), p(r+)] \text{ if } & p(r-) \leq p(r+) \\ [p(r+), p(r-)] \text{ if } & p(r+) \leq p(r-) \end{cases} \quad \textit{for all } r \in \mathbb{R}.$$

(iii) *If $p\colon \mathbb{R} \to \mathbb{R}$ is a continuous function, then $\partial g(r) = p(r)$ for all $r \in \mathbb{R}$.*

Proof. For the proof of (i) and (ii) we refer to [41]. The proof of (iii) is immediate since in this case g is continuously differentiable and, from Proposition 3.32(i), we have $\partial g(r) = \{g'(r)\}$ for all $r \in \mathbb{R}$. □

The filling in a gap procedure described in Lemma 3.50 is illustrated in the following simple case.

Example 3.51. Consider the function $p \in L^\infty(\mathbb{R})$ defined by

$$p(r) = \begin{cases} 1 & \text{if } r \le -4, \\ -2r - 6 & \text{if } r \in (-4, -1), \\ 2r & \text{if } r \in [-1, 1], \\ -2r + 6 & \text{if } r \in (1, 4), \\ -1 & \text{if } r \ge 4. \end{cases}$$

Then the multivalued function $\widehat{p}$ takes the form

$$\widehat{p}(r) = \begin{cases} [1, 2] & \text{if } r = -4, \\ [-4, -2] & \text{if } r = -1, \\ [2, 4] & \text{if } r = 1, \\ [-2, -1] & \text{if } r = 4, \\ p(r) & \text{otherwise.} \end{cases}$$

Moreover, the nonconvex locally Lipschitz function $g\colon \mathbb{R} \to \mathbb{R}$ defined by (3.21) is given by

$$g(r) = \begin{cases} r + 8 & \text{if } r < -4, \\ -r^2 - 6r - 4 & \text{if } r \in [-4, -1], \\ r^2 & \text{if } r \in [-1, 1], \\ -r^2 + 6r - 4 & \text{if } r \in [1, 4], \\ -r + 8 & \text{if } r > 4. \end{cases}$$

Note that it satisfies the inequality $|\partial g(r)| \le 4(1 + |r|)$ for all $r \in \mathbb{R}$.

The following result gives a sufficient condition for the relaxed monotonicity condition.

Lemma 3.52. *Let $p \in L^\infty_{loc}(\mathbb{R})$ be such that*

$$\overline{p}(r_1) \le \underline{p}(r_2) + M(r_2 - r_1) \quad \text{for all } r_1,\ r_2 \in \mathbb{R} \text{ with } M \ge 0$$

and let g be given by (3.21). *Then g satisfies the condition* (3.20) *with constant $m = M$.*

Proof. First observe that $g^0(r;s) \le \overline{p}(r)s$ if $s > 0$ and $g^0(r;s) \le \underline{p}(r)s$ if $s < 0$. Using these facts, for $r_1 \le r_2$, we have

$$g^0(r_1; r_2 - r_1) + g^0(r_2; r_1 - r_2) \le (r_2 - r_1)\,\overline{p}(r_1) + (r_1 - r_2)\,\underline{p}(r_2)$$
$$\le (r_2 - r_1)\,(\overline{p}(r_1) - \underline{p}(r_2)) \le M\,(r_2 - r_1)^2.$$

Analogously, for $r_2 \le r_1$, we get

$$g^0(r_1; r_2 - r_1) + g^0(r_2; r_1 - r_2) \le (r_2 - r_1)\,\underline{p}(r_1) + (r_1 - r_2)\,\overline{p}(r_2)$$
$$\le (r_1 - r_2)\,(\overline{p}(r_2) - \underline{p}(r_1)) \le M\,(r_1 - r_2)^2.$$

On the other hand, we obtain

$$\begin{aligned} g^0(r_1; r_2 - r_1) + g^0(r_2; r_1 - r_2) &= \max\{\zeta\,(r_2 - r_1) \mid \zeta \in \partial g(r_1)\} \\ &\quad + \max\{-\zeta\,(r_2 - r_1) \mid \zeta \in \partial g(r_2)\} \\ &\ge \eta_1(r_2 - r_1) + (-\eta_2)(r_2 - r_1) \\ &= (\eta_1 - \eta_2)(r_2 - r_1) \end{aligned}$$

for all $r_i \in \mathbb{R}$, $\eta_i \in \partial g(r_i)$, $i = 1, 2$. Hence $(\eta_1 - \eta_2)(r_2 - r_1) \le M\,|r_1 - r_2|^2$, which concludes the proof. □

From Lemmas 3.50 and 3.52, we obtain the following.

Corollary 3.53. *Let the function $g: \mathbb{R} \to \mathbb{R}$ be given by* (3.21) *with the integrand p.*

(i) *If $p: \mathbb{R} \to \mathbb{R}$ is a continuous function, then the function g satisfies the condition* (3.20) *with constant $m \ge 0$ if and only if the function $\alpha: \mathbb{R} \to \mathbb{R}$ defined by $\alpha(r) = p(r) + mr$ for $r \in \mathbb{R}$ is nondecreasing.*
(ii) *If $p: \mathbb{R} \to \mathbb{R}$ is Lipschitz continuous, i.e., $|p(r_1) - p(r_2)| \le L|r_1 - r_2|$ for all r_1, $r_2 \in \mathbb{R}$ with $L > 0$, then g satisfies the condition* (3.20) *with constant $m = L$.*

3.4 Operators of Monotone Type

In this section we provide the basic results on operators of monotone type in Banach spaces. Most of the material presented here can be found in standard textbooks such as [67, 109, 264]. The development of a complete theory of operators of monotone type requires the consideration of set-valued operators, called also multivalued operators, as well as the study or single-valued operators.

Multivalued operators. For any multivalued operator $A\colon X \to 2^Y$ between two nonempty sets X and Y, we define

$$D(A) = \{x \in X \mid A(x) \neq \emptyset\} \quad \text{(the domain of } A\text{)},$$
$$R(A) = \cup_{x\in X} A(x) \quad \text{(the range of } A\text{)},$$
$$\mathrm{Gr}\,(A) = \{(x, y) \in X \times Y \mid y \in A(x)\} \quad \text{(the graph of } A\text{)},$$
$$A^{-1}\colon Y \to X,\ A^{-1}(y) = \{x \in X \mid y \in A(x)\} \quad \text{(the inverse of } A\text{)}.$$

If X and Y are linear spaces, A_1, $A_2\colon X \to 2^Y$ and λ, $\mu \in \mathbb{R}$, we define $\lambda A_1 + \mu A_2\colon X \to 2^Y$ by

$$(\lambda A_1 + \mu A_2)(x) = \begin{cases} \lambda A_1(x) + \mu A_2(x) & \text{if } x \in D(A_1) \cap D(A_2) \\ \emptyset & \text{otherwise.} \end{cases}$$

Hence $D(\lambda A_1 + \mu A_2) = D(A_1) \cap D(A_2)$. We also say that $A_2\colon X \to 2^Y$ is an *extension* of $A_1\colon X \to 2^Y$ if and only if $\mathrm{Gr}\,(A_1) \subseteq \mathrm{Gr}\,(A_2)$. We sometimes write Ax instead of $A(x)$.

In what follows X is a real Banach space with a norm $\|\cdot\|_X$, X^* denotes its dual and $\langle\cdot,\cdot\rangle_{X^*\times X}$ is the duality pairing between X^* and X.

Definition 3.54. The multivalued operator $A\colon X \to 2^{X^*}$ is called

(a) *monotone*, if for all (u, u^*), $(v, v^*) \in \mathrm{Gr}\,(A)$, we have

$$\langle u^* - v^*, u - v\rangle_{X^*\times X} \geq 0.$$

(b) *strictly monotone*, if for all (u, u^*), $(v, v^*) \in \mathrm{Gr}\,(A)$, $u \neq v$, we have

$$\langle u^* - v^*, u - v\rangle_{X^*\times X} > 0.$$

(c) *strongly monotone*, if there exist $c > 0$ and $p > 1$ such that for all (u, u^*), $(v, v^*) \in \mathrm{Gr}\,(A)$, we have

$$\langle u^* - v^*, u - v\rangle_{X^*\times X} \geq c\,\|u - v\|_X^p.$$

(d) *uniformly monotone*, if there exists a strictly increasing continuous function $a\colon [0, +\infty) \to [0, +\infty)$ such that $a(0) = 0$, $a(t) \to +\infty$ as $t \to +\infty$ and, for all (u, u^*), $(v, v^*) \in \mathrm{Gr}\,(A)$, we have

$$\langle u^* - v^*, u - v\rangle_{X^*\times X} \geq a(\|u - v\|_X)\|u - v\|_X.$$

(e) *maximal monotone*, if A is monotone and if $(u, u^*) \in X \times X^*$ is such that

$$\langle u^* - v^*, u - v \rangle_{X^* \times X} \geq 0 \text{ for all } (v, v^*) \in \operatorname{Gr}(A)$$

then $(u, u^*) \in \operatorname{Gr}(A)$. The latter is equivalent to saying that $\operatorname{Gr}(A)$ is not properly included in the graph of another monotone multivalued operator.

Remark 3.55. It is clear that the following implications hold: A is strongly monotone $\Longrightarrow$ A is uniformly monotone $\Longrightarrow$ A is strictly monotone $\Longrightarrow$ A is monotone.

Example 3.56. If $X = \mathbb{R}$, then a maximal monotone mapping $f : \mathbb{R} \to 2^{\mathbb{R}}$ is called *maximal monotone graph* in $\mathbb{R}^2$. In particular, every increasing continuous function $f : \mathbb{R} \to \mathbb{R}$ represents a maximal monotone graph in $\mathbb{R}^2$. More generally, if $f : \mathbb{R} \to \mathbb{R}$ is an increasing function which has one-sided limits in every point $s \in \mathbb{R}$, denoted by $f(s\pm)$, then it generates a maximal monotone graph in $\mathbb{R}^2$, defined by $f(s) = [f(s-), f(s+)]$ for all $s \in \mathbb{R}$. Note that this graph is obtained by filling in the gaps at the points of discontinuity of f.

We recall some useful extensions of the notion of monotonicity. The main results of the theory of pseudomonotone multivalued operators are developed in [35].

Definition 3.57. Let X be a reflexive Banach space and $A : X \to 2^{X^*}$ be a multivalued operator. We say that

(a) A is *pseudomonotone*, if

(1) A has values which are nonempty, bounded, closed, and convex.
(2) A is usc from each finite-dimensional subspace of X to X^* endowed with the weak topology.
(3) If $\{u_n\} \subset X$ with $u_n \to u$ weakly in X, and $u_n^* \in Au_n$ is such that

$$\limsup \langle u_n^*, u_n - u \rangle_{X^* \times X} \leq 0,$$

then for every $y \in X$, there exists $u^*(y) \in Au$ such that

$$\langle u^*(y), u - y \rangle_{X^* \times X} \leq \liminf \langle u_n^*, u_n - y \rangle_{X^* \times X}.$$

(b) A is *generalized pseudomonotone*, if for any sequences $\{u_n\} \subset X$, $\{u_n^*\} \subset X^*$ with $u_n^* \in Au_n$, $u_n \to u$ weakly in X, $u_n^* \to u^*$ weakly in X^*, and

$$\limsup \langle u_n^*, u_n - u \rangle_{X^* \times X} \leq 0,$$

we have $u^* \in Au$ and $\langle u_n^*, u_n \rangle_{X^* \times X} \to \langle u^*, u \rangle_{X^* \times X}$.

The next result shows that every pseudomonotone operator is generalized pseudomonotone, while the converse holds under an additional boundedness condition.

Proposition 3.58. *Let X be a reflexive Banach space and $A : X \to 2^{X^*}$ be an operator.*

(i) *If A is a pseudomonotone operator, then A is generalized pseudomonotone.*
(ii) *If A is a generalized pseudomonotone operator which is bounded (i.e., maps bounded sets into bounded ones) and for each $u \in X$, Au is nonempty, closed a and convex subset of X^*, then A is pseudomonotone.*

The properties of pseudomonotone operators listed in the proposition below are essential in the applications.

Proposition 3.59. *Let X be a reflexive Banach space.*

(i) *If $A: X \to 2^{X^*}$ is a maximal monotone operator with $D(A) = X$, then A is pseudomonotone.*
(ii) *If $A_1, A_2: X \to 2^{X^*}$ are pseudomonotone operators, then $A_1 + A_2$ is pseudomonotone.*

In what follows we introduce the notion of coercivity.

Definition 3.60. Let X be a Banach space and $A: X \to 2^{X^*}$ be an operator. We say that A is *coercive* if either $D(A)$ is bounded or $D(A)$ is unbounded and

$$\lim_{\|u\|_X \to \infty,\, u \in D(A)} \frac{\inf\{\langle u^*, u\rangle_{X^* \times X} \mid u^* \in Au\}}{\|u\|_X} = +\infty.$$

The following is the main surjectivity result for pseudomonotone and coercive operators.

Theorem 3.61. *Let X be a reflexive Banach space and $A: X \to 2^{X^*}$ be pseudomonotone and coercive. Then A is surjective, i.e., $R(A) = X^*$.*

Following Definition 1.3.72 and Theorem 1.3.73 of [67], we recall in what follows a useful version of the notion of pseudomonotonicity of multivalued operators together with the corresponding surjectivity result.

Definition 3.62. Let X be a reflexive Banach space, let $L: D(L) \subset X \to X^*$ be a linear maximal monotone operator, and let $A: X \to 2^{X^*}$ be an operator. We say that A is *pseudomonotone with respect to* $D(L)$ or L-*pseudomonotone*, if the following conditions hold:

(1) A has values which are nonempty, bounded, closed, and convex sets.
(2) A is usc from each finite-dimensional subspace of X to X^* endowed with the weak topology.
(3) If $\{u_n\} \subset D(L)$ with $u_n \to u$ weakly in X, $Lu_n \to Lu$ weakly in X^*, $u_n^* \in Au_n$ is such that $u_n^* \to u^*$ weakly in X^* and

$$\limsup \langle u_n^*, u_n - u\rangle_{X^* \times X} \le 0,$$

then $u^* \in Au$ and $\langle u_n^*, u_n\rangle_{X^* \times X} \to \langle u^*, u\rangle_{X^* \times X}$.

Theorem 3.63. *Let X be a reflexive Banach space which is strictly convex, let $L: D(L) \subset X \to X^*$ be a linear, densely defined, and maximal monotone operator.*

If $A: X \to 2^{X^*}$ is bounded, coercive, and pseudomonotone with respect to $D(L)$, then $L + A$ is surjective, i.e., $(L + A)(D(L)) = X^*$.

To provide an example of the operator L which satisfies the properties in this theorem we assume in what follows that (V, H, V^*) is an evolution triple, $0 < T < \infty$, $1 < p < \infty$, and $1/p+1/q = 1$. We consider the subspace of $\mathcal{V} = L^p(0, T; V)$ defined in (2.11) on page 48, i.e.,

$$\mathcal{W} = \{u \in \mathcal{V} \mid u' \in \mathcal{V}^*\},$$

where $\mathcal{V}^* = L^q(0, T; V^*)$. The distributional derivative $Lu = u'$ restricted to the subset $D(L) = \{u \in \mathcal{W} \mid u(0) = 0\}$ defines a linear operator $L: D(L) \subset \mathcal{V} \to \mathcal{V}^*$ by

$$\langle Lu, v\rangle_{\mathcal{V}\times\mathcal{V}^*} = \int_0^T \langle u'(t), v(t)\rangle_{V^*\times V}\, dt \quad \text{for all } v \in \mathcal{V}. \tag{3.22}$$

From Proposition 32.10 and Theorem 32L of [264], we deduce the following result.

Lemma 3.64. *Let (V, H, V^*) be an evolution triple of spaces, $0 < T < \infty$, $1 < p < \infty$, and $\mathcal{V} = L^p(0, T; V)$. Then the operator $L: D(L) \subset \mathcal{V} \to \mathcal{V}^*$ defined by* (3.22) *is linear, densely defined, and maximal monotone.*

Note that we use Theorem 3.63 in the study of the evolution problems presented in Chap. 5.

Single-valued operators. We turn now to the case of single-valued operators. First, we recall that a single-valued operator $A: D(A) \subset X \to X^*$ can be understood as a multivalued operator $A: X \to 2^{X^*}$ by setting $Au = \{Au\}$ if $u \in D(A)$ and $Au = \emptyset$ otherwise. Therefore, all the notions presented above in the case of multivalued operators can be easily formulated in the case of single-valued operators. For example, a single-valued operator $A: D(A) \subset X \to X^*$ is *monotone* if

$$\langle Au - Av, u - v\rangle_{X^*\times X} \geq 0 \quad \text{for all } u, v \in D(A),$$

a single-valued operator $A: D(A) \subset X \to X^*$ is *maximal monotone* if A is monotone and from the conditions $(u, u^*) \in X \times X^*$ and

$$\langle u^* - Av, u - v\rangle_{X^*\times X} \geq 0 \quad \text{for all } v \in D(A)$$

it follows that $u \in D(A)$ and $u^* = Au$, and a single-valued operator $A: D(A) \subset X \to X^*$ is *strongly monotone* if there exists $c > 0$ and $p > 1$ such that

$$\langle Au - Av, u - v\rangle_{X^*\times X} \geq c\, \|u - v\|_X^p \quad \text{for all } u, v \in D(A).$$

Below we specialize the definition of pseudomonotonicity for multivalued operators to the single-valued case.

Definition 3.65. Let X be a Banach space. The single-valued operator $A: X \to X^*$ is *pseudomonotone* if it is bounded (i.e., it maps bounded subsets of X into bounded subsets of X^*) and satisfies the inequality

$$\langle Au, u - v\rangle_{X^* \times X} \leq \liminf \langle Au_n, u_n - v\rangle_{X^* \times X} \quad \text{for all } v \in X, \tag{3.23}$$

whenever the sequence $\{u_n\}$ converges weakly in X towards u with

$$\limsup \langle Au_n, u_n - u\rangle_{X^* \times X} \leq 0. \tag{3.24}$$

In some situations we need the following characterization of pseudomonotonicity.

Proposition 3.66. *Let X be a reflexive Banach space and $A: X \to X^*$. The operator A is pseudomonotone if and only if A is bounded and satisfies the following condition*

$$\left.\begin{aligned} &\text{if } u_n \to u \text{ weakly in } X \text{ and } \limsup \langle Au_n, u_n - u\rangle_{X^* \times X} \leq 0, \\ &\text{then } Au_n \to Au \text{ weakly in } X^* \text{ and } \lim \langle Au_n, u_n - u\rangle_{X^* \times X} = 0. \end{aligned}\right\} \tag{3.25}$$

Proof. Assume the condition (3.25). Let $u_n \to u$ weakly in X be such that $\limsup \langle Au_n, u_n - u\rangle_{X^* \times X} \leq 0$. Then, for every $v \in X$, we have

$$\begin{aligned} \liminf \langle Au_n, u_n - v\rangle_{X^* \times X} &\geq \liminf \langle Au_n, u_n - u\rangle_{X^* \times X} \\ &\quad + \liminf \langle Au_n, u - v\rangle_{X^* \times X} \\ &= \langle Au, u - v\rangle_{X^* \times X}, \end{aligned}$$

which implies that A is pseudomonotone.

Conversely, assume that A is pseudomonotone and let $u_n \to u$ weakly in X such that (3.24) holds. We take $v = u$ in (3.23) and use (3.24) to obtain

$$\begin{aligned} 0 &\leq \liminf \langle Au_n, u_n - u\rangle_{X^* \times X} \\ &\leq \limsup \langle Au_n, u_n - u\rangle_{X^* \times X} \leq 0. \end{aligned}$$

Therefore, it follows that $\lim \langle Au_n, u_n - u\rangle_{X^* \times X} = 0$. Moreover, for any $v = u - \alpha w$ with $w \in X$ and $\alpha \in \mathbb{R}$, by (3.23) we have

$$\begin{aligned} \langle Au, \alpha w\rangle_{X^* \times X} &\leq \liminf \langle Au_n, u_n - u + \alpha w\rangle_{X^* \times X} \\ &= \lim \langle Au_n, u_n - u\rangle_{X^* \times X} + \liminf \langle Au_n, \alpha w\rangle_{X^* \times X} \\ &= \liminf \langle Au_n, \alpha w\rangle_{X^* \times X}. \end{aligned}$$

Taking first $\alpha > 0$ and next $\alpha < 0$, we easily deduce that

$$\lim \langle Au_n, w \rangle_{X^* \times X} = \langle Au, w \rangle_{X^* \times X}$$

for all $w \in X$ and, therefore, condition (3.25) is satisfied. □

We note that when X is finite dimensional, then A is pseudomonotone if it is continuous and vice versa (see Proposition 3.67(ii)). Moreover, it is useful to recall that when X is a Hilbert space and $A\colon X \to X^*$ is a linear and bounded operator, then the weak (strong) convergence in X of $\{u_n\}$ to u implies the weak (strong) convergence of $\{Au_n\}$ to Au.

For pseudomonotone operators on a reflexive Banach space we have the following result.

Proposition 3.67. *Let X be a reflexive Banach space and $A\colon X \to X^*$ be a pseudomonotone operator. The following properties hold:*

(i) *If $u_n \to u$ weakly in X and $Au_n \to z$ weakly in X^* with*

$$\limsup \langle Au_n, u_n \rangle_{X^* \times X} \le \langle z, u \rangle_{X^* \times X},$$

then $z = Au$.

(ii) *If $u_n \to u$ in X, then $Au_n \to Au$ weakly in X^*.*

We recall in what follows the various continuity modes for nonlinear single-valued operators as well as the relationship between them and the pseudomonotonicity.

Definition 3.68. Let X be a Banach space and $A\colon X \to X^*$. We say that

(a) A is *demicontinuous*, if for all $w \in X$ the functional $u \mapsto \langle Au, w \rangle_{X^* \times X}$ is continuous, i.e., A is continuous as a mapping from X to $(w^*\text{–}X^*)$.

(b) A is *hemicontinuous*, if for all $u, v, w \in X$ the functional

$$t \mapsto \langle A(u + tv), w \rangle_{X^* \times X}$$

is continuous on $[0, 1]$, i.e., A is directionally weakly continuous.

(c) A is *radially continuous*, if for all $u, v \in X$ the functional

$$t \mapsto \langle A(u + tv), v \rangle_{X^* \times X}$$

is continuous on $[0, 1]$, i.e., A satisfies the condition (b) only for $w = v$.

(d) A is *weakly continuous*, if for all $w \in X$ the functional $u \mapsto \langle Au, w \rangle_{X^* \times X}$ is weakly continuous, i.e., A is continuous as a mapping from $(w\text{–}X)$ to $(w^*\text{–}X^*)$.

(e) A is *totally continuous*, if A is continuous as a mapping from $(w\text{–}X)$ to X^*.

It is easy to see that if $A\colon X \to X^*$ is continuous, then it is demicontinuous which, in turn, implies that A is hemicontinuous and, therefore, it is radially continuous. If $A\colon X \to X^*$ is linear and demicontinuous, then it is continuous. It can be shown that for monotone operators $A\colon X \to X^*$ with $D(A) = X$, the notions of demicontinuity and hemicontinuity coincide (see Exercise I.9 in Sect. 1.9 of [67]). Moreover, from Proposition 27.6 of [264], we have the following result.

Theorem 3.69. *Let X be a Banach space and $A: X \to X^*$.*

(i) *If the operator A is a bounded, hemicontinuous, and monotone, then A is pseudomonotone.*
(ii) *If A is pseudomonotone, then A is demicontinuous.*

In the following we collect some additional properties of pseudomonotone operators.

Proposition 3.70. *Let X be a reflexive Banach space. Then*

(i) *A radially continuous monotone operator $A: X \to X^*$ satisfies condition (3.23) whenever the sequence $\{u_n\}$ converges weakly in X towards u and (3.24) holds. In particular, a bounded radially continuous monotone mapping is pseudomonotone.*
(ii) *The sum of two pseudomonotone operators remains pseudomonotone, i.e., if $A_1: X \to X^*$ and $A_2: X \to X^*$ are pseudomonotone, then the operator $X \ni u \mapsto (A_1 + A_2)u \in X^*$ is pseudomonotone.*
(iii) *A shift of a pseudomonotone operator remains pseudomonotone, i.e., if $A: X \to X^*$ is pseudomonotone, then $X \ni u \mapsto A(u + v) \in X^*$ is pseudomonotone for any $v \in X$.*
(iv) *A perturbation of a pseudomonotone operator by a totally continuous operator is pseudomonotone, i.e., if $A: X \to X^*$ is pseudomonotone and $B: X \to X^*$ is totally continuous then $X \ni u \mapsto (A + B)u \in X^*$ is pseudomonotone.*

Note that the monotonicity assumption plays a crucial role in the result (i). Indeed, the example which follows shows that even if the operator is linear and continuous, it is not necessarily pseudomonotone.

Example 3.71. Let X be a Hilbert space and let $A: X \to X^*$ be given by $\langle Au, v\rangle_{X^* \times X} = \langle -u, v\rangle_X$. Then A fails to be pseudomonotone.

Proof. Indeed, suppose that the condition in the definition of pseudomonotonicity is valid, i.e., for every $\{u_n\}$ which converges weakly in X towards u with $\limsup \langle Au_n, u_n - u\rangle_{X^* \times X} \le 0$, we have

$$\langle Au, u - v\rangle_{X^* \times X} \le \liminf \langle Au_n, u_n - v\rangle_{X^* \times X} \quad \text{for all } v \in X. \tag{3.26}$$

Consider $\{u_n\}_{n=1}^{\infty}$ an orthonormal set of vectors in X. Then, for every $v \in X$, by the Bessel inequality

$$\sum_{n=1}^{\infty} \langle u_n, v\rangle_X^2 \le \|v\|_X^2$$

we know that $\lim \langle u_n, v\rangle_X = 0$. Hence $u_n \to 0$ weakly in X and, by the definition of A, we have

$$\limsup \langle Au_n, u_n \rangle_{X^* \times X} = \limsup (-\|u_n\|_X^2) = -1. \tag{3.27}$$

We take now $u = v = 0$ in (3.26) to obtain

$$0 \leq \liminf \langle Au_n, u_n \rangle_{X^* \times X}. \tag{3.28}$$

Inequalities (3.27) and (3.28) lead to a contradiction which concludes the proof. □

We reformulate now Definition 3.60 in the case of single-valued operators defined on X with values in X^*.

Definition 3.72. Let X be a Banach space. An operator $A\colon X \to X^*$ is said to be *coercive* if

$$\lim_{\|u\|_X \to \infty} \frac{\langle Au, u \rangle_{X^* \times X}}{\|u\|_X} = +\infty. \tag{3.29}$$

Note that the coercivity condition (3.29) is satisfied if we assume that there exists a function $\alpha\colon \mathbb{R} \to \mathbb{R}$ such that $\lim_{t \to +\infty} \alpha(t) = +\infty$ and

$$\langle Au, u \rangle_{X^* \times X} \geq \alpha(\|u\|_X)\, \|u\|_X \quad \text{for all} \quad u \in X.$$

And, in particular, it is satisfied if there exists $\alpha > 0$ such that

$$\langle Au, u \rangle_{X^* \times X} \geq \alpha\, \|u\|_X^2 \quad \text{for all} \quad u \in X. \tag{3.30}$$

This remark justifies the following definition.

Definition 3.73. Let X be a Banach space. An operator $A\colon X \to X^*$ is said to be *coercive with constant* α if there exists $\alpha > 0$ such that (3.30) holds.

The following surjectivity result shows the importance of the class of pseudomonotone coercive single-valued operators.

Theorem 3.74. *Let X be a Banach space and $A\colon X \to X^*$ be pseudomonotone and coercive. Then A is surjective, i.e., for any $f \in X^*$, there is at least one solution to the equation $Au = f$.*

Bibliographical Notes

Most of the prerequisite material presented in Chap. 1 can be found in standard textbooks on functional analysis such as Brézis [33, 34], Deimling [59], Dunford and Schwartz [73], Hu and Papageorgiou [109], Yosida [259], and Zeidler [260], [263–266].

Details on metric and normed spaces, Banach spaces, and weak topologies presented in Sects. 1.1 and 1.2 can be found in most of the books on real analysis such as Hewitt and Stromberg [105] and Royden [226]. An excellent reference in the subject is Larsen [143], which contains the most widely used results from functional analysis. Details on the results on measure theory resumed in Sect. 1.3 can be found in Halmos [98] and Cohn [55], for instance.

Concerning Definition 1.63 in Sect. 1.3, we note that in the literature one can find an alternative definition of the integral that does not require the separation of positive and negative parts of a function f. This alternative definition introduces the integral of a function f as the limit of integrals of simple functions approximating the function f, in the context of Banach space-valued functions; based on Corollary 2.2.7 of [66], it represents the starting point in the construction of the Bochner integral provided in Sect. 2.4 of the book.

For further properties of uniformly convex, locally uniformly convex and strictly convex Banach spaces, we refer to Chapter 21 of [263]. For the proofs and generalizations of Fatou's lemma (Theorem 1.64) and Lebesgue-dominated convergence theorem (Theorem 1.65), we refer to Sect. 2.2 of [66]. The proofs of Lemma 1.66 and Fubini's theorem (Theorem 1.69) can be found, for instance, in Proposition 2.4.3 and Theorem 2.4.10 of [66]. For additional properties of Carathéodory functions, we refer to Sect. 2.5 of [66].

The standard material of Sects. 2.1–2.3 on Hölder continuous and smooth functions, Lebesgue and Sobolev spaces can be found in Adams [1], Adams and Fournier [2], Brézis [33, 34], Denkowski et al. [66, 67], Evans [80], Gasinski and Papageorgiou [83], Grisvard [92], Kufner et al. [136], Lions and Magenes [150], Maz'ja [159], Nečas [196], Tartar [245], Triebel [249], and Wloka [252]. Results related to abstract analysis, including characterization of the duals of L^p spaces and various properties of reflexive Banach spaces, can be found in the celebrated monographs on functional analysis Dunford and Schwartz [73] and Yosida [259].

In Sect. 2.2 we have restricted ourselves to the spaces defined on an open subset of $\mathbb{R}^d$ although the theory is well developed in an abstract measure space setting, see Sects. 2.2 and 3.8 of [66]. The Jensen inequality and the Riesz representation theorem for L^p spaces (i.e., Theorems 2.5 and 2.8) correspond to Theorems 2.2.51 and 3.8.2 in [66], respectively, where the proof of these results is provided. For the proof of the variational lemma (Lemma 2.9), we refer to Exercise III.26 of [66]. Details on transpose or dual operators can be found in Sect. 3.7 of [66].

We refer to [1, 12, 245, 252] for a more complete list of various embedding theorems for Sobolev spaces presented in Sect. 2.3. In particular, a proof of Theorem 2.16 can be found in [1, p. 144] and Theorem 2.33 corresponds to Theorem 3.10.16 of [66]. There have been several nice proofs of the Korn inequality in the literature, see for instance [44, 77, 197] and the references therein.

The introduction of fractional order spaces in Sect. 2.3 can be accomplished by at least three distinct techniques: the Fourier transform methods (as done in Sect. 5.8 of [80], Sect. 46.2 of [139]), the interpolation space methods (as done in Sect. 50 of [139]), and using the definition of Sobolev-Slobodeckij spaces, as done on page 33 of this book. Also, note that the Sobolev–Slobodeckij spaces introduced in Definition 2.18 form a scale of embeddings with respect to their fractional order of regularity.

More details on the spaces of vector-valued functions presented in Sect. 2.4 can be found in Barbu [22], Barbu and Precupanu [24], Brézis [32], Cazenave and Haraux [40], Diestel and Uhl [68], Dinculeanu [69], Lions and Magenes [150], Schwartz [229], and Zeidler [263]. The result in Proposition 2.50 is due to Komura [135]. For various additional compactness results for Bochner–Sobolev spaces we refer to Simon [236].

For the material on nonlinear analysis presented in Chap. 3, the books of Zeidler [260–266] represent an outstanding reference. Various results on set-valued mappings introduced in Sect. 3.1 can be found in Aubin and Cellina [14], Castaing and Valadier [39], and Kisielewicz [132]. For the analysis and the applications of multifunctions we send the reader to the books of Hu and Papageorgiou [109], Aubin and Frankowska [15], and Deimling [58], which represent well-known recognized references in the field.

For an impressive list of criteria of measurability and semicontinuity of multifunctions, we refer to Castaing and Valadier [39], Chapter 4 of Denkowski et al. [66], and Chapter 2 of Hu and Papageorgiou [109]. In particular, a stronger version of Theorem 3.5 can be found in Theorem 4.3.4 of Denkowski et al. [66]. A proof of Theorem 3.6 concerning the scalar measurability of multifunctions can be obtained by using Propositions 4.3.12 and 4.3.16 of Denkowski et al. [66], and for the proof of Proposition 3.12 we refer to Propositions 4.1.9 and 4.1.11 of Denkowski et al. [66].

Theorem 3.13 represents a particular case of a result obtained in Aubin and Cellina [14, p. 60] in the more general context of upper hemicontinuous maps.

The main idea of the proof is to use the fact that any usc map from X to Y endowed with the weak topology is upper hemicontinuous. Theorems 3.17 and 3.18 are particular cases of the Kuratowski-Ryll-Nardzewski selection theorem (see Theorem 4.3.1 of Denkowski et al. [66]) and the Yankov-von Neumann-Aumann selection theorem (see Theorem 4.3.7 of Denkowski et al. [66]).

The basic material on the subdifferential theory of locally Lipschitz functions presented in Sect. 3.2 can be found in Clarke [46–48], Clarke et al. [49], Aubin [13], and Denkowski et al [66]. The development of a different generalized differentiation theory for nonsmooth functions together with the corresponding variational analysis has been presented in two-volume monograph by Mordukhovich [190, 191]. For the proofs of Propositions 3.34 and 3.36 we send the reader to Clarke [48] and Denkowski et al. [66]. Proposition 3.37 follows from the chain rule for the generalized subdifferential, as shown in Theorem 2.3.10 of Clarke [48], Proposition 5.2.26 of Denkowski et al. [66], and Lemma 4.2 of Migórski et al. [184]. A complete treatment of the general theory of convex functions can be found for instance in Barbu and Precupanu [24], Denkowski et al. [66], Ekeland and Temam [79], Hiriart-Urruty and Lemaréchal [107], Rockafellar [222], and Zeidler [261].

Theorem 3.47 on the subdifferential of integral functionals defined on the Bochner–Lebesgue spaces, stated and proved in Sect. 3.3, is new. It represents a generalization of Theorem 2.7.5 in Clarke [48]; there, the particular case when the integrand j in (3.11) does not depend on its second and third variables was considered.

Most of the material presented in Sect. 3.4 can be found in many books and textbooks in the literature. For instance, classical references for mappings of monotone type include Brézis [32], Pascali and Sburlan [214], and Showalter [234]. The main results of the theory of pseudomonotone multivalued operators are developed by Browder and Hess in [35] and they were first applied in the study of variational and hemivariational inequalities by Naniewicz and Panagiotopoulos [195]. Proposition 3.58 corresponds to Propositions 1.3.65 and 1.3.66 of Denkowski et al. [67]. A proof of Theorem 3.61 on the surjectivity of pseudomonotone and coercive operators can be found in Denkowski et al. [67]. A thorough exposition of pseudomonotone multivalued operators is the handbook by Hu and Papageorgiou [109].

Part II
Nonlinear Inclusions and Hemivariational Inequalities

Chapter 4
Stationary Inclusions and Hemivariational Inequalities

In this chapter we study stationary operator inclusions, i.e., inclusions in which the derivatives of the unknown with respect to the time variable are not involved. We start with a basic existence result for abstract operator inclusions. Then we use it in order to prove the existence of solutions for various operator inclusions of subdifferential type. We also prove that, under additional assumptions, the solution of the corresponding inclusions is unique. Finally, we specialize our existence and uniqueness results in the study of stationary hemivariational inequalities. The theorems presented in this chapter will be applied in the study of static frictional contact problems in Chap. 7.

4.1 A Basic Existence Result

In this section we establish the existence of solutions to an abstract operator inclusion. Given a normed space X, by X^* we denote its (topological) dual and by $\|\cdot\|_X$ its norm. For the duality brackets for the pair (X, X^*) we use the notation $\langle\cdot,\cdot\rangle_{X^*\times X}$. We consider two evolution triples of spaces (V, H, V^*) and (Z, H, Z^*). We recall that this means that V and Z are separable, reflexive Banach spaces, H is a Hilbert space and $V \subset H \subset V^*$, $Z \subset H \subset Z^*$ with continuous and dense embeddings. Moreover, we suppose that $V \subset Z$ with compact embedding. Let $A: V \to V^*$, $B: V \to 2^{Z^*}$ be given operators and $f \in V^*$. The operator inclusion under consideration is as follows.

Problem 4.1. *Find $u \in V$ such that $Au + Bu \ni f$.*

We complete the statement of Problem 4.1 with the following definition.

Definition 4.2. An element $u \in V$ is a solution to Problem 4.1 if and only if there exists $\zeta \in Z^*$ such that $Au + \zeta = f$ and $\zeta \in Bu$.

In the study of Problem 4.1 we consider the following hypotheses.

S. Migórski et al., *Nonlinear Inclusions and Hemivariational Inequalities*, Advances in Mechanics and Mathematics 26, DOI 10.1007/978-1-4614-4232-5_4,

$$A: V \to V^* \text{ is pseudomonotone and coercive with constant } \alpha > 0. \tag{4.1}$$

$$\left.\begin{array}{l} B: V \to 2^{Z^*} \text{ is such that} \\ \text{(a) } \|Bv\|_{Z^*} \le b_0(1 + \|v\|_V) \text{ for all } v \in V \text{ with } b_0 > 0. \\ \text{(b) For all } v \in V,\ Bv \text{ is nonempty, convex,} \\ \quad \text{weakly compact in } Z^*. \\ \text{(c) } \langle Bv, v\rangle_{V^*\times V} \ge -b_1\|v\|_V^2 - b_2\|v\|_V - b_3 \text{ for all } v \in V \\ \quad \text{with } b_1, b_2, b_3 \ge 0. \\ \text{(d) } \mathrm{Gr}\,(B) \subset V \times Z^* \text{ is closed in } Z \times (w\text{–}Z^*) \text{ topology,} \\ \quad \text{i.e., if } \zeta_n \in Bv_n \text{ with } v_n, v \in V,\ v_n \to v \text{ in } Z \text{ and} \\ \quad \zeta_n, \zeta \in Z^*, \zeta_n \to \zeta \text{ weakly in } Z^*, \text{ then } \zeta \in Bv. \end{array}\right\} \tag{4.2}$$

Recall that, according to Definition 3.73, an operator $A: V \to V^*$ is called coercive with constant $\alpha > 0$, if $\langle Av, v\rangle_{V^*\times V} \ge \alpha\,\|v\|_V^2$ for all $v \in V$; the notion of pseudomonotonicity for a single-valued operator is given in Definition 3.65; in (4.2) and below, the notation w–Z^* stands for the space Z^* endowed with the weak topology; and, finally, sufficient conditions for pseudomonotonicity are provided in Theorem 3.69(i).

Note that since B is a multivalued operator, then for each $v \in V$, Bv represents a set in Z^* and therefore notations $\|Bv\|_{Z^*}$ and $\langle Bv, v\rangle_{V^*\times V}$ are not a priori defined. Nevertheless, we specify that the inequality (4.2)(a) is understood in the sense that $\|v^*\|_{Z^*} \le b_0(1 + \|v\|_V)$ for all $v^* \in Bv$ and, similarly, (4.2)(c) means that $\langle v^*, v\rangle_{V^*\times V} \ge -b_1\|v\|_V^2 - b_2\|v\|_V - b_3$ for all $v^* \in Bv$. For the convenience of the reader we shall use such notation for multivalued operators everywhere in the rest of the book.

Our main existence result in the study of Problem 4.1 is the following.

Theorem 4.3. *Assume that* (4.1) *and* (4.2) *hold,* $f \in V^*$ *and* $\alpha > b_1$. *Then Problem* 4.1 *has at least one solution* $u \in V$ *and, moreover,*

$$\|u\|_V \le c\,(1 + \|f\|_{V^*}) \tag{4.3}$$

with a positive constant c.

Proof. We define the multivalued map $\mathcal{F}: V \to 2^{V^*}$ by $\mathcal{F}v = (A + B)v$ for $v \in V$. We show that $\mathcal{F}$ is pseudomonotone and coercive. First, we prove the pseudomonotonicity of $\mathcal{F}$ and, to this end, we use Proposition 3.58 which states that a generalized pseudomonotone operator which is bounded and has nonempty, closed, and convex values is pseudomonotone.

From the property (4.2)(b), it is clear that $\mathcal{F}$ has nonempty, convex, and closed values in V^*. From the boundedness of A (guaranteed by (4.1) and Definition 3.65) and (4.2)(a), it follows that $\mathcal{F}$ is a bounded map, i.e., it maps bounded subsets of V into bounded subsets of V^*.

We show that $\mathcal{F}$ is a generalized pseudomonotone operator. To this end, let v_n, $v \in V$, $v_n \to v$ weakly in V, v_n^*, $v^* \in V^*$, $v_n^* \to v^*$ weakly in V^*, $v_n^* \in \mathcal{F}v_n$ and assume that

$$\limsup \langle v_n^*, v_n - v\rangle_{V^*\times V} \le 0.$$

We prove that $v^* \in \mathcal{F}v$ and

$$\langle v_n^*, v_n\rangle_{V^*\times V} \to \langle v^*, v\rangle_{V^*\times V}.$$

We have $v_n^* = Av_n + \zeta_n$ with $\zeta_n \in Bv_n$. From the compactness of the embedding $V \subset Z$ it follows that

$$v_n \to v \ \text{ in } Z. \tag{4.4}$$

By the boundedness of B, guaranteed by (4.2)(a), passing to a subsequence, if necessary, we have

$$\zeta_n \to \zeta \ \text{ weakly in } Z^* \text{ with some } \zeta \in Z^*. \tag{4.5}$$

From (4.2)(d), (4.4), and (4.5), since $\zeta_n \in Bv_n$, we infer immediately that $\zeta \in Bv$. Furthermore, from the equality

$$\langle v_n^*, v_n - v\rangle_{V^*\times V} = \langle Av_n, v_n - v\rangle_{V^*\times V} + \langle \zeta_n, v_n - v\rangle_{Z^*\times Z},$$

we obtain

$$\limsup \langle Av_n, v_n - v\rangle_{V^*\times V} = \limsup \langle v_n^*, v_n - v\rangle_{V^*\times V} \le 0.$$

Exploiting now the pseudomonotonicity of A, by Proposition 3.66 we deduce that

$$Av_n \to Av \ \text{ weakly in } V^* \tag{4.6}$$

and

$$\lim \langle Av_n, v_n - v\rangle_{V^*\times V} = 0. \tag{4.7}$$

Therefore, passing to the limit in the equation $v_n^* = Av_n + \zeta_n$, we have $v^* = Av + \zeta$ which, together with $\zeta \in Bv$, implies $v^* \in Av + Bv = \mathcal{F}v$. Next, from the convergences (4.4)–(4.7) we obtain

$$\begin{aligned}\lim \ \langle v_n^*, v_n\rangle_{V^*\times V} &= \lim \langle Av_n, v_n - v\rangle_{V^*\times V} + \lim \langle Av_n, v\rangle_{V^*\times V} + \lim \langle \zeta_n, v_n\rangle_{Z^*\times Z} \\ &= \langle Av, v\rangle_{V^*\times V} + \langle \zeta, v\rangle_{Z^*\times Z} = \langle v^*, v\rangle_{V^*\times V}\end{aligned}$$

which, according to Definition 3.57, shows that $\mathcal{F}$ is a generalized pseudomonotone operator.

Next, by the hypotheses on the operators A and B we have

$$\begin{aligned}\langle \mathcal{F}v, v\rangle_{V^*\times V} &= \langle Av, v\rangle_{V^*\times V} + \langle Bv, v\rangle_{Z^*\times Z}\\ &\geq (\alpha - b_1)\|v\|_V^2 - b_2\|v\|_V - b_3 = \beta(\|v\|_V)\,\|v\|_V\end{aligned}$$

for all $v \in V$, where $\beta(t) = (\alpha - b_1)t - b_2 - \frac{b_3}{t}$ and $\beta(t) \to +\infty$, as $t \to +\infty$. Hence

$$\lim_{\|v\|_V\to\infty} \frac{\inf\{\, \langle v^*, v\rangle_{V^*\times V} \mid v^* \in \mathcal{F}v\,\}}{\|v\|_V} = +\infty,$$

which, by Definition 3.60, means that the operator $\mathcal{F}$ is coercive.

We are now in a position to apply Theorem 3.61 to the multivalued operator $\mathcal{F}$. We deduce that $\mathcal{F}$ is surjective, which implies that Problem 4.1 has a solution $u \in V$. Moreover, from the coercivity of $\mathcal{F}$, we have

$$(\alpha - b_1)\|u\|_V^2 - b_2\|u\|_V - b_3 \leq \|f\|_{V^*}\|u\|_V$$

which implies that the estimate (4.3) holds with a positive constant c depending on b_1, b_2, b_3, and α. This completes the proof of the theorem. □

4.2 Inclusions of Subdifferential Type

In this section we use the result of Sect. 4.1 to show the existence of solutions to various general operator inclusion of the subdifferential type. Then, we complete these existence results with various uniqueness results. To this end, we introduce some additional notation.

Let (V, H, V^*) and (Z, H, Z^*) be evolution triples of spaces such that the embedding $V \subset Z$ is compact, as introduced in Sect. 4.1. We denote by $c_e > 0$ the embedding constant of V into Z. Let X be a reflexive Banach space and $M: Z \to X$ be a given linear continuous operator. We denote by $\|M\|$ the norm of the operator M in $\mathcal{L}(Z, X)$ and by $M^*: X^* \to Z^*$ the adjoint operator to M.

General existence and uniqueness results. Let $A: V \to V^*$ be an operator, $J: X \times X \to \mathbb{R}$ be a given functional and $f \in V^*$. We consider the following inclusion of subdifferential type.

Problem 4.4. *Find $u \in V$ such that*

$$Au + M^*\,\partial J(Mu, Mu) \ni f.$$

Recall that a solution of Problem 4.4 is understood in the sense of Definition 4.2. In the study of Problem 4.4 we need the following hypotheses.

$$\left.\begin{array}{l} J\colon X \times X \to \mathbb{R} \text{ is such that} \\ \text{(a) } J(w,\cdot) \text{ is locally Lipschitz on } X \text{ for all } w \in X. \\ \text{(b) } \|\partial J(w,u)\|_{X^*} \le c_0 + c_1\|u\|_X + c_2\|w\|_X \\ \quad \text{for all } w, u \in X \text{ with } c_0, c_1, c_2 \ge 0. \\ \text{(c) } \partial J\colon X \times X \to 2^{X^*} \text{ has a closed graph} \\ \quad \text{in } X \times X \times (w\text{–}X^*) \text{ topology.} \\ \text{(d) } J^0(w,u;-u) \le d_0\,(1 + \|u\|_X + \|w\|_X) \text{ for all} \\ \quad w, u \in X \text{ with } d_0 \ge 0. \end{array}\right\} \tag{4.8}$$

$$M \in \mathcal{L}(Z, X). \tag{4.9}$$

Note that in (4.8) the symbols ∂J and J^0 denote the Clarke subdifferential and the generalized directional derivative of $J(w,\cdot)$, respectively.

Under the notation of this section, we have the following existence result.

Theorem 4.5. *Assume that* (4.1) *and* (4.9) *hold,* $f \in V^*$ *and one of the following hypotheses:*

(i) (4.8) (a)–(c) *and* $\alpha > (c_1 + c_2)\, c_e^2\, \|M\|^2$.
(ii) (4.8).

is satisfied. Then Problem 4.4 *has at least one solution* $u \in V$ *for which the estimate* (4.3) *holds.*

Proof. We apply Theorem 4.3 to the operator $B\colon V \to 2^{Z^*}$ defined by

$$Bv = M^*\, \partial J(Mv, Mv) \quad \text{for } v \in V.$$

To this end, we show that under hypothesis (4.8)(a)–(c) the operator B satisfies (4.2).

First, using (4.8)(b), (4.9), and the continuity of the embedding $V \subset Z$, we have

$$\begin{aligned} \|Bv\|_{Z^*} &\le \|M^*\|\, \|\partial J(Mv, Mv)\|_{X^*} \\ &\le \|M^*\|\, (c_0 + (c_1 + c_2)\, \|M\|\, \|v\|_Z) \\ &\le \|M^*\|\, (c_0 + (c_1 + c_2)\, c_e\, \|M\|\, \|v\|_V), \end{aligned} \tag{4.10}$$

which proves (4.2)(a).

In order to establish (4.2)(b), we recall that the values of $\partial J(w,\cdot)$ are nonempty, convex, and weakly compact subsets of X^* for all $w \in X$, see Proposition 3.23(iv). Let $v \in V$. Then, it follows from above that Bv is a nonempty and convex subset in Z^*. To show that Bv is weakly compact in Z^*, we prove that it is closed in Z^*. Indeed, let $\{\zeta_n\} \subset Bv$ be such that $\zeta_n \to \zeta$ in Z^*. Since $\zeta_n \in M^*\, \partial J(Mv, Mv)$ and the latter is a closed subset of Z^*, we obtain $\zeta \in M^*\, \partial J(Mv, Mv)$ which implies that $\zeta \in Bv$. Therefore, the set Bv is closed in Z^* and convex, so it is also weakly closed in Z^*. Since Bv is a bounded set in a reflexive Banach space Z^*, we get that Bv is weakly compact in Z^*. This proves (4.2)(b).

For the proof of (4.2)(c), let $v \in V$. Using (4.10), we have

$$\begin{aligned} |\langle Bv, v\rangle_{V^*\times V}| &\le \|Bv\|_{V^*}\|v\|_V \le c_e \|Bv\|_{Z^*}\|v\|_V \\ &\le c_e \|v\|_V \left(c_0 \|M\| + (c_1 + c_2)\, c_e \|M\|^2 \|v\|_V\right) \\ &\le (c_1 + c_2)\, c_e^2 \|M\|^2 \|v\|_V^2 + c_0\, c_e \|M\| \|v\|_V. \end{aligned}$$

Hence

$$\langle Bv, v\rangle_{V^*\times V} \ge -(c_1 + c_2)\, c_e^2 \|M\|^2 \|v\|_V^2 - c_0\, c_e \|M\| \|v\|_V$$

and (4.2)(c) holds with $b_1 = (c_1 + c_2)\, c_e^2 \|M\|^2$.

For the proof of (4.2)(d), let $\zeta_n \in Bv_n$, where $v_n, v \in V$, $v_n \to v$ in Z, ζ_n, $\zeta \in Z^*$ and $\zeta_n \to \zeta$ weakly in Z^*. Then $\zeta_n = M^* z_n$ and $z_n \in \partial J(Mv_n, Mv_n)$. The continuity of the operator M implies $Mv_n \to Mv$ in X and the bound (4.8)(b) implies that, at least for a subsequence, we have $z_n \to z$ weakly in X^* with some $z \in X^*$. Using equality $\zeta_n = M^* z_n$ we easily get $\zeta = M^* z$. Exploiting (4.8)(c), from $z_n \in \partial J(Mv_n, Mv_n)$ we obtain $z \in \partial J(Mv, Mv)$ and, subsequently, $\zeta \in M^* \partial J(Mv, Mv)$, i.e., $\zeta \in Bv$. The proof of all conditions in (4.2) is now complete.

Finally, we need to verify that $\alpha > b_1$. In case (i), this condition holds since $\alpha > (c_1 + c_2)\, c_e^2 \|M\|^2 = b_1$. In case (ii), this condition is satisfied with $b_1 = 0$. Indeed, let $v \in V$ and $\zeta \in Bv$. So $\zeta = M^* z$ and $z \in \partial J(Mv, Mv)$. Using (4.8)(d) we have

$$\begin{aligned} -\langle z, Mv\rangle_{X^*\times X} &\le J^0(Mv, Mv; -Mv) \le d_0 (1 + \|Mv\|_X + \|Mv\|_X) \\ &\le d_0 (1 + 2\|M\| \|v\|_Z) \le d_0 (1 + 2\, c_e \|M\| \|v\|_V). \end{aligned}$$

Hence, it follows that

$$\begin{aligned} \langle \zeta, v\rangle_{V^*\times V} &= \langle \zeta, v\rangle_{Z^*\times Z} = \langle M^* z, v\rangle_{Z^*\times Z} \\ &= \langle z, Mv\rangle_{X^*\times X} \ge -d_0 (1 + 2\, c_e \|M\| \|v\|_V) \end{aligned}$$

which implies that (4.2)(c) holds with $b_1 = 0$ and completes the proof. □

We are now in a position to formulate a corollary of Theorem 4.5 which will be useful in the study of the hemivariational inequalities we present in Sect. 4.3. Let $A\colon V \to V^*$ be an operator, $J\colon X \to \mathbb{R}$ be a given functional, and let $f \in V^*$. We consider the following inclusion.

Problem 4.6. *Find $u \in V$ such that*

$$Au + M^* \partial J(Mu) \ni f.$$

For this problem we need the following modification of the previous hypothesis (4.8).

$$
\left.\begin{array}{l}
J\colon X \to \mathbb{R} \text{ is such that} \\
\text{(a) } J \text{ is locally Lipschitz on } X. \\
\text{(b) } \|\partial J(u)\|_{X^*} \le c_0 + c_1 \|u\|_X \text{ for all } u \in X \\
\qquad \text{with } c_0, c_1 \ge 0. \\
\text{(c) } J^0(u; -u) \le d_0 (1 + \|u\|_X) \text{ for all } u \in X \\
\qquad \text{with } d_0 \ge 0.
\end{array}\right\} \tag{4.11}
$$

Under the notation of this section, we have the following existence result.

Corollary 4.7. *Assume that* (4.1) *and* (4.9) *hold,* $f \in V^*$, *and one of the following hypotheses:*

(i) (4.11) (a), (b) *and* $\alpha > c_1 c_e^2 \|M\|^2$.
(ii) (4.11).

is satisfied. Then Problem 4.6 *has at least one solution* $u \in V$ *which satisfies the estimate* (4.3).

Proof. We apply Theorem 4.5 to a functional J which is independent of the variable $w \in X$. Since the graph of ∂J is closed in $X \times (w\text{–}X^*)$, see Proposition 3.23(v) on page 56, the condition (4.8)(c) is easily satisfied. Now, it is obvious to see that the hypotheses (4.8)(a), (b) (and (4.8), respectively) follow from (4.11)(a), (b) (and (4.11), respectively). The application of Theorem 4.5 concludes the proof. □

We complete now the existence result of Corollary 4.7 with a uniqueness result concerning Problem 4.6. This will be done under stronger hypotheses on the data that we present below.

$$
\left.\begin{array}{l}
A\colon V \to V^* \text{ is such that} \\
\text{(a) } A \text{ is pseudomonotone and coercive with} \\
\qquad \text{constant } \alpha > 0. \\
\text{(b) } A \text{ is strongly monotone, i.e.,} \\
\qquad \langle Av_1 - Av_2, v_1 - v_2\rangle_{V^* \times V} \ge m_1 \|v_1 - v_2\|_V^2 \\
\qquad \text{for all } v_1, v_2 \in V \text{ with } m_1 > 0.
\end{array}\right\} \tag{4.12}
$$

$$
\left.\begin{array}{l}
J\colon X \to \mathbb{R} \text{ is such that} \\
\qquad \langle z_1 - z_2, u_1 - u_2\rangle_{X^* \times X} \ge -m_2 \|u_1 - u_2\|_X^2 \\
\qquad \text{for all } z_i \in \partial J(u_i),\ z_i \in X^*,\ u_i \in X,\ i = 1, 2 \text{ with } m_2 \ge 0.
\end{array}\right\} \tag{4.13}
$$

$$
m_1 > m_2 c_e^2 \|M\|^2. \tag{4.14}
$$

Concerning the assumption (4.12), we note that if $A: V \to V^*$ is strongly monotone with a constant $m_1 > 0$ and $A0 = 0$, then A is coercive with constant $\alpha = m_1$. Therefore, using Theorem 3.69(i), it follows that the hypothesis (4.12) is satisfied if, for instance, A is strongly monotone, bounded, hemicontinuous, and $A0 = 0$. Moreover, we remark that the inequality condition which appears in (4.13) represents the relaxed monotonicity condition introduced in Definition 3.49. And, we recall that for convex functionals, this condition holds with $m_2 = 0$.

Theorem 4.8. *Assume that* (4.9), (4.12)–(4.14) *hold and* $f \in V^*$. *If one of the following hypotheses:*

(i) (4.11) (a), (b) *and* $\alpha > c_1 c_e^2 \|M\|^2$.
(ii) (4.11).

is satisfied, then Problem 4.6 *has a unique solution* $u \in V$ *which satisfies the estimate* (4.3).

Proof. The existence of solutions to Problem 4.6 follows from Corollary 4.7. We prove the uniqueness. Let $u_1, u_2 \in V$ be solutions to Problem 4.6. Then, there exist $z_i \in X^*$ and $z_i \in \partial J(M u_i)$ such that

$$A u_i + M^* z_i = f \quad \text{for } i = 1, 2. \tag{4.15}$$

Subtracting the above two equations, multiplying the result by $u_1 - u_2$, and using the strong monotonicity of A, we have

$$m_1 \|u_1 - u_2\|_V^2 + \langle M^* z_1 - M^* z_2, u_1 - u_2 \rangle_{V^* \times V} \le 0. \tag{4.16}$$

Next, by (4.13), we obtain

$$\begin{aligned} \langle M^* z_1 - M^* z_2, u_1 - u_2 \rangle_{V^* \times V} &= \langle z_1 - z_2, M u_1 - M u_2 \rangle_{X^* \times X} \\ &\ge -m_2 \|M u_1 - M u_2\|_X^2 \end{aligned}$$

and, therefore,

$$\langle M^* z_1 - M^* z_2, u_1 - u_2 \rangle_{V^* \times V} \ge -m_2 c_e^2 \|M\|^2 \|u_1 - u_2\|_V^2. \tag{4.17}$$

We combine (4.16) and (4.17) to obtain

$$m_1 \|u_1 - u_2\|_V^2 - m_2 c_e^2 \|M\|^2 \|u_1 - u_2\|_V^2 \le 0,$$

which, in view of (4.14), implies $u_1 = u_2$. Subsequently, from (4.15) we deduce that $z_1 = z_2$ which completes the proof of the theorem. □

Time-dependent subdifferential inclusions. In what follows we study a stationary time-dependent version of Problem 4.6. We provide a result on its unique solvability which will be needed latter in this section, in the study of abstract inclusions with Volterra integral term. To this end, we introduce the following spaces

$$\mathcal{V} = L^2(0,T;V), \quad \mathcal{Z} = L^2(0,T;Z), \quad \text{and} \quad \widehat{\mathcal{H}} = L^2(0,T;H),$$

where $0 < T < +\infty$. Since the embeddings $V \subset Z \subset H \subset Z^* \subset V^*$ are continuous, from Theorems 2.37 and 2.41(vi), it is known that the embeddings $\mathcal{V} \subset \mathcal{Z} \subset \widehat{\mathcal{H}} \subset \mathcal{Z}^* \subset \mathcal{V}^*$ are also continuous, where $\mathcal{Z}^* = L^2(0,T;Z^*)$ and $\mathcal{V}^* = L^2(0,T;V^*)$.

Let $A\colon (0,T) \times V \to V^*$, $J\colon (0,T) \times X \to \mathbb{R}$ and $f\colon (0,T) \to V^*$ be given. Then, we consider the following time-dependent inclusion, in which the time variable plays the role of a parameter.

Problem 4.9. *Find $u \in \mathcal{V}$ such that*

$$A(t,u(t)) + M^* \partial J(t, Mu(t)) \ni f(t) \ \text{ a.e. } t \in (0,T).$$

In the study of Problem 4.9, we specify Definition 4.2 of the solution to include the time-dependent case.

Definition 4.10. A function $u \in \mathcal{V}$ is called a solution to Problem 4.9 if and only if there exists $\zeta \in \mathcal{Z}^*$ such that

$$\left.\begin{aligned} &A(t,u(t)) + \zeta(t) = f(t) \ \text{ a.e. } t \in (0,T),\\ &\zeta(t) \in M^*\partial J(t, Mu(t)) \ \text{ a.e. } t \in (0,T). \end{aligned}\right\}$$

In order to provide the solvability of Problem 4.9 we need the following hypotheses on the data:

$$\left.\begin{aligned} &A\colon (0,T) \times V \to V^* \text{ is such that}\\ &\text{(a) } A(\cdot, v) \text{ is measurable on } (0,T) \text{ for all } v \in V.\\ &\text{(b) } A(t,\cdot) \text{ is pseudomonotone and coercive with}\\ &\qquad \text{constant } \alpha > 0, \text{ for a.e. } t \in (0,T).\\ &\text{(c) } A(t,\cdot) \text{ is strongly monotone for a.e. } t \in (0,T), \text{ i.e.,}\\ &\qquad \langle A(t,v_1) - A(t,v_2), v_1 - v_2\rangle_{V^*\times V} \ge m_1 \|v_1 - v_2\|_V^2\\ &\qquad \text{for all } v_1, v_2 \in V, \text{ a.e. } t \in (0,T) \text{ with } m_1 > 0. \end{aligned}\right\} \qquad (4.18)$$

$$\left.\begin{array}{l} J\colon (0,T)\times X \to \mathbb{R} \text{ is such that} \\ \text{(a) } J(\cdot,u) \text{ is measurable on } (0,T) \text{ for all } u\in X. \\ \text{(b) } J(t,\cdot) \text{ is locally Lipschitz on } X \text{ for a.e. } t\in(0,T). \\ \text{(c) } \|\partial J(t,u)\|_{X^*} \le c_0 + c_1\,\|u\|_X \text{ for all } u\in X, \\ \quad\ \text{a.e. } t\in(0,T) \text{ with } c_0, c_1\ge 0. \\ \text{(d) } \langle z_1 - z_2, u_1-u_2\rangle_{X^*\times X} \ge -m_2\|u_1-u_2\|_X^2 \\ \quad\ \text{for all } z_i\in\partial J(t,u_i), z_i\in X^*,\ u_i\in X,\ i=1,2, \\ \quad\ \text{a.e. } t\in(0,T) \text{ with } m_2\ge 0. \\ \text{(e) } J^0(t,u;-u)\le d_0\,(1+\|u\|_X) \text{ for all } u\in X, \\ \quad\ \text{a.e. } t\in(0,T) \text{ with } d_0\ge 0. \end{array}\right\} \tag{4.19}$$

We are now in a position to state and prove the following existence and uniqueness result.

Theorem 4.11. *Assume that* (4.9), (4.18) *hold and* $f \in \mathcal{V}^*$. *If one of the following hypotheses:*

(i) (4.19) (a)–(d) *and* $\alpha > c_1\,c_e^2\,\|M\|^2$.
(ii) (4.19).

is satisfied and the smallness assumption (4.14) *holds, then Problem* 4.9 *has a unique solution* $u \in \mathcal{V}$. *Moreover, the solution satisfies*

$$\|u\|_{\mathcal{V}} \le c\ (1+\|f\|_{\mathcal{V}^*}) \tag{4.20}$$

with some constant $c > 0$.

Proof. We use Theorem 4.8 for $t \in (0,T)$ fixed. From the hypotheses (4.18)(b), (c) it follows that the operator $A(t,\cdot)$ satisfies (4.12) for a.e. $t \in (0,T)$. It is obvious that the hypothesis (i) (and (ii), respectively) implies the assumption (i) (and (ii), respectively) of Theorem 4.8 and that, for a.e. $t \in (0,T)$, $J(t,\cdot)$ satisfies (4.11) and (4.13). Hence, exploiting Theorem 4.8, we deduce that for a.e. $t \in (0,T)$ Problem 4.9 has a unique solution $u(t) \in V$ and, moreover,

$$\|u(t)\|_V \le c\ (1+\|f(t)\|_{V^*}) \quad \text{a.e. } t\in(0,T) \tag{4.21}$$

with $c > 0$. We point out that the constant c in (4.21) is independent of the parameter t.

We prove that the function $t \mapsto u(t)$ defined above is measurable on $(0,T)$. To this end, given $g \in V^*$ we denote by $w \in V$ the unique solution of the inclusion

$$A(t,w) + M^*\,\partial J(t,Mw) \ni g \quad \text{a.e. } t\in(0,T). \tag{4.22}$$

Since A and J depend on the parameter t, the solution w is also a function of t, i.e., $w = w(t)$. We claim that the solution w depends continuously on the right-hand side g, for a.e. $t \in (0, T)$. Indeed, let $g_1, g_2 \in V^*$ and $w_1(t), w_2(t) \in V$ be the corresponding solutions to (4.22). Using Definition 4.10 we have

$$A(t, w_1(t)) + \zeta_1(t) = g_1 \text{ a.e. } t \in (0, T), \tag{4.23}$$

$$A(t, w_2(t)) + \zeta_2(t) = g_2 \text{ a.e. } t \in (0, T), \tag{4.24}$$

$$\zeta_1(t) \in M^* \partial J(t, M w_1(t)),\ \zeta_2(t) \in M^* \partial J(t, M w_2(t)) \text{ a.e. } t \in (0, T).$$

Subtracting (4.24) from (4.23), multiplying the result by $w_1(t) - w_2(t)$, we get

$$\begin{aligned} \langle A(t, w_1(t)) \ - \ & A(t, w_2(t)), w_1(t) - w_2(t)\rangle_{V^* \times V} \\ & + \langle \zeta_1(t) - \zeta_2(t), w_1(t) - w_2(t)\rangle_{Z^* \times Z} \\ & = \langle g_1 - g_2, w_1(t) - w_2(t)\rangle_{V^* \times V} \end{aligned}$$

for a.e. $t \in (0, T)$. Since $\zeta_i(t) = M^* z_i(t)$ with $z_i(t) \in \partial J(t, M w_i(t))$ for a.e. $t \in (0, T)$ and $i = 1, 2$, by (4.18)(b) and (4.19)(d), we obtain

$$\begin{aligned} m_1 \|w_1(t) - w_2(t)\|_V^2 - m_2\, c_e^2\, \|M\|^2\, \|w_1(t) - w_2(t)\|_V^2 \\ \le \|g_1 - g_2\|_{V^*}\, \|w_1(t) - w_2(t)\|_V \end{aligned}$$

for a.e. $t \in (0, T)$. Exploiting (4.14), we get

$$\|w_1(t) - w_2(t)\|_V \le \widetilde{c}\, \|g_1 - g_2\|_{V^*} \text{ for a.e. } t \in (0, T), \tag{4.25}$$

where $\widetilde{c} = (m_1 - m_2\, c_e^2\, \|M\|^2)^{-1}$ is independent of t. Hence, we have that the mapping $V^* \ni g \mapsto w(t) \in V$ is continuous, for a.e. $t \in (0, T)$, which proves the claim. Now, by (4.25) and the measurability of f, we deduce that the solution u of Problem 4.9 is measurable on $(0, T)$. Since $f \in \mathcal{V}^*$, from the estimate (4.21), we conclude that $u \in \mathcal{V}$ and, moreover, (4.20) holds. □

Subdifferential inclusions with Volterra integral term. We conclude this section with a result on the unique solvability of abstract inclusions with Volterra-type integral term. To this end, let $C(\cdot)$ be a family of linear bounded operators which satisfy

$$C \in L^2(0, T; \mathcal{L}(V, V^*)). \tag{4.26}$$

We consider the following subdifferential inclusion.

Problem 4.12. *Find $u \in \mathcal{V}$ such that*

$$A(t, u(t)) + \int_0^t C(t-s)u(s)\,ds + M^*\,\partial J(t, Mu(t)) \ni f(t) \ \text{ a.e. } t \in (0,T).$$

Under the assumption (4.26) we remark that if $v \in \mathcal{V}$, then the function

$$t \mapsto \int_0^t C(t-s)v(s)\,ds \tag{4.27}$$

belongs to the space $\mathcal{V}^*$. The integral term in the previous inclusion is called the *Volterra integral term*. Moreover, the operator defined by (4.27) is called the *Volterra operator*. For this reason we refer to Problem 4.12 as to a subdifferential inclusion with Volterra integral term. And, as in the case of Problem 4.9, we remark that no derivatives of the unknown are involved in Problem 4.12 and, therefore, in this problem the time variable plays the role of a parameter. Finally, we recall that the solution to Problem 4.12 is understood in the sense of Definition 4.10.

We have the following existence and uniqueness result.

Theorem 4.13. *Assume that* (4.9), (4.18), (4.26) *hold and $f \in \mathcal{V}^*$. If one of the following hypotheses:*

(i) (4.19) (a) – (d) *and* $\alpha > c_1\, c_e^2\, \|M\|^2$.
(ii) (4.19).

is satisfied and (4.14) *holds, then Problem* 4.12 *has a unique solution.*

Proof. We use a fixed point argument. Let $\eta \in \mathcal{V}^*$. We denote by $u_\eta \in \mathcal{V}$ the solution of the inclusion

$$A(t, u_\eta(t)) + M^*\,\partial J(t, Mu_\eta(t)) \ni f(t) - \eta(t) \ \text{ a.e. } t \in (0,T), \tag{4.28}$$

guaranteed by Theorem 4.11. We know that $u_\eta \in \mathcal{V}$ is unique and it satisfies

$$\|u_\eta\|_{\mathcal{V}} \le c\,(1 + \|f\|_{\mathcal{V}^*} + \|\eta\|_{\mathcal{V}^*}) \tag{4.29}$$

with $c > 0$. We consider the operator $\Lambda\colon \mathcal{V}^* \to \mathcal{V}^*$ defined by

$$(\Lambda\eta)(t) = \int_0^t C(t-s)u_\eta(s)\,ds \ \text{ for all } \eta \in \mathcal{V}^*, \ \text{a.e. } t \in (0,T). \tag{4.30}$$

It is easy to check that the operator Λ is well defined. Indeed, for $\eta \in \mathcal{V}^*$, using (4.26) and the Hölder inequality, we have

$$\left\| \int_0^t C(t-s)u_\eta(s)\,ds \right\|_{V^*} \le \int_0^t \|C(t-s)\|_{\mathcal{L}(V,V^*)} \|u_\eta(s)\|_V \,ds$$

$$\le \left(\int_0^t \|C(\tau)\|^2_{\mathcal{L}(V,V^*)}\,d\tau \right)^{1/2} \left(\int_0^t \|u_\eta(\tau)\|^2_V\,d\tau \right)^{1/2}$$

$$\le \|C\|_{L^2(0,t;\mathcal{L}(V,V^*))} \|u_\eta\|_{L^2(0,t;V)}$$

for a.e. $t \in (0,T)$. Hence

$$\|\Lambda\eta\|^2_{\mathcal{V}^*} = \int_0^T \left(\left\| \int_0^t C(t-s)u_\eta(s)\,ds \right\|_{V^*} \right)^2 dt \le T\,\|C\|^2\,\|u_\eta\|^2_{\mathcal{V}},$$

where, here and below, we use the notation $\|C\| = \|C\|_{L^2(0,T;\mathcal{L}(V,V^*))}$. Keeping in mind (4.29), we obtain that the integral in (4.30) is well defined and the operator Λ takes values in $\mathcal{V}^*$.

Next, we show that the operator Λ has a unique fixed point. To this end, in what follows we denote by Λ^k the kth power of the operator Λ. Let $\eta_1, \eta_2 \in \mathcal{V}^*$ and let $u_1 = u_{\eta_1}$ and $u_2 = u_{\eta_2}$ be the corresponding solutions to (4.28). We have $u_1, u_2 \in \mathcal{V}$ and

$$A(t,u_1(t)) + \zeta_1(t) = f(t) - \eta_1(t) \quad \text{a.e. } t \in (0,T), \tag{4.31}$$

$$A(t,u_2(t)) + \zeta_2(t) = f(t) - \eta_2(t) \quad \text{a.e. } t \in (0,T), \tag{4.32}$$

$$\zeta_1(t) \in M^*\partial J(t, Mu_1(t)), \quad \zeta_2(t) \in M^*\partial J(t, Mu_2(t)) \quad \text{a.e. } t \in (0,T).$$

Subtracting (4.32) from (4.31), multiplying the result by $u_1(t) - u_2(t)$ and using (4.18)(b) and (4.19)(d), we obtain

$$\|u_1(t) - u_2(t)\|_V \le \widetilde{c}\,\|\eta_1(t) - \eta_2(t)\|_{V^*} \quad \text{for a.e. } t \in (0,T) \tag{4.33}$$

with $\widetilde{c} > 0$. Using (4.26), from (4.33), we infer

$$\|(\Lambda\eta_1)(t) - (\Lambda\eta_2)(t)\|^2_{V^*} \le \left(\int_0^t \|C(t-s)(u_1(s) - u_2(s))\|_{V^*}\,ds \right)^2$$

$$\le \|C\|^2 \int_0^t \|u_1(s) - u_2(s)\|^2_V\,ds$$

$$\le \widetilde{c}^{\,2}\,\|C\|^2 \int_0^t \|\eta_1(s) - \eta_2(s)\|^2_{V^*}\,ds$$

for a.e. $t \in (0, T)$ and, consequently,

$$\begin{aligned}
\|(\Lambda^2\eta_1)(t) - (\Lambda^2\eta_2)(t)\|_{V^*}^2 &= \|\Lambda(\Lambda\eta_1)(t) - \Lambda(\Lambda\eta_2)(t)\|_{V^*}^2 \\
&\le \widetilde{c}^2\, \|C\|^2 \int_0^t \|(\Lambda\eta_1)(s) - (\Lambda\eta_2)(s)\|_{V^*}^2\, ds \\
&\le \widetilde{c}^4\, \|C\|^4 \int_0^t \left(\int_0^s \|\eta_1(\tau) - \eta_2(\tau)\|_{V^*}^2\, d\tau \right) ds \\
&\le \widetilde{c}^4\, \|C\|^4 \left(\int_0^t \|\eta_1(s) - \eta_2(s)\|_{V^*}^2\, ds \right) \left(\int_0^t ds \right) \\
&= \widetilde{c}^4\, \|C\|^4\, t \int_0^t \|\eta_1(s) - \eta_2(s)\|_{V^*}^2\, ds
\end{aligned}$$

for a.e. $t \in (0, T)$. Reiterating this inequality k times leads to

$$\|(\Lambda^k\eta_1)(t) - (\Lambda^k\eta_2)(t)\|_{V^*}^2 \le \widetilde{c}^{2k}\, \|C\|^{2k}\, \frac{t^{k-1}}{(k-1)!} \int_0^t \|\eta_1(s) - \eta_2(s)\|_{V^*}^2\, ds$$

for a.e. $t \in (0, T)$. This implies that

$$\|\Lambda^k\eta_1 - \Lambda^k\eta_2\|_{\mathcal{V}^*} \le \frac{(\widetilde{c}\, \|C\|\, \sqrt{T})^k}{\sqrt{k!}} \|\eta_1 - \eta_2\|_{\mathcal{V}^*}.$$

Since

$$\lim_{k\to\infty} \frac{a^k}{\sqrt{k!}} = 0 \quad \text{for all } a > 0,$$

from the last inequality we deduce that for k sufficiently large Λ^k is a contraction on $\mathcal{V}^*$. Therefore, the Banach contraction principle implies that there exists a unique $\eta^* \in \mathcal{V}^*$ such that $\eta^* = \Lambda^k\eta^*$. It is clear that $\Lambda^k(\Lambda\eta^*) = \Lambda(\Lambda^k\eta^*) = \Lambda\eta^*$, so $\Lambda\eta^*$ is also a fixed point of Λ^k. By the uniqueness of the fixed point of Λ^k, we have $\eta^* = \Lambda\eta^*$. So $\eta^* \in \mathcal{V}^*$ is the unique fixed point of Λ. Then u_{η^*} is a solution to Problem 4.12, which concludes the existence part of the theorem.

The uniqueness part follows from the uniqueness of the fixed point of Λ. Namely, let $u \in \mathcal{V}$ be a solution to Problem 4.12 and define the element $\eta \in \mathcal{V}^*$ by

$$\eta(t) = \int_0^t C(t-s)u(s)\, ds \quad \text{for all } t \in [0, T].$$

It follows that u is the solution to the problem (4.28) and, by the uniqueness of solutions to (4.28), we obtain $u = u_\eta$. This implies $\Lambda\eta = \eta$ and by the uniqueness of the fixed point of Λ we have $\eta = \eta^*$, so $u = u_{\eta^*}$, which concludes the proof. □

4.3 Hemivariational Inequalities

In this section we use the results of Sect. 4.2 to provide existence and uniqueness results of solutions to hemivariational inequalities. To this end we introduce the following notation.

Let $\Omega \subset \mathbb{R}^d$ be an open bounded subset of $\mathbb{R}^d$ with a Lipschitz boundary $\partial\Omega = \Gamma$ and let $\Gamma_C \subseteq \Gamma$ be any measurable part of $\partial\Omega$. Also, let V be a closed subspace of $H^1(\Omega; \mathbb{R}^s)$, $s \in \mathbb{N}$, $H = L^2(\Omega; \mathbb{R}^s)$, and $Z = H^\delta(\Omega; \mathbb{R}^s)$ with a fixed $\delta \in (\frac{1}{2}, 1)$. Denoting by $i: V \to Z$ the embedding and by $\gamma: Z \to L^2(\Gamma_C; \mathbb{R}^s)$ and $\gamma_0: H^1(\Omega; \mathbb{R}^s) \to H^{1/2}(\Gamma_C; \mathbb{R}^s) \subset L^2(\Gamma_C; \mathbb{R}^s)$ the trace operators, we get $\gamma_0 v = \gamma(iv)$ for all $v \in V$. For simplicity, in what follows we omit the embedding i and we write $\gamma_0 v = \gamma v$ for all $v \in V$. From the theory of Sobolev spaces we know that (V, H, V^*) and (Z, H, Z^*) form evolution triples of spaces and $V \subset Z$ with compact embedding, see Example 2.20 on page 33. We denote by $c_e > 0$ the embedding constant of V into Z. It follows from Theorem 2.22 that the trace operator $\gamma: Z \to L^2(\Gamma_C; \mathbb{R}^s)$ is linear and continuous. We denote by $\|\gamma\|$ the norm of the trace in $\mathcal{L}(Z, L^2(\Gamma_C; \mathbb{R}^s))$ and by $\gamma^*: L^2(\Gamma_C; \mathbb{R}^s) \to Z^*$ the adjoint operator to γ.

A general existence result. Let $A: V \to V^*$ be an operator, $j: \Gamma_C \times \mathbb{R}^s \times \mathbb{R}^s \to \mathbb{R}$ be a prescribed functional and $f \in V^*$. Then we consider the following problem.

Problem 4.14. *Find $u \in V$ such that*

$$\langle Au, v\rangle_{V^*\times V} + \int_{\Gamma_C} j^0(\gamma u, \gamma u; \gamma v)\, d\Gamma \geq \langle f, v\rangle_{V^*\times V} \quad \text{for all } v \in V.$$

An inequality as above is called a *hemivariational inequality*. Here, the notation j^0 stands for the generalized directional derivative of $j(x, \eta, \cdot)$. In what follows sometimes we skip the dependence of various functions on the variable $x \in \Omega \cup \Gamma$ and omit the symbol γ of the trace operator.

For this hemivariational inequality we make the following hypotheses.

$$\left.\begin{array}{l} j\colon \Gamma_C \times \mathbb{R}^s \times \mathbb{R}^s \to \mathbb{R} \text{ is such that} \\ \quad \text{(a) } j(\cdot,\eta,\xi) \text{ is measurable on } \Gamma_C \text{ for all } \eta,\xi \in \mathbb{R}^s \text{ and there} \\ \qquad \text{exists } e \in L^2(\Gamma_C;\mathbb{R}^s) \text{ such that for all } w \in L^2(\Gamma_C;\mathbb{R}^s), \\ \qquad \text{we have } j(\cdot,w(\cdot),e(\cdot)) \in L^1(\Gamma_C). \\ \quad \text{(b) } j(x,\cdot,\xi) \text{ is continuous on } \mathbb{R}^s \text{ for all } \xi \in \mathbb{R}^s, \text{ a.e. } x \in \Gamma_C \\ \qquad \text{and } j(x,\eta,\cdot) \text{ is locally Lipschitz on } \mathbb{R}^s \text{ for all } \eta \in \mathbb{R}^s, \\ \qquad \text{a.e. } x \in \Gamma_C. \\ \quad \text{(c) } \|\partial j(x,\eta,\xi)\|_{\mathbb{R}^s} \le \overline{c}_0 + \overline{c}_1\,\|\xi\|_{\mathbb{R}^s} + \overline{c}_2\,\|\eta\|_{\mathbb{R}^s} \text{ for all } \eta,\xi \in \mathbb{R}^s, \\ \qquad \text{a.e. } x \in \Gamma_C \text{ with } \overline{c}_0,\overline{c}_1,\overline{c}_2 \ge 0. \\ \quad \text{(d) Either } j(x,\eta,\cdot) \text{ or } -j(x,\eta,\cdot) \text{ is regular on } \mathbb{R}^s \text{ for all} \\ \qquad \eta \in \mathbb{R}^s, \text{a.e. } x \in \Gamma_C. \\ \quad \text{(e) } j^0(x,\cdot,\cdot;\rho) \text{ is upper semicontinuous on } \mathbb{R}^s \times \mathbb{R}^s \text{ for all} \\ \qquad \rho \in \mathbb{R}^s, \text{a.e. } x \in \Gamma_C. \\ \quad \text{(f) } j^0(x,\eta,\xi;-\xi) \le \overline{d}_0\,(1 + \|\xi\|_{\mathbb{R}^s} + \|\eta\|_{\mathbb{R}^s}) \text{ for all } \eta,\xi \in \mathbb{R}^s, \\ \qquad \text{a.e. } x \in \Gamma_C \text{ with } \overline{d}_0 \ge 0. \end{array}\right\} \tag{4.34}$$

In order to establish the existence of solutions to Problem 4.14, we associate to this problem an operator inclusion already studied in Sect. 4.2. To this end, we introduce the functional $J\colon L^2(\Gamma_C;\mathbb{R}^s) \times L^2(\Gamma_C;\mathbb{R}^s) \to \mathbb{R}$ defined by

$$J(w,u) = \int_{\Gamma_C} j(x,w(x),u(x))\,d\Gamma \quad \text{for } w,u \in L^2(\Gamma_C;\mathbb{R}^s). \tag{4.35}$$

The following result on the properties of the functional (4.35) represents a direct consequence of Theorem 3.47.

Corollary 4.15. *Assume that* (4.34)(a)–(c) *hold. Then the functional J defined by* (4.35) *satisfies*

(i) *J is well defined and finite on $L^2(\Gamma_C;\mathbb{R}^s) \times L^2(\Gamma_C;\mathbb{R}^s)$.*
(ii) *$J(w,\cdot)$ is Lipschitz continuous on bounded subsets of $L^2(\Gamma_C;\mathbb{R}^s)$ for all $w \in L^2(\Gamma_C;\mathbb{R}^s)$.*
(iii) *For all w, u, $v \in L^2(\Gamma_C;\mathbb{R}^s)$, we have*

$$J^0(w,u;v) \le \int_{\Gamma_C} j^0(x,w(x),u(x);v(x))\,d\Gamma. \tag{4.36}$$

(iv) *For all w, $u \in L^2(\Gamma_C;\mathbb{R}^s)$, we have*

$$\partial J(w,u) \subseteq \int_{\Gamma_C} \partial j(x,w(x),u(x))\,d\Gamma.$$

(v) *For all* $w, u \in L^2(\Gamma_C;\mathbb{R}^s)$, *we have*

$$\|\partial J(w,u)\|_{L^2(\Gamma_C;\mathbb{R}^s)} \le c_0 + c_1\,\|u\|_{L^2(\Gamma_C;\mathbb{R}^s)} + c_2\,\|w\|_{L^2(\Gamma_C;\mathbb{R}^s)}$$

with $c_0 = \sqrt{3\,\mathrm{meas}(\Gamma_C)}\,\overline{c}_0$, $c_1 = \sqrt{3}\,\overline{c}_1$, *and* $c_2 = \sqrt{3}\,\overline{c}_2$.

(vi) *If, in addition,* (4.34)(d) *is satisfied, then* (iii) *and* (iv) *hold with equalities.*

(vii) *If, in addition,* (4.34)(d) *is satisfied, then* $J(w,\cdot)$ *or* $-J(w,\cdot)$ *is regular on* $L^2(\Gamma_C;\mathbb{R}^s)$ *for all* $w \in L^2(\Gamma_C;\mathbb{R}^s)$, *respectively.*

(viii) *If, in addition,* (4.34)(d), (e) *are satisfied, then the multifunction*

$$\partial J\colon L^2(\Gamma_C;\mathbb{R}^s)\times L^2(\Gamma_C;\mathbb{R}^s)\to 2^{L^2(\Gamma_C;\mathbb{R}^s)}$$

has a closed graph in $L^2(\Gamma_C;\mathbb{R}^s)\times L^2(\Gamma_C;\mathbb{R}^s)\times(w\text{–}L^2(\Gamma_C;\mathbb{R}^s))$ *topology.*

(ix) *If, in addition,* (4.34)(f) *holds, then*

$$J^0(w,u;-u) \le d_0(1+\|u\|_{L^2(\Gamma_C;\mathbb{R}^s)} + \|w\|_{L^2(\Gamma_C;\mathbb{R}^s)})$$

for all $w, u \in L^2(\Gamma_C;\mathbb{R}^s)$ *with* $d_0 \ge 0$.

We are now in a position to state and prove the following result.

Theorem 4.16. *Assume that* (4.1) *holds and* $f \in V^*$. *If one of the following hypotheses:*

(i) (4.34) (a) – (e) *and* $\alpha > \sqrt{3}\,(\overline{c}_1+\overline{c}_2)\,c_e^2\,\|\gamma\|^2$.
(ii) (4.34).

is satisfied, then Problem 4.14 *has at least one solution* $u \in V$. *Moreover, the solution satisfies*

$$\|u\|_V \le c\ (1+\|f\|_{V^*}) \tag{4.37}$$

with a constant $c > 0$.

Proof. We apply Theorem 4.5. To this end, we consider the space $X = L^2(\Gamma_C;\mathbb{R}^s)$, the operator $M\colon Z\to X$ given by $Mz = \gamma z$ for all $z \in Z$, and the functional J defined by (4.35). It is clear that $M \in \mathcal{L}(Z,X)$, i.e., (4.9) holds. From Theorem 4.5 combined with Corollary 4.15, we know that Problem 4.4 admits a solution $u \in V$. According to the definition of the solution, there exists $z \in L^2(\Gamma_C;\mathbb{R}^s)$ such that

$$Au + \gamma^* z = f \tag{4.38}$$

with $z \in \partial J(\gamma u, \gamma u)$. The last inclusion is equivalent to

$$\langle z, \widetilde{v}\rangle_{L^2(\Gamma_C;\mathbb{R}^s)} \le J^0(\gamma u, \gamma u; \widetilde{v}) \tag{4.39}$$

for all $\widetilde{v} \in L^2(\Gamma_C;\mathbb{R}^s)$. We combine now (4.38), (4.39), and (4.36) to obtain

$$\begin{aligned}\langle f - Au, v\rangle_{V^*\times V} &= \langle \gamma^* z, v\rangle_{V^*\times V} = \langle z, \gamma v\rangle_{L^2(\Gamma_C;\mathbb{R}^s)}\\ &\le J^0(\gamma u, \gamma u; \gamma v) \le \int_{\Gamma_C} j^0(\gamma u, \gamma u; \gamma v)\, d\Gamma\end{aligned}$$

for all $v \in V$. Hence we deduce that $u \in V$ is a solution to Problem 4.14. Moreover, the estimate (4.37) follows from (4.3), which ends the proof. □

Particular cases. Let $A\colon V \to V^*$ be an operator, $f \in V^*$ and $h_i, j_i\colon \Gamma_C \times \mathbb{R}^s \to \mathbb{R}$, $i = 1,\ldots,k$ be given functions, $k \in \mathbb{N}$. In what follows we study the following particular case of Problem 4.14 which will be applied in the study of a static contact problem we present in Chap. 7.

Problem 4.17. *Find $u \in V$ such that*

$$\langle Au, v\rangle_{V^*\times V} + \int_{\Gamma_C} \sum_{i=1}^{k} h_i(\gamma u)\, j_i^0(\gamma u; \gamma v)\, d\Gamma \ge \langle f, v\rangle_{V^*\times V} \quad \text{for all } v \in V.$$

The hypotheses on the integrands are the following, for $i = 1,\ldots,k$.

$$\left.\begin{array}{l} h_i\colon \Gamma_C \times \mathbb{R}^s \to \mathbb{R} \text{ is such that}\\ \quad \text{(a) } h_i(\cdot,\eta) \text{ is measurable on } \Gamma_C \text{ for all } \eta \in \mathbb{R}^s.\\ \quad \text{(b) } h_i(x,\cdot) \text{ is continuous on } \mathbb{R}^s \text{ for a.e. } x \in \Gamma_C.\\ \quad \text{(c) } 0 \le h_i(x,\eta) \le \overline{h}_i \text{ for all } \eta \in \mathbb{R}^s, \text{ a.e. } x \in \Gamma_C \text{ with } \overline{h}_i > 0. \end{array}\right\} \quad (4.40)$$

$$\left.\begin{array}{l} j_i\colon \Gamma_C \times \mathbb{R}^s \to \mathbb{R} \text{ is such that}\\ \quad \text{(a) } j_i(\cdot,\xi) \text{ is measurable on } \Gamma_C \text{ for all } \xi \in \mathbb{R}^s \text{ and there exists}\\ \qquad e_i \in L^2(\Gamma_C;\mathbb{R}^s) \text{ such that } j_i(\cdot, e_i(\cdot)) \in L^1(\Gamma_C).\\ \quad \text{(b) } j_i(x,\cdot) \text{ is locally Lipschitz on } \mathbb{R}^s \text{ for a.e. } x \in \Gamma_C.\\ \quad \text{(c) } \|\partial j_i(x,\xi)\|_{\mathbb{R}^s} \le c_{0i} + c_{1i}\, \|\xi\|_{\mathbb{R}^s} \text{ for all } \xi \in \mathbb{R}^s, \text{ a.e. } x \in \Gamma_C\\ \qquad \text{with } c_{0i}, c_{1i} \ge 0.\\ \quad \text{(d) Either } j_i(x,\cdot) \text{ or } -j_i(x,\cdot) \text{ is regular on } \mathbb{R}^s \text{ for a.e. } x \in \Gamma_C.\\ \quad \text{(e) } j_i^0(x,\xi;-\xi) \le d_{0i}\,(1 + \|\xi\|_{\mathbb{R}^s}) \text{ for all } \xi \in \mathbb{R}^s, \text{ a.e. } x \in \Gamma_C\\ \qquad \text{with } d_{0i} \ge 0. \end{array}\right\} \quad (4.41)$$

We have the following existence result.

Corollary 4.18. *Assume that* (4.1), (4.40) *are satisfied and* $f \in V^*$. *If one of the following hypotheses:*

(i) (4.41) (a) – (d) *and* $\alpha > \sqrt{3}\left(\sum_{i=1}^{k} c_{1i}\, \overline{h}_i \right) c_e^2\, \|\gamma\|^2$.

(ii) (4.41).

holds, then Problem 4.17 *has at least one solution* $u \in V$. *Moreover, the solution satisfies the estimate* (4.37).

Proof. It is enough to check that the function $j: \Gamma_C \times \mathbb{R}^s \times \mathbb{R}^s \to \mathbb{R}$ defined by

$$j(x, \eta, \xi) = \sum_{i=1}^{k} h_i(x, \eta)\, j_i(x, \xi) \quad \text{for a.e. } x \in \Gamma_C, \text{ all } \eta, \xi \in \mathbb{R}^s$$

satisfies the hypotheses of Theorem 4.16.

First, suppose that hypothesis (i) holds. Then, it is clear that $j(\cdot, \eta, \xi)$ is measurable on Γ_C for all $\eta\ \xi \in \mathbb{R}^s$, $j(x, \cdot, \xi)$ is continuous on $\mathbb{R}^s$ for all $\xi \in \mathbb{R}^s$, a.e. $x \in \Gamma_C$ $j(x, \eta, \cdot)$ is locally Lipschitz on $\mathbb{R}^s$ for all $\eta \in \mathbb{R}^s$, a.e. $x \in \Gamma_C$. If, either $j_i(x, \cdot)$ or $-j_i(x, \cdot)$ are regular on $\mathbb{R}^s$, for a.e. $x \in \Gamma_C$, for all $i = 1, \ldots, k$, then either $j(x, \eta, \cdot)$ or $-j(x, \eta, \cdot)$ is regular on $\mathbb{R}^s$ for all $\eta \in \mathbb{R}^s$, a.e. $x \in \Gamma_C$, respectively. From Corollary 3.48 applied to the functions j_i for $i = 1, \ldots, k$, we have $j_i(\cdot, e(\cdot)) \in L^1(\Gamma_C)$ for all $e \in L^2(\Gamma_C; \mathbb{R}^s)$. Hence, from the inequality

$$\begin{aligned} \int_{\Gamma_C} |j(x, w(x), e(x))|\, d\Gamma &= \int_{\Gamma_C} |\sum_{i=1}^{k} h_i(x, w(x))\, j_i(x, e(x))|\, d\Gamma \\ &\le \sum_{i=1}^{k} \overline{h}_i \int_{\Gamma_C} |j_i(x, e(x))|\, d\Gamma, \end{aligned}$$

which is valid for all $e, w \in L^2(\Gamma_C; \mathbb{R}^s)$, we deduce that $j(\cdot, w(\cdot), e(\cdot)) \in L^1(\Gamma_C)$ for all $e, w \in L^2(\Gamma_C; \mathbb{R}^s)$. It is also clear that j satisfies (4.34)(c) with

$$\overline{c}_0 = \sum_{i=1}^{k} \overline{h}_i\, c_{0i}, \quad \overline{c}_1 = \sum_{i=1}^{k} \overline{h}_i\, c_{1i} \quad \text{and} \quad \overline{c}_2 = 0.$$

We conclude from above that the function j satisfies conditions (4.34)(a)–(d). Moreover, by the hypothesis (i), we have $\alpha > \sqrt{3}\,(\overline{c}_1 + \overline{c}_2)\, c_e^2\, \|\gamma\|^2$.

In order to show (4.34)(e), let $(\eta_n, \xi_n) \in \mathbb{R}^s \times \mathbb{R}^s$, $(\eta_n, \xi_n) \to (\eta, \xi)$, $(\eta, \xi) \in \mathbb{R}^s \times \mathbb{R}^s$, and $\rho \in \mathbb{R}^s$. We have

$$\begin{aligned}
\limsup j^0(x, \eta_n, \xi_n; \rho) &= \limsup \sum_{i=1}^{k} h_i(x, \eta_n)\, j_i^0(x, \xi_n; \rho) \\
&\le \sum_{i=1}^{k} \limsup \Big((h_i(x, \eta_n) - h_i(x, \eta))\, j_i^0(x, \xi_n; \rho) \\
&\quad + h_i(x, \eta)\, j_i^0(x, \xi_n; \rho) \Big) \\
&\le \sum_{i=1}^{k} \Big(\|\rho\|_{\mathbb{R}^s} (c_{0i} + c_{1i} \|\xi_n\|_{\mathbb{R}^s}) \limsup |h_i(x, \eta_n) - h_i(x, \eta)| \\
&\quad + h_i(x, \eta) \limsup j_i^0(x, \xi_n; \rho) \Big) \\
&\le \sum_{i=1}^{k} h_i(x, \eta)\, j_i^0(x, \xi; \rho) = j^0(x, \eta, \xi; \rho)
\end{aligned}$$

for a.e. $x \in \Gamma_C$, which proves the upper semicontinuity of the function $j^0(x, \cdot, \cdot; \rho)$ for all $\rho \in \mathbb{R}^s$ and a.e. $x \in \Gamma_C$. We conclude from here that (4.34)(e) holds and, therefore, we infer that the hypothesis (i) of Theorem 4.16 is satisfied.

Secondly, under the hypothesis (ii), we have

$$j^0(x, \eta, \xi; -\xi) = \sum_{i=1}^{k} h_i(x, \eta)\, j_i^0(x, \xi; -\xi) \le \max_{i=1,\dots,k} \{ \overline{h}_i\, d_{0i} \} (1 + \|\xi\|_{\mathbb{R}^s})$$

for all $\eta, \xi \in \mathbb{R}^s$ and a.e. $x \in \Gamma_C$, and so in this case the assumption (ii) of Theorem 4.16 is satisfied.

Finally, we apply Theorem 4.16 to complete the proof of the corollary. □

Next, we consider a particular form of Problem 4.14 for which we provide a result on its unique solvability and which will be applied, again, to a static contact problem in Chap. 7.

Problem 4.19. *Find $u \in V$ such that*

$$\langle Au, v\rangle_{V^* \times V} + \int_{\Gamma_C} j^0(\gamma u; \gamma v)\, d\Gamma \ge \langle f, v\rangle_{V^* \times V} \quad \text{for all } v \in V.$$

In the study of Problem 4.19, besides (4.12), we need the following hypotheses on the function j.

$$
\left.\begin{array}{l}
j\colon \Gamma_C \times \mathbb{R}^s \to \mathbb{R} \text{ is such that} \\
\text{(a) } j(\cdot,\xi) \text{ is measurable on } \Gamma_C \text{ for all } \xi \in \mathbb{R}^s \text{ and there} \\
\qquad \text{exists } e \in L^2(\Gamma_C;\mathbb{R}^s) \text{ such that } j(\cdot,e(\cdot)) \in L^1(\Gamma_C). \\
\text{(b) } j(x,\cdot) \text{ is locally Lipschitz on } \mathbb{R}^s \text{ for a.e. } x \in \Gamma_C. \\
\text{(c) } \|\partial j(x,\xi)\|_{\mathbb{R}^s} \le \overline{c}_0 + \overline{c}_1\, \|\xi\|_{\mathbb{R}^s} \text{ for all } \xi \in \mathbb{R}^s, \text{ a.e. } x \in \Gamma_C \\
\qquad \text{with } \overline{c}_0, \overline{c}_1 \ge 0. \\
\text{(d) } (\zeta_1 - \zeta_2)\cdot(\xi_1 - \xi_2) \ge -m_2\|\xi_1 - \xi_2\|^2_{\mathbb{R}^s} \text{ for all } \zeta_i,\, \xi_i \in \mathbb{R}^s, \\
\qquad \zeta_i \in \partial j(x,\xi_i),\, i = 1,2, \text{ a.e. } x \in \Gamma_C \text{ with } m_2 \ge 0. \\
\text{(e) } j^0(x,\xi;-\xi) \le \overline{d}_0\,(1 + \|\xi\|_{\mathbb{R}^s}) \text{ for all } \xi \in \mathbb{R}^s, \text{ a.e. } x \in \Gamma_C \\
\qquad \text{with } \overline{d}_0 \ge 0.
\end{array}\right\} \tag{4.42}
$$

We also need the smallness condition

$$m_1 > m_2\, c_e^2\, \|\gamma\|^2 \tag{4.43}$$

where, recall, m_1 and m_2 are the constants in (4.12) and (4.42), respectively.

We have the following existence and uniqueness result.

Theorem 4.20. *Assume that* (4.12) *holds and* $f \in V^*$. *If one of the following hypotheses:*

(i) (4.42) (a) – (d) *and* $\alpha > \sqrt{3}\,\overline{c}_1\, c_e^2\, \|\gamma\|^2$.
(ii) (4.42).

is satisfied and (4.43) *holds, then Problem* 4.19 *has a solution* $u \in V$ *which satisfies the estimate* (4.37). *If, in addition, the regularity condition*

$$\textit{either } j(x,\cdot) \textit{ or } -j(x,\cdot) \textit{ is regular on } \mathbb{R}^s \textit{ for a.e. } x \in \Gamma_C$$

holds, then the solution of Problem 4.19 *is unique, and denoting by* u_i *the unique solution corresponding to* $f = f_i$, $i = 1, 2$, *there exists* $c > 0$ *such that*

$$\|u_1 - u_2\|_V \le c\, \|f_1 - f_2\|_{V^*}. \tag{4.44}$$

Proof. We apply Theorem 4.8. To this end, we observe that (4.43) implies (4.14) and consider the functional $J\colon L^2(\Gamma_C;\mathbb{R}^s) \to \mathbb{R}$ defined by

$$J(u) = \int_{\Gamma_C} j(x,u(x))\, d\Gamma \quad \text{for } u \in L^2(\Gamma_C;\mathbb{R}^s). \tag{4.45}$$

First, let us assume the hypothesis (i). Due to (4.42)(a)–(c), we are able to apply Corollary 4.15 to the functional J given by (4.45). From conditions (i)–(v) of

Corollary 4.15, we infer that (4.11)(a), (b) are satisfied with $c_0 = \sqrt{3\,\text{meas}(\Gamma_C)}\,\overline{c}_0$ and $c_1 = \sqrt{3}\,\overline{c}_1$,

$$\partial J(u) \subseteq \int_{\Gamma_C} \partial j(x, u(x))\, d\Gamma \quad \text{for all } \; u \in L^2(\Gamma_C; \mathbb{R}^s) \tag{4.46}$$

and, moreover,

$$J^0(u; v) \leq \int_{\Gamma_C} j^0(x, u(x); v(x))\, d\Gamma \quad \text{for all } \; u, v \in L^2(\Gamma_C; \mathbb{R}^s). \tag{4.47}$$

Subsequently, under the hypotheses (4.42)(a)–(d), we show that the functional J satisfies condition (4.13). Indeed, let u_i, $z_i \in L^2(\Gamma_C; \mathbb{R}^s)$, $z_i \in \partial J(u_i)$, $i = 1, 2$. From (4.46), we deduce that there exist $\zeta_i \in L^2(\Gamma_C; \mathbb{R}^s)$ such that $\zeta_i(x) \in \partial j(x, u_i(x))$ for a.e. $x \in \Gamma_C$ and

$$\langle z_i, v \rangle_{L^2(\Gamma_C; \mathbb{R}^s)} = \int_{\Gamma_C} \zeta_i(x) \cdot v(x)\, d\Gamma \quad \text{for all } \; v \in L^2(\Gamma_C; \mathbb{R}^s)$$

for $i = 1, 2$. By (4.42)(d), we have

$$\begin{aligned} \langle z_1 - z_2, u_1 - u_2 \rangle_{L^2(\Gamma_C; \mathbb{R}^s)} &= \int_{\Gamma_C} (\zeta_1(x) - \zeta_2(x)) \cdot (u_1(x) - u_2(x))\, d\Gamma \\ &\geq -m_2 \int_{\Gamma_C} \|u_1(x) - u_2(x)\|^2_{\mathbb{R}^s}\, d\Gamma \\ &= -m_2 \|u_1 - u_2\|^2_{L^2(\Gamma_C; \mathbb{R}^s)}. \end{aligned}$$

We conclude from above that condition (4.13) is satisfied. Moreover, we observe that hypothesis (i) implies condition $\alpha > c_1\, c_e^2\, \|\gamma\|^2$ and, therefore, the hypothesis (i) of Theorem 4.8 is verified.

Secondly, we assume the hypothesis (ii). Then, applying again Corollary 4.15(ix), we obtain (4.11). Therefore, it is easy to see that the hypothesis (ii) of Theorem 4.8 is satisfied.

From Theorem 4.8 we deduce that there exists a unique solution $u \in V$ to the problem

$$Au + \gamma^* \,\partial J(\gamma u) \ni f \tag{4.48}$$

which satisfies the estimate (4.37). We proceed our proof with the following step.

Claim: every solution to (4.48) solves Problem 4.19. It follows from (4.48) that there exists $z \in \partial J(\gamma u)$, $z \in L^2(\Gamma_C; \mathbb{R}^s)$ such that $Au + \gamma^* z = f$. Multiplying the latter by $v \in V$, we have

$$\langle Au, v \rangle_{V^* \times V} + \langle z, \gamma v \rangle_{L^2(\Gamma_C; \mathbb{R}^s)} = \langle f, v \rangle_{V^* \times V}$$

while (4.47), together with the definition of the subdifferential, implies

$$\langle z, \gamma v\rangle_{L^2(\Gamma_C;\mathbb{R}^s)} \le J^0(\gamma u; \gamma v) \le \int_{\Gamma_C} j^0(x, \gamma u(x); \gamma v(x))\, d\Gamma.$$

It follows from above that u is a solution to Problem 4.19 and this proves the claim.

Finally, we assume the regularity hypothesis either on j or $-j$. In order to prove that, under this hypothesis, the solution of Problem 4.19 is unique, we show that $u \in V$ solves (4.48) if and only if $u \in V$ solves Problem 4.19. Due to the previous claim, it is enough to prove the "if" part. So let $u \in V$ be a solution to Problem 4.19, i.e.,

$$\langle Au, v\rangle_{V^*\times V} + \int_{\Gamma_C} j^0(x, \gamma u(x); \gamma v(x))\, d\Gamma \ge \langle f, v\rangle_{V^*\times V} \quad \text{for all } v \in V.$$

Then, by Corollary 4.15(vi), we know that in (4.47) we have the equality. Hence

$$\langle Au, v\rangle_{V^*\times V} + J^0(\gamma u; \gamma v) \ge \langle f, v\rangle_{V^*\times V} \quad \text{for all } v \in V.$$

From the latter, by exploiting the equalities

$$J^0(\gamma u; \gamma v) = (J \circ \gamma)^0(u; v) \quad \text{and} \quad \partial(J \circ \gamma)(u) = \gamma^*\, \partial J(\gamma u)$$

(which represent a consequence of Proposition 3.37), we have

$$\langle f - Au, v\rangle_{V^*\times V} \le J^0(\gamma u; \gamma v) = (J \circ \gamma)^0(u; v) \quad \text{for all } v \in V$$

and

$$f - Au \in \partial(J \circ \gamma)(u) = \gamma^*\, \partial J(\gamma u).$$

We deduce from here that $u \in V$ is a solution to (4.48).

It remains to prove the inequality (4.44). Let $f_i \in V^*$ and u_i be the unique solution to Problem 4.19 corresponding to f_i, $i = 1, 2$. Since Problem 4.19 is equivalent to (4.48), we have $Au_i + \gamma^*\zeta_i = f_i$ and $\zeta_i \in \partial J(\gamma u_i)$, $i = 1, 2$. Hence

$$Au_1 - Au_2 + \gamma^*\zeta_1 - \gamma^*\zeta_2 = f_1 - f_2$$

and by (4.12)(b), we have

$$m_1 \|u_1 - u_2\|_V^2 + \langle \zeta_1 - \zeta_2, \gamma(u_1 - u_2)\rangle_{L^2(\Gamma_C;\mathbb{R}^s)} \le \langle f_1 - f_2, u_1 - u_2\rangle_{V^*\times V}.$$

Since J satisfies the relaxed monotonicity condition, we get

$$\langle \zeta_1 - \zeta_2, \gamma(u_1 - u_2)\rangle_{L^2(\Gamma_C;\mathbb{R}^s)} \ge -m_2\, c_e^2\, \|\gamma\|^2\, \|u_1 - u_2\|_V^2.$$

Therefore, from the previous two inequalities we obtain that

$$\left(m_1 - m_2\, c_e^2\, \|\gamma\|^2\right) \|u_1 - u_2\|_V^2 \le \|f_1 - f_2\|_{V^*} \|u_1 - u_2\|_V.$$

Now, by (4.43) we deduce that inequality (4.44) is satisfied, which concludes the proof. □

Hemivariational inequalities with Volterra integral term. The arguments presented above in this section can be used in order to study time-dependent hemivariational inequalities, i.e., versions of Problem 4.14 in which both A, j, and f depend on time. Nevertheless, in what follows we skip this study and pass directly to an important class of inequalities with Volterra integral operators. To present an existence and uniqueness result of solutions for such inequalities we use the notation introduced on page 103 in the study of time-dependent subdifferential inclusions.

The problem under consideration reads as follows.

Problem 4.21. *Find $u \in \mathcal{V}$ such that*

$$\langle A(t, u(t)), v \rangle_{V^* \times V} + \left\langle \int_0^t C(t-s)u(s)\, ds, v \right\rangle_{V^* \times V} + \int_{\Gamma_C} j^0(t, \gamma u(t); \gamma v)\, d\Gamma \geq \langle f(t), v \rangle_{V^* \times V}$$

for all $v \in V$ and a.e. $t \in (0, T)$.

Using the terminology introduced on page 106 we refer to the inequality in Problem 4.21 as a *hemivariational inequality with Volterra integral term*. To provide the analysis of such inequality we consider the following assumption on the superpotential j.

$$\left.\begin{array}{l}
j\colon \Gamma_C \times (0,T) \times \mathbb{R}^s \to \mathbb{R} \text{ is such that} \\
\text{(a) } j(\cdot,\cdot,\xi) \text{ is measurable on } \Gamma_C \times (0,T) \text{ for all } \xi \in \mathbb{R}^s \\
\quad \text{and there exists } e \in L^2(\Gamma_C;\mathbb{R}^s) \text{ such that} \\
\quad j(\cdot,\cdot,e(\cdot)) \in L^1(\Gamma_C \times (0,T)). \\
\text{(b) } j(x,t,\cdot) \text{ is locally Lipschitz on } \mathbb{R}^s \text{ for} \\
\quad \text{a.e. } (x,t) \in \Gamma_C \times (0,T). \\
\text{(c) } \|\partial j(x,t,\xi)\|_{\mathbb{R}^s} \le \overline{c}_0 + \overline{c}_1\, \|\xi\|_{\mathbb{R}^s} \text{ for all } \xi \in \mathbb{R}^s, \\
\quad \text{a.e. } (x,t) \in \Gamma_C \times (0,T) \text{ with } \overline{c}_0, \overline{c}_1 \ge 0. \\
\text{(d) } (\zeta_1 - \zeta_2)\cdot(\xi_1 - \xi_2) \ge -m_2 \|\xi_1 - \xi_2\|^2_{\mathbb{R}^s} \text{ for all } \zeta_i,\ \xi_i \in \mathbb{R}^s, \\
\quad \zeta_i \in \partial j(x,t,\xi_i),\ i = 1,2, \text{ a.e. } (x,t) \in \Gamma_C \times (0,T) \text{ with} \\
\quad m_2 \ge 0. \\
\text{(e) } j^0(x,t,\xi;-\xi) \le \overline{d}_0\,(1 + \|\xi\|_{\mathbb{R}^s}) \text{ for all } \xi \in \mathbb{R}^s, \\
\quad \text{a.e. } (x,t) \in \Gamma_C \times (0,T) \text{ with } \overline{d}_0 \ge 0.
\end{array}\right\} \qquad (4.49)$$

We are now in a position to state and prove the following existence and uniqueness result.

Theorem 4.22. *Assume that* (4.18), (4.26) *hold and* $f \in \mathcal{V}^*$. *If one of the following hypotheses:*

(i) (4.49) (a) – (d) *and* $\alpha > \sqrt{3}\,\overline{c}_1\, c_e^2\, \|\gamma\|^2$.
(ii) (4.49).

is satisfied and (4.43) *holds, then Problem* 4.21 *has a solution* $u \in \mathcal{V}$. *If, in addition, the regularity condition*

$$\textit{either } j(x,t,\cdot) \textit{ or } -j(x,t,\cdot) \textit{ is regular on } \mathbb{R}^s \textit{ for a.e. } (x,t) \in \Gamma_C \times (0,T)$$

is satisfied, then the solution of Problem 4.21 *is unique.*

Note that in the statement of Theorem 4.22 the constants m_1 and m_2 represent the constants in (4.18) and (4.49), respectively.

Proof. We apply Theorem 4.13. We consider the space $X = L^2(\Gamma_C;\mathbb{R}^s)$, the operator $M = \gamma$ being the trace operator from Z into X and the functional $J\colon (0,T) \times L^2(\Gamma_C;\mathbb{R}^s) \to \mathbb{R}$ defined by

$$J(t,u) = \int_{\Gamma_C} j(x,t,u(x))\, d\Gamma \quad \text{for a.e. } t \in (0,T),\ \text{all } u \in L^2(\Gamma_C;\mathbb{R}^s).$$

First, let us assume the hypothesis (i). From (4.49)(a)–(c), by Theorem 3.47, we have

(i1) $J(\cdot,u)$ is measurable on $(0,T)$ for all $u \in L^2(\Gamma_C;\mathbb{R}^s)$.
(i2) $J(t,\cdot)$ is locally Lipschitz on $L^2(\Gamma_C;\mathbb{R}^s)$ for a.e. $t \in (0,T)$.
(i3) $\|\partial J(t,u)\|_{L^2(\Gamma_C;\mathbb{R}^s)} \le \sqrt{3\,\mathrm{meas}(\Gamma_C)}\,\overline{c}_0 + \sqrt{3}\,\overline{c}_1\, \|u\|_{L^2(\Gamma_C;\mathbb{R}^s)}$ for all $u \in L^2(\Gamma_C;\mathbb{R}^s)$, a.e. $t \in (0,T)$.
(i4) $J^0(t,u;v) \le \displaystyle\int_{\Gamma_C} j^0(x,t,u(x);v(x))\, d\Gamma$ for all $u, v \in L^2(\Gamma_C;\mathbb{R}^s)$, a.e. $t \in (0,T)$.
(i5) $\partial J(t,u) \subseteq \displaystyle\int_{\Gamma_C} \partial j(x,t,u(x))\, d\Gamma$ for all $u \in L^2(\Gamma_C;\mathbb{R}^s)$, a.e. $t \in (0,T)$.

Next, since (4.49)(d) holds then, using arguments similar to those in the proof of Theorem 4.20, we obtain

(i6) $\langle z_1(t) - z_2(t), u_1 - u_2\rangle_{L^2(\Gamma_C;\mathbb{R}^s)} \ge -m_2\, \|u_1 - u_2\|^2_{L^2(\Gamma_C;\mathbb{R}^s)}$ for all $z_i(t) \in \partial J(t,u_i)$, $u_i \in L^2(\Gamma_C;\mathbb{R}^s)$, $z_i \in L^\infty(0,T;L^2(\Gamma_C;\mathbb{R}^s))$, $i = 1, 2$, a.e. $t \in (0,T)$.

Moreover, the condition (4.14) holds due to the smallness condition (4.43). Hence we conclude that the hypothesis (i) of Theorem 4.13 is verified.

Second, we assume the hypothesis (ii). Using (4.49)(e), by Theorem 3.47(x), we get (i7) $J^0(t, u; -u) \leq d_0(1 + \|u\|_{L^2(\Gamma_C;\mathbb{R}^s)})$ for all $u \in L^2(\Gamma_C; \mathbb{R}^s)$, a.e. $t \in (0, T)$ with $d_0 \geq 0$.

It follows from here that the hypothesis (ii) of Theorem 4.13 is satisfied in this case.

We are now in a position to apply Theorem 4.13 to obtain a unique solution $u \in \mathcal{V}$ of the inclusion

$$A(t, u(t)) + \int_0^t C(t-s)u(s)\, ds + \gamma^* \partial J(t, \gamma u(t)) \ni f(t) \ \text{ a.e. } t \in (0, T). \quad (4.50)$$

Moreover, as in the proof of Theorem 4.20, using (i5) and (i6), it follows that $u \in \mathcal{V}$ is a solution to Problem 4.21.

If, in addition, the regularity hypothesis is assumed, then (i4) and (i5) hold with equalities. The argument used in the proof of Theorem 4.20 shows that $u \in \mathcal{V}$ is a solution to (4.50) if and only if $u \in \mathcal{V}$ is a solution to Problem 4.21, which completes the proof of the theorem. □

We end this chapter with a general remark concerning the assumptions we use to provide the solvability of the problems in Sects. 4.2 and 4.3. To present this remark we shall consider in what follows the particular case of Problem 4.21 but a careful analysis shows that similar comments can be formulated for all the problems mentioned above.

Recall that the solvability of Problem 4.21 is provided by Theorem 4.22 and, in the statement of the theorem, there is the possibility to choose one of the following assumptions:

(i) (4.49) (a)–(d) and $\alpha > \sqrt{3}\,\overline{c}_1\, c_e^2\, \|\gamma\|^2$.
(ii) (4.49).

Note that assumption (ii) concerns only the functional j, which represents one of the data of Problem 4.21. For this reason, we refer to this assumption as to an intrinsic assumption in the study of this problem. In contrast, besides the assumption (4.49)(a)–(d) on j, assumption (i) above contains the smallness assumption $\alpha > \sqrt{3}\,\overline{c}_1\, c_e^2\, \|\gamma\|^2$ which involves the embedding constant of V into Z and the norm of the trace operator $\gamma: Z \to L^2(\Gamma_C; \mathbb{R}^s)$. Or, recall that the space Z represents only an auxiliary space, related to our mathematical tools, and it is not related to the statement of Problem 4.21. For this reason, it follows that assumption (i) is not an intrinsic assumption in the study of this problem. Moreover, note that the efficiency of its use is determined by a good estimation of the various constants involved in the smallness assumption.

To conclude, the solvability of Problem 4.21, guaranteed by Theorem 4.22, is obtained either by considering an intrinsic assumption or by considering an assumption depending on our mathematical tools. The question of knowing which of these assumption is more useful in the study of a given hemivariational inequality with Volterra integral term remains widely open. Clearly, it deserves more investigation in the future.

Chapter 5
Evolutionary Inclusions and Hemivariational Inequalities

In this chapter we study evolutionary inclusions of second order. These are multivalued relations which involve the second-order time derivative of the unknown. We start with a basic existence result for such inclusions. Then we provide results on existence and uniqueness of solutions to evolutionary inclusions of the subdifferential type, i.e., inclusions involving the Clarke subdifferential operator of locally Lipschitz functionals. We also prove an existence and uniqueness result for integro-differential evolutionary inclusions. Next, we consider a class of hyperbolic hemivariational inequalities for which we provide a theorem on existence of solutions and, under stronger hypotheses, their uniqueness. We conclude this chapter with a result on existence and uniqueness of solutions to the evolutionary integro-differential hemivariational inequality with the Volterra integral term. The results provided below represent the dynamic counterparts of theorems presented in Chap. 4 and will be used in the study of the dynamic frictional contact problems in Chap. 8.

5.1 A Basic Existence Result

We begin by recalling the notation we need for the statement of the problem. Given a normed space X, by X^* we denote its (topological) dual and by $\|\cdot\|_X$ its norm. For the duality brackets for the pair (X, X^*) we use the notation $\langle\cdot,\cdot\rangle_{X^*\times X}$. Let V and Z be separable and reflexive Banach spaces, H be a separable Hilbert space such that

$$V \subset Z \subset H \subset Z^* \subset V^*$$

with continuous embeddings. We assume that the embedding $V \subset Z$ is compact and we denote by $c_e > 0$ the embedding constant of V into Z. Given $0 < T < \infty$, we introduce the spaces

$$\mathcal{V} = L^2(0,T;V), \quad \mathcal{Z} = L^2(0,T;Z), \quad \widehat{\mathcal{H}} = L^2(0,T;H),$$

$$\mathcal{Z}^* = L^2(0,T;Z^*), \quad \mathcal{V}^* = L^2(0,T;V^*), \quad \mathcal{W} = \{\, v \in \mathcal{V} \mid v' \in \mathcal{V}^* \,\}.$$

S. Migórski et al., *Nonlinear Inclusions and Hemivariational Inequalities*, Advances in Mechanics and Mathematics 26, DOI 10.1007/978-1-4614-4232-5_5,

The duality pairing between $\mathcal{V}^*$ and $\mathcal{V}$ is given by

$$\langle v, w\rangle_{\mathcal{V}^*\times\mathcal{V}} = \int_0^T \langle v(t), w(t)\rangle_{V^*\times V}\, dt \ \text{ for } v \in \mathcal{V}^*,\ w \in \mathcal{V}.$$

Also, recall that, as stated in Proposition 2.54, $\mathcal{W}$ is a Banach space with the norm given by (2.12).

Let $A\colon (0,T)\times V \to V^*$, $B\colon V \to V^*$, $F\colon (0,T)\times V\times V \to 2^{Z^*}$ be given operators and let $f\colon (0,T) \to V^*$. Also, let u_0 and v_0 be prescribed initial data. Then, the nonlinear evolutionary inclusion under consideration is as follows.

Problem 5.1. *Find $u \in \mathcal{V}$ such that $u' \in \mathcal{W}$ and*

$$\left.\begin{aligned} & u''(t) + A(t,u'(t)) + Bu(t) + F(t,u(t),u'(t)) \ni f(t) \ \text{ a.e. } t \in (0,T),\\ & u(0) = u_0,\ \ u'(0) = v_0. \end{aligned}\right\}$$

We note that the initial conditions in Problem 5.1 have sense in V and H, respectively, since the embeddings $\{\, v \in \mathcal{V} \mid v' \in \mathcal{W}\,\} \subset C(0,T;V)$ and $\mathcal{W} \subset C(0,T;H)$ hold, see Propositions 2.46 and 2.54. A solution to Problem 5.1 is understood as follows.

Definition 5.2. A function $u \in \mathcal{V}$ is a solution of Problem 5.1 if and only if $u' \in \mathcal{W}$ and there exists $\zeta \in \mathcal{Z}^*$ such that

$$\left.\begin{aligned} & u''(t) + A(t,u'(t)) + Bu(t) + \zeta(t) = f(t) \ \text{ a.e. } t \in (0,T),\\ & \zeta(t) \in F(t,u(t),u'(t)) \ \text{ a.e. } t \in (0,T),\\ & u(0) = u_0,\ \ u'(0) = v_0. \end{aligned}\right\}$$

We remark that if u is a solution of Problem 5.1, then it has the regularity $u \in C(0,T;V)$, $u' \in C(0,T;H)$ and $u'' \in \mathcal{V}^*$.

We need the following hypotheses on the data.

$$\left.\begin{aligned} & A\colon (0,T)\times V \to V^* \text{ is such that}\\ & \text{(a) } A(\cdot,v) \text{ is measurable on } (0,T) \text{ for all } v \in V.\\ & \text{(b) } A(t,\cdot) \text{ is pseudomonotone on } V \text{ for a.e. } t \in (0,T).\\ & \text{(c) } \|A(t,v)\|_{V^*} \le a_0(t) + a_1\|v\|_V \text{ for all } v \in V,\ \text{a.e. } t \in (0,T)\\ & \quad\ \text{with } a_0 \in L^2(0,T),\ a_0 \ge 0 \text{ and } a_1 > 0.\\ & \text{(d) } \langle A(t,v), v\rangle_{V^*\times V} \ge \alpha\|v\|_V^2 \text{ for all } v \in V,\ \text{a.e. } t \in (0,T)\\ & \quad\ \text{with } \alpha > 0. \end{aligned}\right\} \qquad (5.1)$$

$$B \in \mathcal{L}(V,V^*) \text{ is symmetric and monotone.} \qquad (5.2)$$

$$\left.\begin{array}{l} F\colon (0,T)\times V\times V\to \mathcal{P}_{fc}(Z^*) \text{ is such that} \\ \quad \text{(a) } F(\cdot,u,v) \text{ is measurable on } (0,T) \text{ for all } u,v\in V. \\ \quad \text{(b) } F(t,\cdot,\cdot) \text{ is upper semicontinuous from } V\times V \text{ into } w\text{–}Z^* \\ \qquad \text{for a.e. } t\in(0,T), \text{ where } V\times V \text{ is endowed with} \\ \qquad (Z\times Z) \text{ topology.} \\ \quad \text{(c) } \|F(t,u,v)\|_{Z^*}\le d_0(t)+d_1\|u\|_V+d_2\|v\|_V \text{ for all } u,v\in V, \\ \qquad \text{a.e. } t\in(0,T) \text{ with } d_0\in L^2(0,T) \text{ and } d_0,d_1,d_2\ge 0. \end{array}\right\} \tag{5.3}$$

$$f\in\mathcal{V}^*,\ u_0\in V,\ v_0\in H. \tag{5.4}$$

$$\alpha > 2\sqrt{3}\,c_e\,(d_1T+d_2). \tag{5.5}$$

We recall that an operator $B\colon V\to V^*$ is symmetric if $\langle Bu,v\rangle_{V^*\times V} = \langle Bv,u\rangle_{V^*\times V}$, for all $u,\ v\in V$. It is easy to see that every symmetric operator $B\colon V\to V^*$ is a linear operator. Using this result it follows that a symmetric operator $B\colon V\to V^*$ is monotone if $\langle Bv,v\rangle_{V^*\times V}\ge 0$ for all $v\in V$. Note also that in (5.3)(c) we adopt the convention introduced on page 96. More precisely, this inequality is understood in the following sense: given $u,\ v\in V$, we have $\|\zeta\|_{Z^*}\le d_0(t)+d_1\|u\|_V+d_2\|v\|_V$ for all $\zeta\in F(t,u,v)$, a.e. $t\in(0,T)$.

We underline that the condition (5.5) gives a restriction on the length of time interval T unless $d_1 = 0$. It means that under (5.5) the existence result of Theorem 5.4 is local and holds for a sufficiently small time interval. On the other hand, if the condition (5.5) is satisfied with $d_1 = 0$, then the existence result is global in time. For example, if the multifunction F is independent of u, i.e., $F(t,u,v) = F(t,v)$ for all $u,\ v\in V$, a.e. $t\in(0,T)$, then we may choose $d_1 = 0$ in (5.3)(c) and in this case the hypothesis (5.5) gives no restriction on the length of time interval T.

In the following, we justify the existence of $\mathcal{Z}^*$ selections of the multifunction F which appears in Definition 5.2. It is known that given a measurable space $(\mathcal{O},\Sigma)$, a separable metric space X and a metric space Y, a multifunction $\mathcal{F}\colon\mathcal{O}\times X\to\mathcal{P}(Y)$ which is measurable in $\omega\in\mathcal{O}$ and upper semicontinuous in $x\in X$ is not necessarily jointly measurable (see Example 7.2 in Chap. 2 of [109]). As a consequence, the theorems on the existence of measurable selections of measurable multifunctions, presented e.g. in Chap. 4 of [66], are not directly applicable in this case. Therefore, it is not immediately clear that, under the hypothesis (5.3), the multifunction $t\mapsto F(t,u(t),u'(t))$ has a measurable selection. The following lemma deals with this issue. To present it, we define a multifunction $S_F\colon W^{1,2}(0,T;V)\to 2^{\mathcal{Z}^*}$ by

$$S_F(u) = \{\,\zeta\in\mathcal{Z}^*\mid \zeta(t)\in F(t,u(t),u'(t))\ \text{ a.e. } t\in(0,T)\,\}$$

for all $u\in W^{1,2}(0,T;V)$.

Lemma 5.3. *If* $F\colon (0,T) \times V \times V \to \mathcal{P}_{fc}(Z^*)$ *satisfies* (5.3), *then* S_F *is* $\mathcal{P}_{wkc}(\mathcal{Z}^*)$*–valued.*

Proof. It is easy to see that S_F has convex and weakly compact values. We show that its values are nonempty. Let $u \in W^{1,2}(0,T;V)$. Then, by Theorem 2.35 (ii), there exist two sequences $\{s_n\}, \{r_n\} \subset \mathcal{V}$ of simple functions such that

$$s_n(t) \to u(t), \;\; r_n(t) \to u'(t) \text{ in } V, \text{ a.e. } t \in (0,T). \tag{5.6}$$

From hypothesis (5.3)(a), the multifunction $t \mapsto F(t, s_n(t), r_n(t))$ is measurable from $(0,T)$ into $\mathcal{P}_{fc}(Z^*)$. Applying Theorem 3.18, for every $n \geq 1$, there exists a measurable function $\zeta_n\colon (0,T) \to Z^*$ such that $\zeta_n(t) \in F(t, s_n(t), r_n(t))$ a.e. $t \in (0,T)$. Next, from (5.3)(c), we have

$$\|\zeta_n\|_{\mathcal{Z}^*} \leq \sqrt{3}\left(\|d_0\|_{L^2(0,T)} + d_1\|s_n\|_{\mathcal{V}} + d_2\|r_n\|_{\mathcal{V}}\right).$$

Hence, $\{\zeta_n\}$ remains in a bounded subset of $\mathcal{Z}^*$. Thus, by passing to a subsequence, if necessary, we may suppose, by Theorem 1.36, that $\zeta_n \to \zeta$ weakly in $\mathcal{Z}^*$ with $\zeta \in \mathcal{Z}^*$. From Proposition 3.16 it follows that

$$\zeta(t) \in \overline{\text{conv}}\left((w\text{–}Z^*)\text{- }\limsup\{\zeta_n(t)\}_{n\geq 1}\right) \text{ a.e. } t \in (0,T), \tag{5.7}$$

where $\overline{\text{conv}}$ denotes the closed convex hull of a set. Recalling that the graph of an upper semicontinuous multifunction with closed values is closed (see Proposition 3.12), from (5.3)(b) we get for a.e. $t \in (0,T)$: if $w_n \in F(t, \xi_n, \eta_n)$, $w_n \in Z^*$, $w_n \to w$ weakly in Z^*, $\xi_n, \eta_n \in V$, $\xi_n \to \xi$, $\eta_n \to \eta$ in Z, then $w \in F(t, \xi, \eta)$. Therefore, by (5.6), we have

$$(w\text{–}Z^*)\text{- }\limsup F(t, s_n(t), r_n(t)) \subset F(t, u(t), u'(t)) \text{ a.e. } t \in (0,T), \tag{5.8}$$

where the Kuratowski upper limit of sets is introduced in Definition 3.14. So, from (5.7) and (5.8), we deduce that

$$\begin{aligned}
\zeta(t) &\in \overline{\text{conv}}\left((w\text{–}Z^*)\text{- }\limsup\{\zeta_n(t)\}_{n\geq 1}\right)\\
&\subset \overline{\text{conv}}\left((w\text{–}Z^*)\text{- }\limsup F(t, s_n(t), r_n(t))\right)\\
&\subset F(t, u(t), u'(t)) \text{ a.e. } t \in (0,T).
\end{aligned}$$

Since $\zeta \in \mathcal{Z}^*$ and $\zeta(t) \in F(t, u(t), u'(t))$ a.e. $t \in (0,T)$, it is clear that $\zeta \in S_F(u)$. This proves that S_F has nonempty values and completes the proof of the lemma. □

The main existence result for Problem 5.1 reads as follows.

Theorem 5.4. *Assume that* (5.1)–(5.2) *hold. Then Problem* 5.1 *has at least one solution.*

Before providing a proof we need some preliminaries. First, we define the operator $K: \mathcal{V} \to C(0,T;V)$ by equality

$$Kv(t) = \int_0^t v(s)\,ds + u_0 \tag{5.9}$$

for all $v \in \mathcal{V}$. Then, Problem 5.1 can be formulated as follows

$$\left.\begin{array}{l} \text{find } z \in \mathcal{W} \text{ such that} \\ z'(t) + A(t,z(t)) + B(Kz(t)) + F(t,Kz(t),z(t)) \ni f(t) \\ \hfill \text{a.e. } t \in (0,T), \\ z(0) = v_0. \end{array}\right\} \tag{5.10}$$

We note that a function $z \in \mathcal{W}$ solves (5.10) if and only if $u = Kz$ is a solution to Problem 5.1.

Next, for $v_0 \in V$, we define the operators $\mathcal{A}_0: \mathcal{V} \to \mathcal{V}^*$, $\mathcal{B}_0: \mathcal{V} \to \mathcal{V}^*$, and $\mathcal{F}_0: \mathcal{V} \to 2^{\mathcal{Z}^*}$ by

$$(\mathcal{A}_0 v)(t) = A(t, v(t) + v_0), \tag{5.11}$$

$$(\mathcal{B}_0 v)(t) = B(K(v + v_0)(t)), \tag{5.12}$$

$$(\mathcal{F}_0 v)(t) = F(t, K(v + v_0)(t), v(t) + v_0) \tag{5.13}$$

for $v \in \mathcal{V}$ and a.e. $t \in (0,T)$. We observe that

$$\mathcal{A}_0 v = \mathcal{A}(v + v_0), \quad \mathcal{B}_0 v = \mathcal{B}K(v + v_0), \quad \mathcal{F}_0 v = \mathcal{F}(v + v_0),$$

where $\mathcal{A}$, $\mathcal{B}$, and $\mathcal{F}$ are the Nemytski operators given by

$$(\mathcal{A}v)(t) = A(t, v(t)), \quad (\mathcal{B}v)(t) = Bv(t), \quad \text{and} \quad (\mathcal{F}v)(t) = F(t, Kv(t), v(t))$$

for $v \in \mathcal{V}$ and a.e. $t \in (0,T)$. Also, below we shall use the operator $L : D(L) \subset \mathcal{V} \to \mathcal{V}^*$ defined on page 84.

We collect the properties of the operators $\mathcal{A}_0$, $\mathcal{B}_0$, and $\mathcal{F}_0$ in the following three lemmas.

Lemma 5.5. *If* (5.1) *holds and* $v_0 \in V$, *then the operator* $\mathcal{A}_0: \mathcal{V} \to \mathcal{V}^*$ *defined by* (5.11) *satisfies*

(a) $\|\mathcal{A}_0 v\|_{\mathcal{V}^*} \le \widehat{a}_0 + \widehat{a}_1 \|v\|_{\mathcal{V}}$ *for all* $v \in \mathcal{V}$ *with* $\widehat{a}_0 \ge 0$ *and* $\widehat{a}_1 > 0$.
(b) $\langle \mathcal{A}_0 v, v\rangle_{\mathcal{V}^* \times \mathcal{V}} \ge \frac{\alpha}{2}\|v\|_{\mathcal{V}}^2 - \alpha_1 \|v\|_{\mathcal{V}} - \alpha_2$ *for all* $v \in \mathcal{V}$ *with* $\alpha_1, \alpha_2 \ge 0$.

(c) $\mathcal{A}_0$ *is demicontinuous.*
(d) $\mathcal{A}_0$ *is L-pseudomonotone.*

Also, if (5.1) *holds, then the Nemytski operator* $\mathcal{A}\colon \mathcal{V} \to \mathcal{V}^*$ *corresponding to* A *has the following property:*

(e) *for every sequence* $\{v_n\} \subset \mathcal{W}$ *with* $v_n \to v$ *weakly in* $\mathcal{W}$ *and*

$$\limsup \langle \mathcal{A}v_n, v_n - v\rangle_{\mathcal{V}^*\times\mathcal{V}} \le 0,$$

it follows that $\langle \mathcal{A}v_n, v_n\rangle_{\mathcal{V}^*\times\mathcal{V}} \to \langle \mathcal{A}v, v\rangle_{\mathcal{V}^*\times\mathcal{V}}$ *and* $\mathcal{A}v_n \to \mathcal{A}v$ *weakly in* $\mathcal{V}^*$.

Lemma 5.6. *If* (5.2) *holds and* $v_0 \in V$, *then the operator* $\mathcal{B}_0\colon \mathcal{V} \to \mathcal{V}^*$ *defined by* (5.12) *satisfies*

(a) $\|\mathcal{B}_0 v\|_{\mathcal{V}^*} \le b_1(1 + \|v\|_{\mathcal{V}})$ *for all* $v \in \mathcal{V}$ *with* $b_1 > 0$.
(b) $\langle \mathcal{B}_0 v, v\rangle_{\mathcal{V}^*\times\mathcal{V}} \ge -b_2\|v\|_{\mathcal{V}} - b_3$ *for all* $v \in \mathcal{V}$ *with* $b_2, b_3 \ge 0$.
(c) $\|\mathcal{B}_0 v - \mathcal{B}_0 w\|_{\mathcal{V}^*} \le b_4\|v - w\|_{\mathcal{V}}$ *for all* $v, w \in \mathcal{V}$ *with* $b_4 > 0$.
(d) $\mathcal{B}_0$ *is monotone.*
(e) $\mathcal{B}_0$ *is weakly continuous.*

Also, if (5.2) *holds, then the Nemytski operator* $\mathcal{B}$ *corresponding to* B *is such that*

(f) $\langle \mathcal{B}v, v'\rangle_{\mathcal{V}^*\times\mathcal{V}} \ge 0$ *for all* $v \in \mathcal{W}$ *such that* $v(0) = 0$.

Lemma 5.7. *If* (5.3) *holds and* $v_0 \in V$, *then the operator* $\mathcal{F}_0\colon \mathcal{V} \to 2^{\mathcal{Z}^*}$ *defined by* (5.13) *satisfies*

(a) $\|\mathcal{F}_0 v\|_{\mathcal{Z}^*} \le \sqrt{3}\,(d_1 T + d_2)\|v\|_{\mathcal{V}} + d$ *for all* $v \in \mathcal{V}$ *with* $d \ge 0$.
(b) $\mathcal{F}_0 v$ *has nonempty convex and weakly compact values in* $\mathcal{Z}^*$.
(c) $\langle \zeta, v\rangle_{\mathcal{Z}^*\times\mathcal{Z}} \ge -\sqrt{3}\,c_e\,(d_1 T + d_2)\|v\|^2_{\mathcal{V}} - d\, c_e\|v\|_{\mathcal{V}}$ *for all* $\zeta \in \mathcal{F}_0 v$, $v \in \mathcal{V}$ *with* $d \ge 0$.
(d) *for every* $v_n, v \in \mathcal{V}$ *with* $v_n \to v$ *in* $\mathcal{Z}$ *and every* $\zeta_n, \zeta \in \mathcal{Z}^*$ *with* $\zeta_n \to \zeta$ *weakly in* $\mathcal{Z}^*$, *if* $\zeta_n \in \mathcal{F}_0 v_n$, *then* $\zeta \in \mathcal{F}_0 v$.

In the proof of Theorem 5.4 we also need the following result concerning the a priori estimates on the solutions.

Lemma 5.8. *Assume that hypotheses* (5.1)–(5.5) *hold and let* u *be a solution to Problem* 5.1. *Then there exists a positive constant* $\overline{c}$ *such that*

$$\|u\|_{C(0,T;V)} + \|u'\|_{\mathcal{W}} \le \overline{c}\;(1 + \|u_0\|_V + \|v_0\|_H + \|f\|_{\mathcal{V}^*})\,.$$

For the convenience of the reader, the proofs of Lemmas 5.5–5.8 are postponed to the end of this section.

Proof of Theorem 5.4. In order to prove Theorem 5.4, we solve the first-order evolutionary inclusion (5.10). To this end we proceed in two steps, as follows: in the first step we suppose that $v_0 \in V$ and then, in the second step, we treat the general case $v_0 \in H$.

Step 1. Assume that $v_0 \in V$. Consider the Cauchy problem for the first-order evolutionary inclusion

$$\left.\begin{array}{l}\text{find } z \in \mathcal{W} \text{ such that}\\ z'(t) + (\mathcal{A}_0 z)(t) + (\mathcal{B}_0 z)(t) + (\mathcal{F}_0 z)(t) \ni f(t) \ \text{ a.e. } t \in (0,T),\\ z(0) = 0,\end{array}\right\} \tag{5.14}$$

and note that $z \in \mathcal{W}$ solves (5.10) if and only if $z - v_0 \in \mathcal{W}$ solves (5.14). Let

$$L\colon D(L) \subset \mathcal{V} \to \mathcal{V}^*, \quad D(L) = \{\, v \in \mathcal{W} \mid v(0) = 0\,\}$$

be the operator given by $Lv = v'$ for all $v \in D(L)$, already considered on page 84. Then, problem (5.14) can be written as

$$\text{find } z \in D(L) \ \text{ such that } \ (L + \mathcal{G})z \ni f, \tag{5.15}$$

where $\mathcal{G}\colon \mathcal{V} \to 2^{\mathcal{V}^*}$ is defined by

$$\mathcal{G}v = (\mathcal{A}_0 + \mathcal{B}_0 + \mathcal{F}_0)\, v \tag{5.16}$$

for all $v \in \mathcal{V}$. The existence of solutions to (5.15) will be proved by applying Theorem 3.63 on page 83. It follows from Lemma 3.64 that the operator L is densely defined linear maximal monotone operator. Therefore, in order to apply Theorem 3.63, it is enough to show that $\mathcal{G}$ is bounded, coercive, and pseudomonotone with respect to $D(L)$.

The fact that $\mathcal{G}$ is a bounded operator, i.e., it maps bounded subsets of $\mathcal{V}$ into bounded subsets of $\mathcal{V}^*$, follows from the continuity of the embedding $\mathcal{Z}^* \subset \mathcal{V}^*$, Lemma 5.5(a), Lemma 5.6(a), and Lemma 5.7(a). The coercivity of $\mathcal{G}$ is a consequence of the following inequality: for all $v \in \mathcal{V}$ and $v^* \in \mathcal{G}v$, we have

$$\begin{aligned}\langle v^*, v\rangle_{\mathcal{V}^*\times\mathcal{V}} &= \langle \mathcal{A}_0 v, v\rangle_{\mathcal{V}^*\times\mathcal{V}} + \langle \mathcal{B}_0 v, v\rangle_{\mathcal{V}^*\times\mathcal{V}} + \langle \zeta, v\rangle_{\mathcal{V}^*\times\mathcal{V}}\\ &\ge \frac{\alpha}{2}\,\|v\|_{\mathcal{V}}^2 - \alpha_1\|v\|_{\mathcal{V}} - \alpha_2 - b_2\|v\|_{\mathcal{V}} - b_3\\ &\quad -\sqrt{3}\,c_e\,(d_1\,T + d_2)\|v\|_{\mathcal{V}}^2 - d\,c_e\|v\|_{\mathcal{V}}\\ &\ge \left(\frac{\alpha}{2} - \sqrt{3}\,c_e\,(d_1\,T + d_2)\right)\|v\|_{\mathcal{V}}^2 - b_5\|v\|_{\mathcal{V}} - b_6\end{aligned}$$

with $b_5, b_6 \geq 0$, where $\zeta \in \mathcal{F}_0 v$. This inequality follows from Lemma 5.5(b), Lemma 5.6(c), and Lemma 5.7(c). Due to (5.5) it turns out that the operator $\mathcal{G}$ is coercive.

We prove now that $\mathcal{G}$ is L-pseudomonotone. From Lemma 5.7(b) it follows that for every $v \in \mathcal{V}$, $\mathcal{G}v$ is a nonempty convex and weakly compact subset of $\mathcal{V}^*$. We show that $\mathcal{G}$ is upper semicontinuous in $\mathcal{V} \times (w\text{–}\mathcal{V}^*)$. To this end, using Proposition 3.8 on page 53, it is enough to show that if $\mathcal{K} \subset \mathcal{V}^*$ is weakly closed, then the set

$$\mathcal{G}^-(\mathcal{K}) = \{\, v \in \mathcal{V} \mid \mathcal{G}v \cap \mathcal{K} \neq \emptyset \,\}$$

is closed in $\mathcal{V}$. Let $\{v_n\} \subset \mathcal{G}^-(\mathcal{K})$ and $v_n \to v$ in $\mathcal{V}$. Then, for all $n \in \mathbb{N}$ there is $v_n^* \in \mathcal{G}v_n \cap \mathcal{K}$ such that

$$v_n^* = \mathcal{A}_0 v_n + \mathcal{B}_0 v_n + \zeta_n \ \text{ with } \ \zeta_n \in \mathcal{F}_0 v_n. \tag{5.17}$$

Since $\mathcal{G}$ is a bounded operator, it is clear that $\{v_n^*\}$ belongs to a bounded subset of $\mathcal{V}^*$. So we may suppose, by passing to a subsequence if necessary, that

$$v_n^* \to v^* \ \text{ weakly in } \ \mathcal{V}^* \tag{5.18}$$

and $v^* \in \mathcal{K}$, since $\mathcal{K}$ is weakly closed in $\mathcal{V}^*$. Similarly, from Lemma 5.7(a), it follows that

$$\zeta_n \to \zeta \ \text{ weakly in } \ \mathcal{Z}^*, \tag{5.19}$$

at least for a subsequence, with $\zeta \in \mathcal{Z}^*$. Using the continuity of the embedding $\mathcal{V} \subset \mathcal{Z}$ and Lemma 5.7(d), we obtain that $\zeta \in \mathcal{F}_0 v$. On the other hand, by Lemma 5.5(c) and Lemma 5.6(c), we deduce that

$$\mathcal{A}_0 v_n \to \mathcal{A}_0 v \ \text{ weakly in } \mathcal{V}^*, \tag{5.20}$$

$$\mathcal{B}_0 v_n \to \mathcal{B}_0 v \ \text{ in } \mathcal{V}^*. \tag{5.21}$$

We use the convergences (5.18)–(5.21) to pass to the limit in (5.17). As a result we obtain $v^* = \mathcal{A}_0 v + \mathcal{B}_0 v + \zeta$ with $\zeta \in \mathcal{F}_0 v$ and $v^* \in \mathcal{K}$. Thus, we have $v^* \in \mathcal{G}v \cap \mathcal{K}$, i.e., $v^* \in \mathcal{G}^-(\mathcal{K})$. This shows that $\mathcal{G}^-(\mathcal{K})$ is closed in $\mathcal{V}$ and, therefore, $\mathcal{G}$ is upper semicontinuous from $\mathcal{V}$ into $w\text{–}\mathcal{V}^*$.

To conclude the proof that $\mathcal{G}$ is L-pseudomonotone, it is enough to show the condition (3) in Definition 3.62. To this end we assume that $\{v_n\} \subset D(L)$, $v_n \to v$ weakly in $\mathcal{W}$, $v_n^* \in \mathcal{G}v_n$, $v_n^* \to v^*$ weakly in $\mathcal{V}^*$ with $v^* \in \mathcal{V}^*$ and we suppose that

$$\limsup \, \langle v_n^*, v_n - v \rangle_{\mathcal{V}^* \times \mathcal{V}} \leq 0. \tag{5.22}$$

Thus, $v_n^* = \mathcal{A}_0 v_n + \mathcal{B}_0 v_n + \zeta_n$ with $\zeta_n \in \mathcal{F}_0 v_n$, for all $n \in \mathbb{N}$. From the boundedness of $\mathcal{F}_0$ (guaranteed by Lemma 5.7(a)) we know that $\{\zeta_n\}$ lies in a bounded subset of $\mathcal{Z}^*$. By passing to a subsequence, if necessary, we may assume that

$$\zeta_n \to \zeta \ \text{ weakly in } \mathcal{Z}^* \tag{5.23}$$

for some $\zeta \in \mathcal{Z}^*$. Since $\mathcal{W} \subset \mathcal{Z}$ compactly (see Theorem 2.56 on page 49), we also assume that

$$v_n \to v \ \text{in } \mathcal{Z}. \tag{5.24}$$

Next, we use Lemma 5.7(d) to see that (5.23) and (5.24) imply that $\zeta \in \mathcal{F}_0 v$. And, using again Lemma 5.7(a) and (5.24), we have

$$|\langle \zeta_n, v_n - v\rangle_{\mathcal{Z}^*\times\mathcal{Z}}| \le \|\zeta_n\|_{\mathcal{Z}^*}\|v_n - v\|_{\mathcal{Z}} \le c\,(1 + \|v_n\|_{\mathcal{V}})\,\|v_n - v\|_{\mathcal{Z}} \to 0 \tag{5.25}$$

as $n \to +\infty$. Next, by the monotonicity of $\mathcal{B}_0$ (guaranteed by Lemma 5.6(d)) and the convergence $v_n \to v$ weakly in $\mathcal{V}$, we deduce

$$\limsup\,\langle \mathcal{B}_0 v_n, v - v_n\rangle_{\mathcal{V}^*\times\mathcal{V}} \le \limsup\,\langle \mathcal{B}_0 v, v - v_n\rangle_{\mathcal{V}^*\times\mathcal{V}} = 0. \tag{5.26}$$

Combining (5.23), (5.25), and (5.26), we have

$$\begin{aligned}\limsup\,\langle \mathcal{A}_0 v_n, v_n - v\rangle_{\mathcal{V}^*\times\mathcal{V}} \le{}& \limsup\,\langle v_n^*, v_n - v\rangle_{\mathcal{V}^*\times\mathcal{V}}\\ &+ \limsup\,\langle \mathcal{B}_0 v_n, v - v_n\rangle_{\mathcal{V}^*\times\mathcal{V}}\\ &+ \lim\,\langle \zeta_n, v - v_n\rangle_{\mathcal{V}^*\times\mathcal{V}} \le 0.\end{aligned}$$

Thus, the L-pseudomonotonicity of $\mathcal{A}_0$ (guaranteed by Lemma 5.5(d)) implies that

$$\mathcal{A}_0 v_n \to \mathcal{A}_0 v \ \text{ weakly in } \mathcal{V}^* \tag{5.27}$$

and

$$\langle \mathcal{A}_0 v_n, v_n - v\rangle_{\mathcal{V}^*\times\mathcal{V}} \to 0. \tag{5.28}$$

Due to (5.23), (5.27), and Lemma 5.6(e), we have

$$v_n^* = \mathcal{A}_0 v_n + \mathcal{B}_0 v_n + \zeta_n \to \mathcal{A}_0 v + \mathcal{B}_0 v + \zeta \ \text{ weakly in } \mathcal{V}^*. \tag{5.29}$$

We combine (5.18) and (5.29), then we use the fact that $\zeta \in \mathcal{F}_0 v$ and the definition (5.16) of the operator $\mathcal{G}$ to see that $v^* \in \mathcal{G}v$. Furthermore, from (5.22), (5.25), and (5.28), we infer that

$$\begin{aligned}\limsup\,\langle \mathcal{B}_0 v_n, v_n - v\rangle_{\mathcal{V}^*\times\mathcal{V}} \le{}& \limsup\,\langle v_n^*, v_n - v\rangle_{\mathcal{V}^*\times\mathcal{V}}\\ &- \lim\,\langle \mathcal{A}_0 v_n, v_n - v\rangle_{\mathcal{V}^*\times\mathcal{V}}\\ &- \lim\,\langle \zeta_n, v_n - v\rangle_{\mathcal{V}^*\times\mathcal{V}} \le 0.\end{aligned}$$

This inequality and (5.26) yield $\lim\,\langle \mathcal{B}_0 v_n, v_n - v\rangle_{\mathcal{V}^*\times\mathcal{V}} = 0$ which implies that

$$\lim\,\langle \mathcal{B}_0 v_n, v_n\rangle_{\mathcal{V}^*\times\mathcal{V}} = \langle \mathcal{B}_0 v, v\rangle_{\mathcal{V}^*\times\mathcal{V}}. \tag{5.30}$$

Now we pass to the limit in the equality

$$\langle v_n^*, v_n\rangle_{\mathcal{V}^*\times\mathcal{V}} = \langle \mathcal{A}_0 v_n, v_n\rangle_{\mathcal{V}^*\times\mathcal{V}} + \langle \mathcal{B}_0 v_n, v_n\rangle_{\mathcal{V}^*\times\mathcal{V}} + \langle \zeta_n, v_n\rangle_{\mathcal{V}^*\times\mathcal{V}}$$

and use (5.23), (5.24), (5.27), (5.28), (5.30) to obtain

$$\lim \langle v_n^*, v_n\rangle_{\mathcal{V}^*\times\mathcal{V}} = \langle v^*, v\rangle_{\mathcal{V}^*\times\mathcal{V}}$$

with $v^* \in \mathcal{G}v$. This shows that $\mathcal{G}$ is pseudomonotone with respect to $D(L)$.

We are now in a position to apply Theorem 3.63 to deduce that the problem (5.14) admits a solution $z \in D(L)$. It follows from here that $z + v_0$ is a solution to (5.10) and, therefore, $u = K(z + v_0)$ solves Problem 5.1 in the case $v_0 \in V$.

Step 2. We pass to the second step of the proof in which we suppose that $v_0 \in H$. Since the embedding $V \subset H$ is dense, there exists a sequence $\{v_{0n}\} \subset V$ such that $v_{0n} \to v_0$ in H, as $n \to +\infty$. We denote by u_n a solution of the problem

$$\left.\begin{array}{l} \text{find } u_n \in \mathcal{V} \text{ such that } u_n' \in \mathcal{W} \text{ and} \\ u_n''(t) + A(t, u_n'(t)) + Bu_n(t) + F(t, u_n(t), u_n'(t)) \ni f(t) \\ \hfill \text{a.e. } t \in (0, T), \\ u_n(0) = u_0,\ u_n'(0) = v_{0n}. \end{array}\right\} \tag{5.31}$$

The existence of u_n which solves (5.31), for all $n \in \mathbb{N}$, follows from the first step of the proof. So, we have

$$u_n''(t) + A(t, u_n'(t)) + Bu_n(t) + \zeta_n(t) = f(t) \ \text{ a.e. } t \in (0, T) \tag{5.32}$$

with

$$\zeta_n(t) \in F(t, u_n(t), u_n'(t)) \ \text{ a.e. } t \in (0, T) \tag{5.33}$$

and the corresponding initial conditions. From Lemma 5.8 it follows that there exists a subsequence of $\{u_n\}$, again denoted $\{u_n\}$, such that

$$u_n \to u, \quad u_n' \to u' \quad \text{both weakly in } \mathcal{V},$$

$$u_n'' \to u'' \ \text{ weakly in } \mathcal{V}^*.$$

Our goal is to show that u is a solution to Problem 5.1. To this end, first, we remark that

$$u_n \to u \ \text{ weakly in } W^{1,2}(0, T; V) \ \text{ and } \ u_n' \to u' \ \text{ weakly in } \mathcal{W}. \tag{5.34}$$

From Lemma 2.55 on page 49 we have $u_n(t) \to u(t)$ weakly in V and $u_n'(t) \to u'(t)$ weakly in H, for all $t \in [0, T]$. Hence $u_0 = u_n(0) \to u(0)$ weakly in V, which gives $u(0) = u_0$. Similarly, it results that $v_{0n} = u_n'(0) \to u'(0)$ weakly in H, which implies that $u'(0) = v_0$.

Using the compactness of the embedding $\mathcal{W} \subset \mathcal{Z}$, from (5.34) we have $u_n \to u$, $u_n' \to u'$, both in $\mathcal{Z}$, and, subsequently,

$$u_n(t) \to u(t), \quad u_n'(t) \to u'(t) \quad \text{both in } Z, \text{ for a.e. } t \in (0,T).$$

Exploiting (5.3)(c) and (5.33), we may suppose that

$$\zeta_n \to \zeta \text{ weakly in } \mathcal{Z}^*. \tag{5.35}$$

Therefore, we are in a position to apply the convergence result stated in Theorem 3.13, to the inclusion (5.33). In this way we deduce that

$$\zeta(t) \in F(t, u(t), u'(t)) \text{ for a.e. } t \in (0,T). \tag{5.36}$$

In what follows we prove that

$$\mathcal{A}u_n' \to \mathcal{A}u' \text{ weakly in } \mathcal{V}^*. \tag{5.37}$$

First, since $\langle f, u_n' - u'\rangle_{\mathcal{V}^*\times\mathcal{V}} \to 0$ and $\langle \zeta_n, u_n' - u'\rangle_{\mathcal{Z}^*\times\mathcal{Z}} \to 0$ (recall that $\zeta_n \to \zeta$ weakly in $\mathcal{Z}^*$ and $u_n' \to u'$ in $\mathcal{Z}$), by (5.30), we get

$$\begin{aligned}\limsup \langle \mathcal{A}u_n', u_n' - u'\rangle_{\mathcal{V}^*\times\mathcal{V}} &\le \limsup \langle u_n'', u' - u_n'\rangle_{\mathcal{V}^*\times\mathcal{V}} \\ &\quad + \limsup \langle \mathcal{B}u_n, u' - u_n'\rangle_{\mathcal{V}^*\times\mathcal{V}}. \end{aligned} \tag{5.38}$$

Next, from the integration by parts formula (Proposition 2.54 (iii) on page 49), we have

$$\begin{aligned}\langle u_n'' - u'', u_n' - u'\rangle_{\mathcal{V}^*\times\mathcal{V}} &= \frac{1}{2}\int_0^T \frac{d}{dt}\|u_n'(t) - u'(t)\|_H^2\, dt \\ &= \frac{1}{2}\|u_n'(T) - u'(T)\|_H^2 - \frac{1}{2}\|u_n'(0) - u'(0)\|_H^2,\end{aligned}$$

which implies that

$$\lim \langle u_n'' - u'', u' - u_n'\rangle_{\mathcal{V}^*\times\mathcal{V}} \le 0. \tag{5.39}$$

In addition, taking the property (f) of Lemma 5.6 into consideration, we obtain

$$\begin{aligned}\limsup \langle \mathcal{B}u_n, u' - u_n'\rangle_{\mathcal{V}^*\times\mathcal{V}} &= \limsup \Big(- \langle \mathcal{B}u - \mathcal{B}u_n, u' - u_n'\rangle_{\mathcal{V}^*\times\mathcal{V}} \\ &\quad + \langle \mathcal{B}u, u' - u_n'\rangle_{\mathcal{V}^*\times\mathcal{V}}\Big) \\ &\le \limsup \langle \mathcal{B}u, u' - u_n'\rangle_{\mathcal{V}^*\times\mathcal{V}} = 0.\end{aligned} \tag{5.40}$$

We combine (5.38)–(5.40) to find that

$$\limsup \langle \mathcal{A}u_n', u_n' - u' \rangle_{\mathcal{V}^* \times \mathcal{V}} \leq 0.$$

Since $u_n' \to u'$ weakly in $\mathcal{W}$, from Lemma 5.5(e), we deduce (5.37).

Finally, the convergences (5.35), (5.37), and the weak continuity of $\mathcal{B}$ (which can be proved by using an argument similar to that used in the proof of Lemma 5.6(e)) allow to pass to the limit in (5.32). We obtain

$$u''(t) + A(t, u'(t)) + Bu(t) + F(t, u(t), u'(t)) \ni f(t) \ \text{ a.e. } t \in (0, T),$$

which, together with (5.36) and conditions $u(0) = u_0$, $u'(0) = v_0$, implies that u is a solution to Problem 5.1. This completes the proof of the theorem. □

We conclude this section with the proofs of Lemmas 5.5–5.8.

Proof of Lemma 5.5. The property (a) follows easily from (5.1)(a), (c). The coercivity condition in (b) is a consequence of the inequality

$$\begin{aligned}
\langle \mathcal{A}_0 v, v \rangle_{\mathcal{V}^* \times \mathcal{V}} &= \int_0^T \Big(\langle A(t, v(t) + v_0), v(t) + v_0 \rangle_{V^* \times V} \\
&\qquad - \langle A(t, v(t) + v_0), v_0 \rangle_{V^* \times V} \Big)\, dt \\
&\geq \alpha \int_0^T \left(\frac{1}{2} \|v(t)\|_V^2 - \|v_0\|_V^2 \right) dt - T\, \|v_0\|_V\, \|a_0\|_{L^2(0,T)} \\
&\qquad - a_1 \|v_0\|_V \int_0^T \|v(t) + v_0\|_V \, dt \\
&\geq \frac{\alpha}{2} \|v\|_{\mathcal{V}}^2 - \alpha_1 \|v\|_{\mathcal{V}} - \alpha_2,
\end{aligned}$$

which is valid for all $v \in \mathcal{V}$. Here we have used (5.1)(c), (d), and the inequality $(a+b)^2 \geq \frac{1}{2} a^2 - b^2$, valid for all $a, b \in \mathbb{R}$.

The details on the proof of properties (c)–(e) can be found in Lemma 11 of [165] and, for this reason, we skip them. □

Proof of Lemma 5.6. In the proof we use the following elementary properties of the nonlinear operator $K : \mathcal{V} \to C(0, T; V)$ defined by (5.9):

$$\|Kv\|_{C(0,T;V)} \leq \sqrt{T}\, \|v\|_{\mathcal{V}} + \|u_0\|_V \ \text{ for all } v \in \mathcal{V}, \tag{5.41}$$

$$\|Kv - Kw\|_{C(0,T;V)} \leq \sqrt{T}\, \|v - w\|_{\mathcal{V}} \ \text{ for all } v, w \in \mathcal{V}. \tag{5.42}$$

Let $v \in \mathcal{V}$. Using (5.41), we have

$$\|\mathcal{B}_0 v\|_{\mathcal{V}^*}^2 = \int_0^T \|B(K(v+v_0)(t))\|_{V^*}^2 \, dt$$
$$\leq \|B\|_{\mathcal{L}(V,V^*)}^2 \int_0^T \|K(v+v_0)(t)\|_V^2 \, dt$$
$$\leq T\|B\|_{\mathcal{L}(V,V^*)}^2 \left(\sqrt{T}\|v+v_0\|_{\mathcal{V}} + \|u_0\|_V\right)^2.$$

Hence $\|\mathcal{B}_0 v\|_{\mathcal{V}^*} \leq \sqrt{T}\,\|B\|_{\mathcal{L}(V,V^*)}\left(\sqrt{T}\|v+v_0\|_{\mathcal{V}} + \|u_0\|_V\right)$ and the condition (a) follows.

Next, since B is a monotone symmetric operator and (5.41) guarantees that K is bounded, we have

$$\langle \mathcal{B}_0 v, v\rangle_{\mathcal{V}^*\times\mathcal{V}} = \int_0^T \langle B(K(v+v_0)(t)), (K(v+v_0))'(t) - v_0\rangle_{V^*\times V} \, dt$$
$$= \frac{1}{2}\int_0^T \frac{d}{dt}\langle B(K(v+v_0)(t)), K(v+v_0)(t)\rangle_{V^*\times V} \, dt$$
$$- \int_0^T \langle B(K(v+v_0)(t)), v_0\rangle_{V^*\times V} \, dt$$
$$\geq -\frac{1}{2}\,\|B\|_{\mathcal{L}(V,V^*)}\|u_0\|_V^2 - T\|B\|_{\mathcal{L}(V,V^*)}\|v_0\|_V\,\|K(v+v_0)\|_{C(0,T;V)}$$
$$\geq -\frac{1}{2}\,\|B\|_{\mathcal{L}(V,V^*)}\|u_0\|_V^2$$
$$-T\|B\|_{\mathcal{L}(V,V^*)}\|v_0\|_V\left(\sqrt{T}\,\|v+v_0\|_{\mathcal{V}} + \|u_0\|_V\right)$$

for all $v \in \mathcal{V}$, which proves the property (b).

In order to obtain (c), we use (5.42) to see that

$$\|\mathcal{B}_0 v - \mathcal{B}_0 w\|_{\mathcal{V}^*}^2 = \int_0^T \|B(K(v+v_0)(t)) - B(K(w+v_0)(t))\|_{V^*}^2 \, dt$$
$$\leq \|B\|_{\mathcal{L}(V,V^*)}^2 \int_0^T \|K(v+v_0)(t) - K(w+v_0)(t)\|_V^2 \, dt$$
$$\leq T^2\,\|B\|_{\mathcal{L}(V,V^*)}^2\|v-w\|_{\mathcal{V}}^2$$

for all $v, w \in \mathcal{V}$. It follows from here that the operator $\mathcal{B}_0$ is Lipschitz continuous, i.e., it satisfies condition (c).

Next, using the monotonicity and symmetry of the operator B, we obtain

$$
\begin{aligned}
&\langle \mathcal{B}_0 v - \mathcal{B}_0 w, v - w\rangle_{\mathcal{V}^*\times\mathcal{V}} \\
&\quad = \int_0^T \langle B(K(v+v_0)(t)) - B(K(w+v_0)(t)), \\
&\qquad\qquad (K(v+v_0))'(t) - (K(w+v_0))'(t)\rangle_{V^*\times V}\, dt \\
&\quad = \frac{1}{2}\int_0^T \frac{d}{dt}\langle B(K(v+v_0)(t)) - B(K(w+v_0)(t)), \\
&\qquad\qquad K(v+v_0)(t) - K(w+v_0)(t)\rangle_{V^*\times V}\, dt \\
&\quad = \frac{1}{2}\langle B(K(v+v_0)(T)) - B(K(w+v_0)(T)), \\
&\qquad\qquad K(v+v_0)(T) - K(w+v_0)(T)\rangle_{V^*\times V} \geq 0
\end{aligned}
$$

for all $v, w \in \mathcal{V}$, which proves that $\mathcal{B}_0$ is monotone and, therefore, condition (d) holds.

Next, we show that $\mathcal{B}_0$ is weakly continuous. To this end, we consider a sequence $\{v_n\} \subset \mathcal{V}$ such that $v_n \to v$ weakly in $\mathcal{V}$. Since, for all $t \in [0, T]$, the operator

$$
\mathcal{V} \ni w \mapsto \int_0^t w(s)\, ds \in V
$$

is linear and continuous, we have

$$
\int_0^t v_n(s)\, ds \to \int_0^t v(s)\, ds \quad \text{weakly in } V
$$

for all $t \in [0, T]$. Therefore,

$$
K(v_n + v_0)(t) \to K(v + v_0)(t) \quad \text{weakly in } V
$$

for all $t \in [0, T]$ and, subsequently,

$$
B(K(v_n + v_0)(t)) \to B(K(v + v_0)(t)) \quad \text{weakly in } V^*
$$

for all $t \in [0, T]$. In view of the property (a), we can apply the Lebesgue-dominated convergence theorem (Theorem 2.38 on page 42) to obtain

$$
\begin{aligned}
\langle \mathcal{B}_0 v_n, w\rangle_{\mathcal{V}^*\times\mathcal{V}} &= \int_0^T \langle B(K(v_n+v_0)(t)), w(t)\rangle_{V^*\times V}\, dt \\
&\to \int_0^T \langle B(K(v+v_0)(t)), w(t)\rangle_{V^*\times V}\, dt \\
&= \langle \mathcal{B}_0 v, w\rangle_{\mathcal{V}^*\times\mathcal{V}}
\end{aligned}
$$

for all $w \in \mathcal{V}$. We conclude from here that $\mathcal{B}_0 v_n \to \mathcal{B}_0 v$ weakly in $\mathcal{V}^*$ and, therefore, (e) holds.

Finally, we observe that by the monotonicity and symmetry of B, for all element $v \in \mathcal{W}$ which satisfies $v(0) = 0$ we have

$$\langle \mathcal{B}v, v' \rangle_{\mathcal{V}^* \times \mathcal{V}} = \frac{1}{2} \int_0^T \frac{d}{dt} \langle Bv(t), v(t) \rangle_{V^* \times V}\, dt$$

$$= \frac{1}{2} \langle Bv(T), v(T) \rangle_{V^* \times V} - \frac{1}{2} \langle Bv(0), v(0) \rangle_{V^* \times V} \geq 0.$$

This implies that (f) holds and completes the proof of Lemma 5.6. □

Proof of Lemma 5.7. For the proof of (a) we consider $v \in \mathcal{V}$ and $\zeta \in \mathcal{F}_0 v$. It follows from here that

$$\zeta(t) \in F(t, K(v + v_0)(t), v(t) + v_0) \ \text{ a.e. } t \in (0, T).$$

Moreover, from the estimates (5.41) and (5.3)(c), we have

$$\|\zeta(t)\|_{Z^*} \leq d_0(t) + d_1 \sqrt{T} \|v\|_{\mathcal{V}} + d_1 T \|v_0\|_V$$

$$+ d_1 \|u_0\|_V + d_2 \|v(t)\|_V + d_2 \|v_0\|_V.$$

Hence, using the inequality $(a + b + c)^2 \leq 3\,(a^2 + b^2 + c^2)$ with $a, b, c \geq 0$, we deduce

$$\|\zeta(t)\|_{Z^*}^2 \leq 3\, d_1^2\, T\, \|v\|_{\mathcal{V}}^2 + 3\, d_2^2\, \|v(t)\|_V^2$$

$$+ 3\,(d_0(t) + d_1 T \|v_0\|_V + d_1 \|u_0\|_V + d_2 \|v_0\|_V)^2$$

and

$$\|\zeta\|_{\mathcal{Z}^*}^2 = \int_0^T \|\zeta(t)\|_{Z^*}^2\, dt \leq 3\, d_1^2\, T^2\, \|v\|_{\mathcal{V}}^2 + 3\, d_2^2 \int_0^T \|v(t)\|_V^2\, dt$$

$$+ 3 \int_0^T (d_0(t) + d_1 T \|v_0\|_V + d_1 \|u_0\|_V + d_2 \|v_0\|_V)^2\, dt$$

$$\leq 3\,(d_1^2 T^2 + d_2^2)\, \|v\|_{\mathcal{V}}^2 + d^2 \ \text{ with } d \geq 0.$$

Therefore, we obtain

$$\|\zeta\|_{\mathcal{Z}^*} \leq \sqrt{3} \sqrt{d_1^2 T^2 + d_2^2}\, \|v\|_{\mathcal{V}} + d \leq \sqrt{3}\,(d_1 T + d_2)\, \|v\|_{\mathcal{V}} + d,$$

which shows that (a) holds.

Next, we turn to the proof of the property (b). From (5.3) we know that for $v \in \mathcal{V}$ the set $\mathcal{F}_0 v$ is nonempty and convex in $\mathcal{Z}^*$. In order to show that $\mathcal{F}_0 v$ is weakly compact in $\mathcal{Z}^*$, we shall prove that it is closed in $\mathcal{Z}^*$. Let $v \in \mathcal{V}$, $\{\zeta_n\} \subset \mathcal{F}_0 v$, $\zeta_n \to \zeta$ in $\mathcal{Z}^*$. Passing to a subsequence, again denoted $\{\zeta_n\}$, we have $\zeta_n(t) \to \zeta(t)$ in Z^* for a.e. $t \in (0,T)$. From the relation

$$\zeta_n(t) \in F(t, K(v+v_0)(t), v(t)+v_0) \quad \text{a.e. } t \in (0,T),$$

since the set is closed in Z^*, we get

$$\zeta(t) \in F(t, K(v+v_0)(t), v(t)+v_0) \quad \text{a.e. } t \in (0,T).$$

Hence $\zeta \in \mathcal{F}_0 v$ and thus $\mathcal{F}_0 v$ is closed in $\mathcal{Z}^*$ and convex, so it is also weakly closed in $\mathcal{Z}^*$. Since $\mathcal{F}_0 v$ is a bounded set (see property (a) of this lemma) in the reflexive Banach space $\mathcal{Z}^*$, we obtain that $\mathcal{F}_0 v$ is weakly compact in $\mathcal{Z}^*$, which ends the proof of (b).

For the proof of (c), consider $v \in \mathcal{V}$ and let $\zeta \in \mathcal{F}_0 v$. Using the property (a) we have

$$\begin{aligned} |\langle \zeta, v\rangle_{\mathcal{V}^*\times\mathcal{V}}| &= |\langle \zeta, v\rangle_{\mathcal{Z}^*\times\mathcal{Z}}| \le c_e \|\zeta\|_{\mathcal{Z}^*} \|v\|_{\mathcal{V}} \\ &\le \sqrt{3}\, c_e\, (d_1\, T + d_2) \|v\|_{\mathcal{V}}^2 + d\, c_e\, \|v\|_{\mathcal{V}}, \end{aligned}$$

where $d \ge 0$. Hence we obtain that

$$\langle \zeta, v\rangle_{\mathcal{V}^*\times\mathcal{V}} \ge -\sqrt{3}\, c_e\, (d_1\, T + d_2)\, \|v\|_{\mathcal{V}}^2 - d\, c_e\, \|v\|_{\mathcal{V}}$$

and, therefore, (c) follows.

Finally, we show the property (d). Let $v_n, v \in \mathcal{V}$ with $v_n \to v$ in $\mathcal{Z}$, $\zeta_n, \zeta \in \mathcal{Z}^*$ with $\zeta_n \to \zeta$ weakly in $\mathcal{Z}^*$ and $\zeta_n \in \mathcal{F}_0 v_n$. Hence

$$v_n(t) \to v(t) \;\text{ in } Z \text{ for a.e. } t \in (0,T), \tag{5.43}$$

$$\zeta_n(t) \in F(t, K(v_n+v_0)(t), v_n(t)+v_0) \text{ a.e. } t \in (0,T). \tag{5.44}$$

From the inequality

$$\begin{aligned} &\|K(v_n+v_0) - K(v+v_0)\|_{\mathcal{Z}}^2 \\ &= \int_0^T \left\| \int_0^t v_n(s)\, ds + v_0 t + u_0 - \int_0^t v(s)\, ds - v_0 t - u_0 \right\|_Z^2 dt \\ &\le T\, \|v_n - v\|_{\mathcal{Z}}^2, \end{aligned}$$

we have $K(v_n + v_0) \to K(v + v_0)$ in $\mathcal{Z}$ and for a subsequence, again denoted by $\{v_n\}$, we obtain

$$K(v_n + v_0)(t) \to K(v + v_0)(t) \text{ in } Z, \text{ a.e. } t \in (0, T). \tag{5.45}$$

We use (5.3)(b), (5.43), and (5.45), and apply Theorem 3.13 to the inclusion (5.44) to find that

$$\zeta(t) \in F(t, K(v + v_0)(t), v(t) + v_0) \text{ a.e. } t \in (0, T).$$

This implies that $\zeta \in \mathcal{F}_0 v$ and concludes the proof of the lemma. □

Proof of Lemma 5.8. We use Definition 5.2, take the duality brackets with $u'(s)$ and integrate over $[0, t]$ for all $t \in [0, T]$ to obtain

$$\begin{aligned}
&\int_0^t \langle u''(s), u'(s)\rangle_{V^*\times V}\, ds + \int_0^t \langle A(s, u'(s)), u'(s)\rangle_{V^*\times V}\, ds \\
&\quad + \int_0^t \langle Bu(s), u'(s)\rangle_{V^*\times V}\, ds + \int_0^t \langle \zeta(s), u'(s)\rangle_{Z^*\times Z}\, ds \\
&= \int_0^t \langle f(s), u'(s)\rangle_{V^*\times V}\, ds
\end{aligned} \tag{5.46}$$

with $\zeta(s) \in F(s, u(s), u'(s))$ for a.e. $s \in (0, t)$. Next, from the integration by parts formula (see Proposition 2.54 (iii)), we have

$$\int_0^t \langle u''(s), u'(s)\rangle_{V^*\times V}\, ds = \frac{1}{2}\|u'(t)\|_H^2 - \frac{1}{2}\|v_0\|_H^2 \tag{5.47}$$

and, by the monotonicity and symmetry of B, we deduce

$$\begin{aligned}
\int_0^t \langle Bu(s), u'(s)\rangle_{V^*\times V}\, ds &= \frac{1}{2}\int_0^t \frac{d}{ds}\langle Bu(s), u(s)\rangle_{V^*\times V}\, ds \\
&= \frac{1}{2}\langle Bu(t), u(t)\rangle_{V^*\times V} - \frac{1}{2}\langle Bu_0, u_0\rangle_{V^*\times V} \\
&\geq -\frac{1}{2}\|B\|_{\mathcal{L}(V,V^*)}\|u_0\|_V^2.
\end{aligned} \tag{5.48}$$

We use (5.46)–(5.48) and the coercivity of the operator A to obtain

$$\begin{aligned}
&\frac{1}{2}\|u'(t)\|_H^2 + \alpha\,\|u'\|_{L^2(0,t;V)}^2 \\
&\quad \leq \frac{1}{2}\|B\|_{\mathcal{L}(V,V^*)}\|u_0\|_V^2 + \frac{1}{2}\|v_0\|_H^2 \\
&\quad\quad + \int_0^t \Big(\|f(s)\|_{V^*} + c_e\,\|\zeta(s)\|_{Z^*}\Big)\|u'(s)\|_V\, ds
\end{aligned} \tag{5.49}$$

for all $t \in [0, T]$. Moreover, using the Young inequality (Lemma 2.6 on page 27), we have

$$\int_0^t \Big(\|f(s)\|_{V^*} + c_e \, \|\zeta(s)\|_{Z^*} \Big) \|u'(s)\|_V \, ds$$

$$\le \frac{\alpha}{2} \|u'\|^2_{L^2(0,t;V)} + \frac{1}{\alpha} \int_0^t \Big(\|f(s)\|^2_{V^*} + c_e^2 \, \|\zeta(s)\|^2_{Z^*} \Big) ds \quad (5.50)$$

for all $t \in [0, T]$. Taking into account (5.49) and (5.50) we deduce that

$$\frac{1}{2} \, \|u'(t)\|^2_H + \frac{\alpha}{2} \, \|u'\|^2_{L^2(0,t;V)}$$

$$\le \frac{1}{2} \, \|B\|_{\mathcal{L}(V,V^*)} \|u_0\|^2_V + \frac{1}{2} \, \|v_0\|^2_H$$

$$+ \frac{1}{\alpha} \|f\|^2_{\mathcal{V}^*} + \frac{c_e^2}{\alpha} \int_0^t \|\zeta(s)\|^2_{Z^*} \, ds \quad (5.51)$$

for all $t \in [0, T]$. We estimate now the last term on the right-hand side of (5.51). By the hypothesis (5.3)(c), we have

$$\int_0^t \|\zeta(s)\|^2_{Z^*} \, ds \le \int_0^t \big(d_0(s) + d_1 \, \|u(s)\|_V + d_2 \, \|u'(s)\|_V \big)^2 \, ds$$

$$\le 3 \int_0^t \Big(d_0^2(s) + d_1^2 \, \|u(s)\|^2_V + d_2^2 \, \|u'(s)\|^2_V \Big) \, ds. \quad (5.52)$$

On the other hand, since $u \in W^{1,2}(0, T; V)$ and V is reflexive, by Propositions 2.50 and 2.51, we have

$$u(t) = u(0) + \int_0^t u'(s) \, ds \;\text{ for all }\; t \in [0, T]. \quad (5.53)$$

Combining (5.53) with the Jensen inequality (Theorem 2.5 on page 27), we get

$$\|u(s)\|^2_V \le 2 \, \|u_0\|^2_V + 2 \left(\int_0^s \|u'(\tau)\|_V \, d\tau \right)^2$$

$$\le 2 \, \|u_0\|^2_V + 2T \int_0^s \|u'(\tau)\|^2_V \, d\tau$$

for all $s \in [0, t]$. Hence, (5.52) implies that

$$\|\zeta\|^2_{L^2(0,t;Z^*)} \le 3\,\|d_0\|^2_{L^2(0,t)}$$

$$+3\,d_1^2 \int_0^t \Big(2\|u_0\|_V^2 + 2T \int_0^s \|u'(\tau)\|_V^2 d\tau\Big)\, ds + 3\,d_2^2\,\|u'\|^2_{L^2(0,t;V)}$$

$$\le 3\,\|d_0\|^2_{L^2(0,T)} + 6\,d_1^2\,T\,\|u_0\|_V^2 + (6\,d_1^2\,T^2 + 3\,d_2^2)\,\|u'\|^2_{L^2(0,t;V)} \tag{5.54}$$

for all $t \in [0, T]$. We combine (5.51) and (5.54) to obtain

$$\frac{1}{2}\,\|u'(t)\|_H^2 + \left(\frac{\alpha}{2} - \frac{3c_e^2}{\alpha}\,(2\,d_1^2\,T^2 + d_2^2)\right)\|u'\|^2_{L^2(0,t;V)} \le \widetilde{c} \tag{5.55}$$

for all $t \in [0, T]$, where

$$\widetilde{c} = \frac{1}{2}\,\|B\|_{\mathcal{L}(V,V^*)}\|u_0\|_V^2 + \frac{1}{2}\,\|v_0\|_H^2 + \frac{1}{\alpha}\,\|f\|^2_{\mathcal{V}^*}$$

$$+\frac{3\,c_e^2}{\alpha}\left(\|d_0\|^2_{L^2(0,T)} + 2\,d_1^2\,T\,\|u_0\|_V^2\right).$$

Since the hypothesis (5.5) implies $\alpha^2 > 6\,c_e^2\,(2d_1^2\,T^2 + d_2^2)$, from (5.55) we deduce that

$$\|u'\|_{L^2(0,t;V)} \le c\,\big(1 + \|u_0\|_V + \|v_0\|_H + \|f\|_{\mathcal{V}^*}\big) \tag{5.56}$$

for all $t \in [0, T]$ where, here and below, c represents a positive constant whose value may change from line to line. Next, from (5.53), we have

$$\|u(t)\|_V \le \|u_0\|_V + \int_0^t \|u'(s)\|_V\,ds \le \|u_0\|_V + \sqrt{T}\,\|u'\|_{L^2(0,t;V)}$$

for all $t \in [0, T]$ which, together with (5.56), gives

$$\|u\|_{C(0,T;V)} \le c\,(1 + \|u_0\|_V + \|v_0\|_H + \|f\|_{\mathcal{V}^*})\,. \tag{5.57}$$

To conclude the proof, it is enough to show the bound for $\|u''\|_{\mathcal{V}^*}$. From (5.1)–(5.3), we obtain

$$\|u''\|_{\mathcal{V}^*} \le \|f\|_{\mathcal{V}^*} + c\|u'\|_{\mathcal{V}} + \|B\|_{\mathcal{L}(V,V^*)}\|u\|_{\mathcal{V}} + \|\zeta\|_{\mathcal{V}^*}\,.$$

Moreover, since $\zeta(s) \in F(s, u(s), u'(s))$ for a.e. $s \in (0, T)$, we have

$$\|\zeta\|_{\mathcal{V}^*} \le c_e\|\zeta\|_{\mathcal{Z}^*} \le \sqrt{3}\,c_e\left(\|d_0\|_{L^2(0,T)} + d_1\|u\|_{\mathcal{V}} + d_2\|u'\|_{\mathcal{V}}\right).$$

Combining the above inequalities, we deduce that

$$\|u''\|_{\mathcal{V}^*} \le c\left(1 + \|u\|_{\mathcal{V}} + \|u'\|_{\mathcal{V}} + \|f\|_{\mathcal{V}^*}\right). \tag{5.58}$$

Lemma 5.8 is now a consequence of inequalities (5.56)–(5.58), combined with the definition (2.12) on page 48 of the norm on the space $\mathcal{W}$. □

5.2 Evolutionary Inclusions of Subdifferential Type

In this section we investigate evolutionary inclusions involving the multivalued Clarke subdifferential operator. We use the same functional spaces as in Sect. 5.1, introduced on page 121. Let $A\colon (0,T) \times V \to V^*$, $B\colon V \to V^*$, and $R\colon Z \times Z \to X \times X$ be given operators where, recall, X, V, and Z are assumed to be separable reflexive Banach spaces. We denote by ∂J the Clarke generalized subdifferential of a prescribed functional $J\colon (0,T) \times X \times X \times X \times X \to \mathbb{R}$ with respect to its last two variables, $R^*\colon X^* \times X^* \to Z^* \times Z^*$ stands for the adjoint operator to R and $S\colon Z^* \times Z^* \to Z^*$ is the operator defined by $S(z_1, z_2) = z_1 + z_2$.

We start by considering the following version of Problem 5.1 which will be applied in the study of the hemivariational inequalities in Sect. 5.3.

Problem 5.9. *Find $u \in \mathcal{V}$ such that $u' \in \mathcal{W}$ and*

$$\left.\begin{aligned} & u''(t) + A(t, u'(t)) + Bu(t) + SR^*\partial J(t, R(u(t), u'(t)), R(u(t), u'(t))) \ni f(t) \\ & \qquad\qquad\qquad\qquad\qquad\qquad\qquad\qquad \text{a.e. } t \in (0,T), \\ & u(0) = u_0, \ \ u'(0) = v_0. \end{aligned}\right\}$$

The concept of solution to Problem 5.9 is understood in the sense of Definition 5.2.

In addition to the assumptions on the operators A and B formulated in Sect. 5.1, we need the following hypotheses on the data.

$$\left.\begin{aligned} & J\colon (0,T) \times X \times X \times X \times X \to \mathbb{R} \text{ is such that} \\ & \text{(a) } J(\cdot, w, z, u, v) \text{ is measurable on } (0,T) \text{ for all } w, z, u, v \in X. \\ & \text{(b) } J(t, w, z, \cdot, \cdot) \text{ is locally Lipschitz on } X \times X \text{ for all } w, z \in X, \\ & \quad\ \text{a.e. } t \in (0,T). \\ & \text{(c) } \|\partial J(t, w, z, u, v)\|_{X^* \times X^*} \le c_0(t) + c_1\left(\|u\|_X + \|v\|_X\right) + \\ & \quad\ + c_2\left(\|w\|_X + \|z\|_X\right) \text{ for all } w, z, u, v \in X, \text{ a.e. } t \in (0,T) \\ & \quad\ \text{with } c_0 \in L^2(0,T),\, c_0, c_1, c_2 \ge 0, \text{ where the subdifferential} \\ & \quad\ \text{of } J \text{ is taken with respect to } (u, v). \\ & \text{(d) } \partial J(t, \cdot, \cdot, \cdot, \cdot) \text{ has a closed graph in } X \times X \times X \times X \times \\ & \quad\ \times (w\text{–}(X^* \times X^*)) \text{ topology.} \end{aligned}\right\} \tag{5.59}$$

$$R \in \mathcal{L}(Z \times Z, X \times X). \tag{5.60}$$

In the study of Problem 5.9 we need the following lemma.

Lemma 5.10. *Let $(\mathcal{O}, \Sigma)$ be a measurable space, Y_1, Y_2 be separable Banach spaces, $L \in \mathcal{L}(Y_1, Y_2)$, and let $G: \mathcal{O} \to \mathcal{P}_{wkc}(Y_1)$ be measurable. Then the multifunction $H: \mathcal{O} \to \mathcal{P}_{wkc}(Y_2)$ given by $H(\omega) = LG(\omega)$ for $\omega \in \mathcal{O}$ is measurable.*

Proof. We start by recalling that if $L \in \mathcal{L}(Y_1, Y_2)$, then $L \in \mathcal{L}(w\text{–}Y_1, w\text{–}Y_2)$. Hence, it follows that H is $\mathcal{P}_{wkc}(Y_2)$-valued. Given an open set $U \subset Y_2$, we will show that

$$H^-(U) = \{\omega \in \mathcal{O} \mid H(\omega) \cap U \neq \emptyset\} \in \Sigma.$$

First, from the definition of H, we have

$$H^-(U) = \{\omega \in \mathcal{O} \mid G(\omega) \cap L^{-1}(U) \neq \emptyset\} = G^-(U'),$$

where $U' = L^{-1}(U)$. Next, since the mapping $L: Y_1 \to Y_2$ is continuous, for every open set $U \subset Y_2$, the inverse image $L^{-1}(U) \subset Y_1$ is an open set. Finally, from the definition of measurability of G, we have $G^-(U') \in \Sigma$. Therefore, we deduce that $H^-(U) \in \Sigma$ which implies that H is measurable, as claimed. □

The main result for Problem 5.9 reads as follows.

Theorem 5.11. *Assume that* (5.1), (5.2), (5.4), (5.59), (5.60) *hold and*

$$\alpha > 2\sqrt{3}\, c_e^2\, \|R\|^2\, T\, (c_1 + c_2), \tag{5.61}$$

where $\|R\| = \|R\|_{\mathcal{L}(Z\times Z, X\times X)}$. Then Problem 5.9 *has at least one solution.*

Proof. We apply Theorem 5.4 to the multivalued operator $F: (0,T)\times V \times V \to 2^{Z^*}$ defined by

$$F(t,u,v) = SR^*\partial J(t, R(u,v), R(u,v)) \text{ for } u, v \in V, \text{ a.e. } t \in (0,T).$$

To this end we show that, under the hypotheses (5.59) and (5.60), the operator F satisfies (5.3) with $d_0(t) = c_0(t)\, \|R\|$ and $d_1 = d_2 = (c_1 + c_2)\, c_e\, \|R\|^2$. First, we observe that the mapping F has nonempty and convex values. This follows from the nonemptiness and convexity of values of the Clarke subdifferential of J, guaranteed by Proposition 3.23 on page 56. Since the values of the subdifferential are weakly closed subsets of $X^* \times X^*$ and $S\, R^*: X^* \times X^* \to Z^*$ is a linear continuous operator, we easily obtain that the mapping F has closed values in Z^*. Hence F is $\mathcal{P}_{fc}(Z^*)$–valued.

We show that $F(\cdot, u, v)$ is measurable on $(0,T)$ for all u, $v \in V$. Since, by the hypothesis (5.59)(a), $J(\cdot, w, z, \overline{w}, \overline{z})$ is measurable on $(0,T)$ for all w, z, $\overline{w}$, $\overline{z} \in X$

and $J(t, w, z, \cdot, \cdot)$ is locally Lipschitz on $X \times X$ for all $w, z \in X$, a.e. $t \in (0, T)$, according to Proposition 3.44, we know that the multifunction

$$(0, T) \times X \times X \ni (t, \overline{w}, \overline{z}) \mapsto \partial J(t, w, z, \overline{w}, \overline{z}) \subset X^* \times X^*$$

is measurable. Hence, by Lemma 1.66 on page 21 we infer that also the multifunction

$$(0, T) \ni t \mapsto \partial J(t, w, z, \overline{w}, \overline{z}) \subset X^* \times X^*$$

is measurable for all $w, z, \overline{w}, \overline{z} \in X$ and, clearly, it is $\mathcal{P}_{wkc}(X^* \times X^*)$–valued. These properties, together with the fact that $S\, R^*\colon X^* \times X^* \to Z^*$ is a linear continuous operator, allow to apply Lemma 5.10. So we have that

$$(0, T) \ni t \mapsto S\, R^* \partial J(t, w, z, \overline{w}, \overline{z}) \subset Z^*$$

is measurable for all $w, z, \overline{w}, \overline{z} \in X$. As a consequence the multifunction $F(\cdot, u, v)$ is measurable for all $u, v \in V$.

Next, we prove the upper semicontinuity of $F(t, \cdot, \cdot)$ for a.e. $t \in (0, T)$. Let $t \in (0, T) \setminus N$, meas $(N) = 0$. According to Proposition 3.8 on page 53, we show that for every weakly closed subset K of Z^*, the weak inverse image

$$F^-(K) = \{\, (u, v) \in V \times V \mid F(t, u, v) \cap K \neq \emptyset \,\}$$

is a closed subset of $Z \times Z$. Let $\{(u_n, v_n)\} \subset F^-(K)$ and $(u_n, v_n) \to (u, v)$ in $Z \times Z$. Then, for all $n \in \mathbb{N}$ we can find $\zeta_n \in F(t, u_n, v_n) \cap K$. By the definition of F, we have $\zeta_n = S\, R^*(w_n, z_n)$ with $w_n, z_n \in X^*$ and

$$(w_n, z_n) \in \partial J(t, R(u_n, v_n), R(u_n, v_n)) \text{ for a.e. } t \in (0, T). \tag{5.62}$$

Using the continuity of the operator R, we obtain

$$R(u_n, v_n) \to R(u, v) \text{ in } X \times X.$$

Since by (5.59)(c) the operator $\partial J(t, \cdot, \cdot, \cdot, \cdot)$ is bounded (i.e. it maps bounded sets of $X \times X \times X \times X$ into bounded sets in $X^* \times X^*$), from (5.62) it follows that the sequence $\{(w_n, z_n)\}$ belongs to a bounded subset of $X^* \times X^*$. Also, since X is reflexive, the weak and the weak* topologies for X^* are the same. Thus, by passing to a subsequence, if necessary, we may suppose that

$$(w_n, z_n) \to (w, z) \text{ weakly in } X^* \times X^*$$

for some $(w, z) \in X^* \times X^*$. Now, we use (5.59)(d) to see that the graph of $\partial J(t, \cdot, \cdot, \cdot, \cdot)$ is closed in $X \times X \times X \times X \times (w\text{–}(X^* \times X^*))$ topology, for a.e. $t \in (0, T)$. Therefore, from (5.62), we obtain

$$(w, z) \in \partial J(t, R(u, v), R(u, v)).$$

Furthermore, since $\{\zeta_n\}$ also belongs to a bounded subset of Z^*, we may assume that $\zeta_n \to \zeta$ weakly in Z^*. And, since $\zeta_n \in K$ and K is weakly closed in Z^*, it follows that $\zeta \in K$. By the continuity and linearity of the operator $S\, R^*\colon X^* \times X^* \to Z^*$ we find

$$\zeta_n = S\, R^*(w_n, z_n) \to S\, R^*(w, z) = \zeta \text{ weakly in } Z^*,$$

where $\zeta \in Z^*$ and $(w, z) \in \partial J(t, R(u, v), R(u, v))$. This, by the definition of F, implies that $\zeta \in F(t, u, v)$. As a consequence, once $\zeta \in K$, we see that $F^-(K)$ is closed in $Z \times Z$ and, therefore, (5.3)(b) holds.

Next, we show that F satisfies (5.3)(c). Assume that $t \in (0, T) \setminus N$ with meas $(N) = 0$ and let $u, v \in V$, $\zeta \in Z^*$ be such that $\zeta \in F(t, u, v)$. The latter is equivalent to $\zeta = S\,R^*(w, z)$, where $(w, z) \in X^* \times X^*$ and $(w, z) \in \partial J(t, R(u, v), R(u, v))$. Using the estimate (5.59)(c), we obtain

$$\begin{aligned}
\|\zeta\|_{Z^*} &= \|S\,R^*(w, z)\|_{Z^*} \le \|S\,R^*\|\,\|(w, z)\|_{X^* \times X^*} \\
&\le \|S\,R^*\|\,(c_0(t) + c_1\,\|R(u, v)\|_{X \times X} + c_2\,\|R(u, v)\|_{X \times X}) \\
&\le \|S\,R^*\|\,(c_0(t) + c_1\,\|R\|\,\|(u, v)\|_{Z \times Z} + c_2\,\|R\|\,\|(u, v)\|_{Z \times Z}) \\
&\le c_0(t)\|S\,R^*\| + ((c_1 + c_2)\,c_e\,\|R\|\,\|S\,R^*\|)\|u\|_V \\
&\quad +((c_1 + c_2)\,c_e\,\|R\|\,\|S\,R^*\|)\|v\|_V \\
&\le c_0(t)\|R\| + (c_1 + c_2)\,c_e\,\|R\|^2\,\|u\|_V + (c_1 + c_2)\,c_e\,\|R\|^2\,\|v\|_V.
\end{aligned}$$

Here we used the equality

$$\|R^*\|_{\mathcal{L}(X^* \times X^*, Z^* \times Z^*)} = \|R\|_{\mathcal{L}(Z \times Z, X \times X)}$$

(guaranteed by Proposition 1.51) and equality $\|S\|_{\mathcal{L}(Z^* \times Z^*, Z^*)} = 1$ which easily follows from the definition of S. Also, we recall that $c_e > 0$ is the embedding constant of V into Z. We conclude from above that F satisfies (5.3)(c) with $d_0(t) = c_0(t)\|R\|$ and $d_1 = d_2 = (c_1 + c_2)\,c_e\,\|R\|^2$.

Finally, we observe that the condition (5.61) implies (5.5). Theorem 5.11 is now a consequence of Theorem 5.4. □

We note that the existence result of Theorem 5.11 is local in time, since the various constants related to Problem 5.9 are restricted to the smallness assumption (5.61). This is the characteristic feature of a problem in which the subdifferential of the superpotential is taken with respect to a pair of variables. We will see on page 146 below that, under additional assumptions, for an evolutionary inclusion involving the subdifferential with respect to its last variable only, we can remove the smallness assumption on the length of the interval of time, i.e., we can obtain a global existence result.

Assume now that X and Z are given separable reflexive Banach spaces, $A: (0, T) \times V \to V^*$ is a nonlinear operator, $B: V \to V^*$ and $M: Z \to X$ are linear and continuous operators. Also, denote by ∂J the Clarke generalized subdifferential of a prescribed functional $J: (0, T) \times X \times X \times X \to \mathbb{R}$ with respect to its last variable and let M^* be the adjoint operator to M. With these data we consider the following problem.

Problem 5.12. *Find $u \in \mathcal{V}$ such that $u' \in \mathcal{W}$ and*

$$\left.\begin{aligned} & u''(t) + A(t, u'(t)) + Bu(t) + M^*\partial J(t, Mu(t), Mu'(t), Mu'(t)) \ni f(t) \\ & \qquad\qquad\qquad\qquad\qquad\qquad\qquad\qquad\qquad\qquad \text{a.e. } t \in (0, T), \\ & u(0) = u_0, \quad u'(0) = v_0. \end{aligned}\right\}$$

In addition to the assumptions on the operators A and B formulated in Sect. 5.1, we consider the following hypotheses on the data.

$$\left.\begin{aligned} & J\colon (0,T) \times X \times X \times X \to \mathbb{R} \text{ is such that} \\ & \text{(a) } J(\cdot, w, u, v) \text{ is measurable on } (0,T) \text{ for all } w, u, v \in X. \\ & \text{(b) } J(t, w, u, \cdot) \text{ is locally Lipschitz on } X \text{ for all } w, u \in X, \\ & \quad\ \text{a.e. } t \in (0,T). \\ & \text{(c) } \|\partial J(t, w, u, v)\|_{X^*} \le c_0(t) + c_1\|v\|_X + c_2\|w\|_X + c_3\|u\|_X \\ & \quad\ \text{for all } w, u, v \in X, \text{ a.e. } t \in (0,T) \text{ with } c_0 \in L^2(0,T), \\ & \quad\ c_0, c_1, c_2, c_3 \ge 0, \text{ where the subdifferential of } J \text{ is taken} \\ & \quad\ \text{with respect to the last variable.} \\ & \text{(d) } \partial J(t, \cdot, \cdot, \cdot) \text{ has a closed graph in } X \times X \times X \times (w\text{–}X^*) \\ & \quad\ \text{topology.} \end{aligned}\right\} \tag{5.63}$$

$$M \in \mathcal{L}(Z, X). \tag{5.64}$$

The main existence result for Problem 5.12 is following.

Theorem 5.13. *Assume that* (5.1), (5.2), (5.4), (5.63), (5.64) *hold and*

$$\alpha > 2\sqrt{3}\, c_e^2\, \|M\|^2\, (c_2 T + c_1 + c_3), \tag{5.65}$$

where $\|M\| = \|M\|_{\mathcal{L}(Z,X)}$. Then Problem 5.12 *has at least one solution.*

Proof. It is similar to the proof of Theorem 5.11 however, for the convenience of the reader we provide it in what follows. We apply Theorem 5.4 to the multivalued operator $F\colon (0,T) \times V \times V \to 2^{Z^*}$ given by

$$F(t, u, v) = M^*\partial J(t, Mu, Mv, Mv) \ \text{ for } u, v \in V, \ \text{a.e. } t \in (0,T).$$

To this end, we show that under the hypotheses (5.63) and (5.64), the operator F satisfies the condition (5.3) with $d_0(t) = c_0(t)\, \|M\|$, $d_1 = c_2\, c_e\, \|M\|^2$, and $d_2 = (c_1 + c_3)\, c_e\, \|M\|^2$. First, we observe that the mapping F has nonempty and convex values. This follows from the nonemptiness and convexity of values of the Clarke subdifferential of J, see Proposition 3.23 on page 56. Moreover, it is easy to verify that the mapping F has closed values in Z^*. Indeed, let $t \in (0,T)$, $u, v \in V$,

$\{\zeta_n\} \subset F(t, u, v)$, and $\zeta_n \to \zeta$ in Z^*. Since $\zeta_n \in M^* \partial J(t, Mu, Mv, Mv)$ and the latter is a weakly closed subset of Z^* (recall that the values of the subdifferential are weakly star closed subsets of X^*), we obtain $\zeta \in M^* \partial J(t, Mu, Mv, Mv)$. It follows from here that $\zeta \in F(t, u, v)$ and, therefore, F is $\mathcal{P}_{fc}(Z^*)$-valued.

Next, we prove that $F(\cdot, u, v)$ is measurable on $(0, T)$ for all $u, v \in V$. Let $w, \overline{u}, \overline{v} \in X$. From hypothesis (5.63)(a) it follows that $J(\cdot, w, \overline{u}, \overline{v})$ is measurable on $(0, T)$ for all $w, \overline{u}, \overline{v} \in X$ and, since $J(t, w, \overline{u}, \cdot)$ is locally Lipschitz on X for all $w, \overline{u} \in X$ and a.e. $t \in (0, T)$, by using Proposition 3.44, we deduce that the multifunction

$$(0, T) \times X \ni (t, \overline{v}) \mapsto \partial J(t, w, \overline{u}, \overline{v}) \subset X^*$$

is measurable. Therefore, by Lemma 1.66, we infer that also the multifunction $(0, T) \ni t \mapsto \partial J(t, w, \overline{u}, \overline{v})$ is measurable for all $w, \overline{u}, \overline{v} \in X$ and, clearly, it is $\mathcal{P}_{wkc}(X^*)$-valued. These properties, together with the fact that $M^* \colon X^* \to Z^*$ is a linear continuous operator, allow to apply Lemma 5.10. So we have that $(0, T) \ni t \mapsto M^* \partial J(t, w, \overline{u}, \overline{v})$ is measurable for all $w, \overline{u}, \overline{v} \in X$. As a consequence the multifunction $F(\cdot, u, v)$ is measurable for all $u, v \in V$.

Subsequently, in order to prove that $F(t, \cdot, \cdot)$ is upper semicontinuous for a.e. $t \in (0, T)$, according to Proposition 3.8, it is enough to show that for every weakly closed subset K of Z^*, the weak inverse image

$$F^-(K) = \{\, (u, v) \in V \times V \mid F(t, u, v) \cap K \neq \emptyset \,\}$$

is a closed subset of $Z \times Z$. Let $t \in (0, T) \setminus N$, $\text{meas}(N) = 0$, $\{(u_n, v_n)\} \subset F^-(K)$ and assume that $(u_n, v_n) \to (u, v)$ in $Z \times Z$. Thus, for all $n \in \mathbb{N}$ we can find $\zeta_n \in F(t, u_n, v_n) \cap K$. From the definition of F, we obtain $\zeta_n = M^* z_n$ with $z_n \in X^*$ and

$$z_n \in \partial J(t, M u_n, M v_n, M v_n). \tag{5.66}$$

From the continuity of the operator M, we have

$$(M u_n, M v_n) \to (M u, M v) \text{ in } X \times X.$$

Since by (5.63)(c) the operator $\partial J(t, \cdot, \cdot, \cdot)$ is bounded (i.e., it maps bounded sets of $X \times X \times X$ into bounded sets in X^*), from (5.66) it follows that the sequence $\{z_n\}$ remains in a bounded subset of X^*. Recall also that X is reflexive and, therefore, the weak and the weak* topologies for X^* are the same. Thus, by passing to a subsequence, if necessary, we may suppose that

$$z_n \to z \text{ weakly in } X^*$$

for some $z \in X^*$. Now we use the fact that, for a.e. $t \in (0, T)$, the graph of $\partial J(t, \cdot, \cdot, \cdot)$ is closed in $X \times X \times X \times (w\text{–}X^*)$ topology, see (5.63)(d). Therefore, from (5.66), we find

$$z \in \partial J(t, M u, M v, M v).$$

Also, since $\{\zeta_n\}$ remains in a bounded subset of Z^*, we may assume that $\zeta_n \to \zeta$ weakly in Z^*. And, since $\zeta_n \in K$ and K is weakly closed in Z^*, it follows that $\zeta \in K$. By the continuity and linearity of the operator $M^*\colon X^* \to Z^*$, we obtain

$$\zeta_n = M^* z_n \to M^* z = \zeta \text{ weakly in } Z^*,$$

where $\zeta \in Z^*$ and $z \in \partial J(t, Mu, Mv, Mv)$. This, by the definition of F, implies that $\zeta \in F(t, u, v)$. As a consequence, once $\zeta \in K$, we know that $F^-(K)$ is closed in $Z \times Z$ and, therefore, (5.3)(b) holds.

Next, we show that F satisfies the condition (5.3)(c). Let $t \in (0, T) \setminus N$, $\text{meas}\,(N) = 0$ and let $u, v \in V$, $\zeta \in Z^*$ be such that $\zeta \in F(t, u, v)$. Hence $\zeta = M^* z$, where $z \in X^*$ and $z \in \partial J(t, Mu, Mv, Mv)$. Using the estimate (5.63)(c), we have

$$\begin{aligned}
\|\zeta\|_{Z^*} &= \|M^* z\|_{Z^*} \le \|M^*\| \|z\|_{X^*} \\
&\le \|M^*\| \left(c_0(t) + c_1 \|Mv\|_X + c_2 \|Mu\|_X + c_3 \|Mv\|_X \right) \\
&\le \|M^*\| \left(c_0(t) + (c_1 + c_3) \|M\| \|v\|_Z + c_2 \|M\| \|u\|_Z \right) \\
&\le c_0(t) \|M\| + c_2\, c_e\, \|M\|^2 \|u\|_V + (c_1 + c_3)\, c_e\, \|M\|^2 \|v\|_V .
\end{aligned}$$

Here we used the equality $\|M^*\|_{\mathcal{L}(X^*, Z^*)} = \|M\|_{\mathcal{L}(Z, X)}$, which is a consequence of Proposition 1.51 and, recall, $c_e > 0$ is the embedding constant of V into Z. This implies that F satisfies (5.3)(c) with $d_0(t) = c_0(t)\, \|M\|$, $d_1 = c_2\, c_e\, \|M\|^2$, and $d_2 = (c_1 + c_3)\, c_e\, \|M\|^2$.

Finally, we observe that the condition (5.65) implies (5.5). Theorem 5.13 is now a consequence of Theorem 5.4. □

As in the case of Theorem 5.11, we note that Theorem 5.13 provides the existence of local solutions to the evolutionary inclusion in Problem 5.12. Nevertheless, in the case when the functional J in (5.63) is independent of w, we see that the smallness condition (5.65) is satisfied with $c_2 = 0$. And, therefore, in contrast to Theorem 5.11, in this particular case Theorem 5.13 provides a global existence result.

We are now in a position to formulate a corollary of Theorem 5.13 which concerns the unique solvability of evolutionary inclusions and which will be useful in the study of dynamic contact problems presented in Chap. 8. The problem under consideration is the following.

Problem 5.14. *Find $u \in \mathcal{V}$ such that $u' \in \mathcal{W}$ and*

$$\left.\begin{aligned}
&u''(t) + A(t, u'(t)) + Bu(t) + M^* \partial J(t, Mu'(t)) \ni f(t) \text{ a.e. } t \in (0, T), \\
&u(0) = u_0, \ \ u'(0) = v_0.
\end{aligned}\right\}$$

Note that here and below in this section ∂J represents the Clarke generalized subdifferential of J with respect to its last variable.

In the study of this problem we consider the following modification of the hypotheses (5.1) and (5.63).

$$\left.\begin{array}{l}
A\colon (0,T)\times V\to V^* \text{ is such that}\\
\quad\text{(a) } A(\cdot,v) \text{ is measurable on } (0,T) \text{ for all } v\in V.\\
\quad\text{(b) } A(t,\cdot) \text{ is pseudomonotone on } V \text{ for a.e. } t\in(0,T).\\
\quad\text{(c) } \|A(t,v)\|_{V^*}\le a_0(t)+a_1\|v\|_V \text{ for all } v\in V,\ \text{a.e. } t\in(0,T)\\
\qquad\quad \text{with } a_0\in L^2(0,T), a_0\ge 0 \text{ and } a_1>0.\\
\quad\text{(d) } \langle A(t,v),v\rangle_{V^*\times V}\ge\alpha\|v\|_V^2 \text{ for all } v\in V,\ \text{a.e. } t\in(0,T)\\
\qquad\quad \text{with } \alpha>0.\\
\quad\text{(e) } A(t,\cdot) \text{ is strongly monotone for a.e. } t\in(0,T),\ \text{i.e. there}\\
\qquad\quad \text{is } m_1>0 \text{ such that for all } v_1,v_2\in V,\ \text{a.e. } t\in(0,T)\\
\qquad\quad \langle A(t,v_1)-A(t,v_2),v_1-v_2\rangle_{V^*\times V}\ge m_1\|v_1-v_2\|_V^2.
\end{array}\right\} \tag{5.67}$$

$$\left.\begin{array}{l}
J\colon (0,T)\times X\to\mathbb{R} \text{ is such that}\\
\quad\text{(a) } J(\cdot,v) \text{ is measurable on } (0,T) \text{ for all } v\in X.\\
\quad\text{(b) } J(t,\cdot) \text{ is locally Lipschitz on } X \text{ for a.e. } t\in(0,T).\\
\quad\text{(c) } \|\partial J(t,v)\|_{X^*}\le c_0(t)+c_1\|v\|_X \text{ for all } v\in X,\\
\qquad\quad \text{a.e. } t\in(0,T) \text{ with } c_0\in L^2(0,T), c_0,c_1\ge 0.\\
\quad\text{(d) } \langle z_1-z_2,v_1-v_2\rangle_{X^*\times X}\ge -m_2\|v_1-v_2\|_X^2 \text{ for all}\\
\qquad\quad z_i\in\partial J(t,v_i),\ z_i\in X^*,\ v_i\in X,\ i=1,2,\ \text{a.e. } t\in(0,T)\\
\qquad\quad \text{with } m_2\ge 0.
\end{array}\right\} \tag{5.68}$$

$$m_1\ge m_2\,c_e^2\,\|M\|^2,\ \text{where } \|M\|=\|M\|_{\mathcal{L}(Z,X)}. \tag{5.69}$$

We have the following existence and uniqueness result.

Theorem 5.15. *Assume that hypotheses* (5.2), (5.4), (5.64), (5.67)–(5.69) *hold and*

$$\alpha>2\sqrt{3}\,c_1\,c_e^2\,\|M\|^2. \tag{5.70}$$

Then Problem 5.14 *has a unique solution.*

Proof. For the existence part, we apply Theorem 5.13. Since the functional J is independent of (w,u), it is clear that (5.68) implies (5.63) with $c_2=c_3=0$. Also, the condition (5.70) implies (5.65), and (5.1) is guaranteed by (5.67). Therefore, using Theorem 5.13 we deduce that Problem 5.14 admits at least one solution.

Next, we establish the uniqueness of the solution. Let u_1, u_2 be solutions to Problem 5.14. Then, by Definition 5.2, there exist $\zeta_1, \zeta_2 \in \mathcal{Z}^*$ such that

$$\left.\begin{aligned} & u_i''(s) + A(s, u_i'(s)) + Bu_i(s) + \zeta_i(s) = f(s) \ \text{ a.e. } s \in (0,T), \\ & \zeta_i(s) \in M^* \partial J(s, Mu_i'(s)) \ \text{ a.e. } s \in (0,T), \\ & u_i(0) = u_0, \ u_i'(0) = v_0 \end{aligned}\right\} \tag{5.71}$$

for $i = 1, 2$. Subtracting the two equations in (5.71), taking the result in duality with $u_1'(s) - u_2'(s)$ and integrating by parts, we obtain

$$\begin{aligned} \frac{1}{2} \|u_1'(t) - u_2'(t)\|_H^2 &+ \int_0^t \langle A(s, u_1'(s)) - A(s, u_2'(s)), u_1'(s) - u_2'(s) \rangle_{V^* \times V} \, ds \\ &+ \int_0^t \langle Bu_1(s) - Bu_2(s), u_1'(s) - u_2'(s) \rangle_{V^* \times V} \, ds \\ &+ \int_0^t \langle \zeta_1(s) - \zeta_2(s), u_1'(s) - u_2'(s) \rangle_{Z^* \times Z} \, ds = 0 \end{aligned}$$

for all $t \in [0, T]$. We also have $\zeta_i(s) = M^* z_i(s)$ with $z_i(s) \in \partial J(s, Mu_i'(s))$ for a.e. $s \in (0, T)$ and $i = 1, 2$. Therefore, using (5.68)(d) yields

$$\begin{aligned} & \int_0^t \langle \zeta_1(s) - \zeta_2(s), u_1'(s) - u_2'(s) \rangle_{Z^* \times Z} \, ds \\ & \quad = \int_0^t \langle z_1(s) - z_2(s), Mu_1'(s) - Mu_2'(s) \rangle_{X^* \times X} \, ds \\ & \quad \geq -m_2 \int_0^t \|Mu_1'(s) - Mu_2'(s)\|_X^2 \, ds \\ & \quad \geq -m_2 \, c_e^2 \, \|M\|^2 \int_0^t \|u_1'(s) - u_2'(s)\|_V^2 \, ds \end{aligned} \tag{5.72}$$

for all $t \in [0, T]$. Next,

$$\begin{aligned} & \int_0^t \langle Bu_1(s) - Bu_2(s), u_1'(s) - u_2'(s) \rangle_{V^* \times V} \, ds \\ & \quad = \frac{1}{2} \int_0^t \frac{d}{ds} \langle B(u_1(s) - u_2(s)), u_1(s) - u_2(s) \rangle_{V^* \times V} \, ds \\ & \quad = \frac{1}{2} \langle B(u_1(t) - u_2(t)), u_1(t) - u_2(t) \rangle_{V^* \times V} \geq 0 \end{aligned} \tag{5.73}$$

for all $t \in [0, T]$. We combine now (5.72), (5.73), and use (5.67)(e) to obtain

$$\frac{1}{2}\,\|u_1'(t) - u_2'(t)\|_H^2 + m_1 \int_0^t \|u_1'(s) - u_2'(s)\|_V^2\, ds$$
$$- m_2\, c_e^2\, \|M\|^2 \int_0^t \|u_1'(s) - u_2'(s)\|_V^2\, ds \le 0$$

for all $t \in [0, T]$. From this inequality and (5.69) we deduce that $u_1' = u_2'$ on $[0, T]$ and, since

$$u_i(t) = u_0 + \int_0^t u_i'(s)\, ds \quad \text{for all } t \in [0, T],\ i = 1, 2,$$

we deduce that $u_1 = u_2$ on $[0, T]$, which completes the proof. □

Note that the assumptions of Theorem 5.15 do not include any restriction on the length of the interval of time T. Therefore, in contrast to Theorems 5.11 and 5.13 which provide local existence results of the solution, Theorem 5.15 provides a global existence result of the solution and, in addition, the solution is unique.

Subdifferential inclusions with Volterra integral term. We conclude this section with a result on the unique solvability of evolutionary inclusions with Volterra integral term. The problem under consideration can be formulated as follows.

Problem 5.16. *Find $u \in \mathcal{V}$ such that $u' \in \mathcal{W}$ and*

$$\left.\begin{aligned} & u''(t) + A(t, u'(t)) + Bu(t) + \int_0^t C(t-s)u(s)\, ds + M^*\partial J(t, Mu'(t)) \ni f(t) \\ & \qquad\qquad\qquad\qquad\qquad\qquad\qquad\qquad \text{a.e. } t \in (0, T), \\ & u(0) = u_0,\ \ u'(0) = v_0. \end{aligned}\right\}$$

For this problem we need the additional hypothesis

$$C \in L^2(0, T; \mathcal{L}(V, V^*)). \tag{5.74}$$

The unique solvability of Problem 5.16 is given by the following existence and uniqueness result.

Theorem 5.17. *Assume that* (5.2), (5.4), (5.64)–(5.70), *and* (5.74) *hold. Then Problem* 5.16 *admits a unique solution.*

Proof. The proof is carried out into two steps and is based on Theorem 5.15 (which provides an existence and uniqueness result for evolutionary inclusions without memory term) combined with a fixed-point argument, similar to that already used in the proof of Theorem 4.13.

Let us fix $\eta \in \mathcal{V}^*$. In the first step we consider the following evolutionary inclusion:

$$\left.\begin{aligned} &\text{find } u_\eta \in \mathcal{V} \text{ with } u_\eta' \in \mathcal{W} \text{ such that} \\ &u_\eta''(t) + A(t, u_\eta'(t)) + Bu_\eta(t) + M^* \,\partial J(t, Mu_\eta'(t)) \ni f(t) - \eta(t) \\ &\qquad\qquad\qquad\qquad\qquad\qquad\qquad\qquad \text{a.e. } t \in (0,T), \\ &u_\eta(0) = u_0,\ u_\eta'(0) = v_0. \end{aligned}\right\} \tag{5.75}$$

It follows from Theorem 5.15 that the problem (5.75) has a unique solution. Moreover, from (5.70) and (5.68), it is clear that (5.5) holds with $d_1 = 0$ and $d_2 = c_1\, c_e \|M\|^2$. Therefore, by Lemma 5.8, we deduce the estimate

$$\|u_\eta\|_{C(0,T;V)} + \|u_\eta'\|_{\mathcal{W}} \le \overline{c}\ (1 + \|u_0\|_V + \|v_0\|_H + \|f\|_{\mathcal{V}^*} + \|\eta\|_{\mathcal{V}^*}) \tag{5.76}$$

with a positive constant $\overline{c}$.

Next, in the second step, we consider the operator $\Lambda\colon \mathcal{V}^* \to \mathcal{V}^*$ defined by

$$(\Lambda\eta)(t) = \int_0^t C(t-s)u_\eta(s)\,ds \ \text{ for all } \eta \in \mathcal{V}^*,\ \text{a.e. } t \in (0,T), \tag{5.77}$$

where $u_\eta \in \mathcal{V}$ is the unique solution to (5.75). It is easy to check that the operator Λ is well defined. Indeed, for $\eta \in \mathcal{V}^*$, by using (5.74) and the Hölder inequality, we have

$$\begin{aligned} \left\| \int_0^t C(t-s)u_\eta(s)\,ds \right\|_{V^*} &\le \int_0^t \|C(t-s)\|_{\mathcal{L}(V,V^*)} \|u_\eta(s)\|_V\,ds \\ &\le \left(\int_0^t \|C(\tau)\|^2_{\mathcal{L}(V,V^*)}\,d\tau \right)^{1/2} \left(\int_0^t \|u_\eta(\tau)\|_V^2\,d\tau \right)^{1/2} \\ &\le \|C\|_{L^2(0,t;\mathcal{L}(V,V^*))}\, \|u_\eta\|_{L^2(0,t;V)} \end{aligned}$$

for a.e. $t \in (0,T)$. Therefore,

$$\left\| \int_0^t C(t-s)u_\eta(s)\,ds \right\|_{V^*} \le \|C\|\ \|u_\eta\|_{\mathcal{V}}, \tag{5.78}$$

where $\|C\|$ represents the norm of the operator C in $L^2(0,T;\mathcal{L}(V,V^*))$, i.e., $\|C\| = \|C\|_{L^2(0,T;\mathcal{L}(V,V^*))}$. Hence,

$$
\begin{aligned}
\|\Lambda\eta\|^2_{\mathcal{V}^*} &= \int_0^T \|(\Lambda\eta)(t)\|^2_{V^*}\,dt \\
&= \int_0^T \left(\left\|\int_0^t C(t-s)u_\eta(s)ds\right\|_{V^*}\right)^2 dt \\
&\le T\|C\|^2\|u_\eta\|^2_{\mathcal{V}}.
\end{aligned}
$$

Keeping in mind (5.76), we obtain that the integral in (5.77) is well defined and the operator Λ takes values in $\mathcal{V}^*$.

Subsequently, we show that the operator Λ has a unique fixed point. Let η_1, $\eta_2 \in \mathcal{V}^*$ and let $u_1 = u_{\eta_1}$ and $u_2 = u_{\eta_2}$ be the corresponding solutions to (5.75) such that $u_i \in \mathcal{V}$ and $u_i' \in \mathcal{W}$ for $i = 1, 2$. We have

$$u_1''(s) + A(s,u_1'(s)) + Bu_1(s) + \zeta_1(s) = f(s) - \eta_1(s) \text{ a.e. } s \in (0,T), \tag{5.79}$$

$$u_2''(s) + A(s,u_2'(s)) + Bu_2(s) + \zeta_2(s) = f(s) - \eta_2(s) \text{ a.e. } s \in (0,T), \tag{5.80}$$

$$\zeta_1(s) \in M^*\partial J(s, Mu_1'(s)),\ \zeta_2(s) \in M^*\partial J(s, Mu_2'(s)) \text{ a.e. } s \in (0,T), \tag{5.81}$$

$$u_1(0) = u_2(0) = u_0,\ u_1'(0) = u_2'(0) = v_0. \tag{5.82}$$

Subtracting (5.80) from (5.79), multiplying the result in duality by $u_1'(s) - u_2'(s)$ and integrating by parts with the initial conditions (5.82) we obtain

$$
\begin{aligned}
&\frac{1}{2}\|u_1'(t) - u_2'(t)\|^2_H + \int_0^t \langle A(s,u_1'(s)) - A(s,u_2'(s)), u_1'(s) - u_2'(s)\rangle_{V^*\times V}\,ds \\
&\qquad + \int_0^t \langle Bu_1(s) - Bu_2(s), u_1'(s) - u_2'(s)\rangle_{V^*\times V}\,ds \\
&\qquad + \int_0^t \langle \zeta_1(s) - \zeta_2(s), u_1'(s) - u_2'(s)\rangle_{Z^*\times Z}\,ds \\
&= \int_0^t \langle \eta_2(s) - \eta_1(s), u_1'(s) - u_2'(s)\rangle_{V^*\times V}\,ds
\end{aligned}
\tag{5.83}
$$

for all $t \in [0, T]$. Note also that from (5.81) we have $\zeta_i(s) = M^* z_i(s)$ with $z_i(s) \in \partial J(s, M u_i'(s))$ for a.e. $s \in (0, t)$ and $i = 1, 2$. Therefore, by using (5.68)(d), we obtain

$$\begin{aligned}
&\int_0^t \langle \zeta_1(s) - \zeta_2(s), u_1'(s) - u_2'(s) \rangle_{Z^* \times Z} \, ds \\
&= \int_0^t \langle z_1(s) - z_2(s), M u_1'(s) - M u_2'(s) \rangle_{X^* \times X} \, ds \\
&\geq -m_2 \int_0^t \| M u_1'(s) - M u_2'(s) \|_X^2 \, ds \\
&\geq -m_2 \, c_e^2 \, \| M \|^2 \int_0^t \| u_1'(s) - u_2'(s) \|_V^2 \, ds
\end{aligned} \tag{5.84}$$

for all $t \in [0, T]$. And, finally, using the properties of the operator B we have

$$\begin{aligned}
&\int_0^t \langle B u_1(s) - B u_2(s), u_1'(s) - u_2'(s) \rangle_{V^* \times V} \, ds \\
&= \frac{1}{2} \int_0^t \frac{d}{ds} \langle B(u_1(s) - u_2(s)), u_1(s) - u_2(s) \rangle_{V^* \times V} \, ds \\
&= \frac{1}{2} \langle B(u_1(t) - u_2(t)), u_1(t) - u_2(t) \rangle_{V^* \times V} \geq 0
\end{aligned} \tag{5.85}$$

for all $t \in [0, T]$. We combine now (5.83)–(5.85), and use (5.67)(e) to obtain

$$\begin{aligned}
&\frac{1}{2} \| u_1'(t) - u_2'(t) \|_H^2 + \widetilde{c} \int_0^t \| u_1'(s) - u_2'(s) \|_V^2 \, ds \\
&\leq \int_0^t \| \eta_1(s) - \eta_1(s) \|_{V^*} \| u_1'(s) - u_2'(s) \|_V \, ds
\end{aligned}$$

for all $t \in [0, T]$, with $\widetilde{c} = m_1 - m_2 \, c_e^2 \, \| M \|^2 > 0$. It follows from here that

$$\widetilde{c} \, \| u_1' - u_2' \|_{L^2(0,t;V)}^2 \leq \| \eta_1 - \eta_1 \|_{L^2(0,t;V^*)} \, \| u_1' - u_2' \|_{L^2(0,t;V)}$$

for all $t \in [0, T]$, which implies that

$$\| u_1' - u_2' \|_{L^2(0,t;V)} \leq \frac{1}{\widetilde{c}} \| \eta_1 - \eta_1 \|_{L^2(0,t;V^*)} \tag{5.86}$$

for all $t \in [0, T]$. Recall also that

$$u_i(t) = u_0 + \int_0^t u_i'(s) \, ds \quad \text{for all } t \in [0, T], \; i = 1, 2$$

and, therefore,

$$\|u_1(t) - u_2(t)\|_V \le \int_0^t \|u_1'(s) - u_2'(s)\|_V \, ds$$
$$\le \sqrt{T}\, \|u_1' - u_2'\|_{L^2(0,t;V)} \tag{5.87}$$

for all $t \in [0, T]$. Combining (5.86) and (5.87), we find

$$\|u_1(t) - u_2(t)\|_V \le \frac{\sqrt{T}}{\widetilde{c}} \|\eta_1 - \eta_2\|_{L^2(0,t;V^*)} \tag{5.88}$$

for all $t \in [0, T]$.

On the other hand, by an estimate similar to (5.78), we infer

$$\|(\Lambda\eta_1)(t) - (\Lambda\eta_2)(t)\|_{V^*} \le \int_0^t \|C(t-s)(u_1(s) - u_2(s))\|_{V^*} \, ds$$
$$\le \|C\| \, \|u_1 - u_2\|_{L^2(0,t;V)}$$

for a.e. $t \in (0, T)$. Hence, using (5.88) and the previous inequality, we obtain

$$\|(\Lambda\eta_1)(t) - (\Lambda\eta_2)(t)\|_{V^*}^2 \le \|C\|^2 \, \|u_1 - u_2\|_{L^2(0,t;V)}^2$$
$$\le \frac{T\|C\|^2}{\widetilde{c}^2} \int_0^t \|\eta_1 - \eta_2\|_{L^2(0,s;V^*)}^2 \, ds \le \frac{T\|C\|^2}{\widetilde{c}^2} t \, \|\eta_1 - \eta_2\|_{\mathcal{V}^*}^2$$

and, consequently,

$$\|(\Lambda^2\eta_1)(t) - (\Lambda^2\eta_2)(t)\|_{V^*}^2 = \|\Lambda(\Lambda\eta_1)(t) - \Lambda(\Lambda\eta_2)(t)\|_{V^*}^2$$
$$\le \frac{T\|C\|^2}{\widetilde{c}^2} t \int_0^t \|(\Lambda\eta_1)(s) - (\Lambda\eta_2)(s)\|_{V^*}^2 \, ds$$
$$\le \frac{T^3\|C\|^4}{\widetilde{c}^4} \frac{t^2}{2} \|\eta_1 - \eta_2\|_{\mathcal{V}^*}^2$$

for a.e. $t \in (0, T)$. Reiterating this inequality k times leads to

$$\|(\Lambda^k\eta_1)(t) - (\Lambda^k\eta_2)(t)\|_{V^*}^2 \le \frac{T^{2k-1}\|C\|^{2k}}{\widetilde{c}^{2k}} \frac{t^k}{k!} \|\eta_1 - \eta_2\|_{\mathcal{V}^*}^2$$
$$\le \frac{T^{3k-1}\,\|C\|^{2k}}{\widetilde{c}^{2k}} \frac{1}{k!} \|\eta_1 - \eta_2\|_{\mathcal{V}^*}^2$$

for a.e. $t \in (0, T)$. This implies that

$$\|\Lambda^k\eta_1 - \Lambda^k\eta_2\|_{\mathcal{V}^*} \le \left(\frac{T\sqrt{T}\,\|C\|}{\widetilde{c}} \right)^k \frac{1}{\sqrt{k!}} \|\eta_1 - \eta_2\|_{\mathcal{V}^*}.$$

Since

$$\lim_{k\to\infty} \frac{a^k}{\sqrt{k!}} = 0 \ \text{ for all } \ a > 0,$$

from the previous inequality we deduce that for k sufficiently large Λ^k is a contraction on $\mathcal{V}^*$. Therefore, the Banach contraction principle implies that there exists a unique $\eta^* \in \mathcal{V}^*$ such that $\eta^* = \Lambda^k \eta^*$. Moreover, it follows that $\eta^* \in \mathcal{V}^*$ is the unique fixed point of Λ. Then u_{η^*} is a solution to Problem 5.16, which concludes the proof of the existence part of the theorem.

To prove the uniqueness part let $u \in \mathcal{V}$ with $u' \in \mathcal{W}$ be a solution to Problem 5.16 and define the element $\eta \in \mathcal{V}^*$ by

$$\eta(t) = \int_0^t C(t-s)u(s)\,ds \ \text{ for a.e. } t \in [0,T].$$

It follows that u is the solution to the problem (5.75) and, by the uniqueness of solution to (5.75), we obtain $u = u_\eta$. This implies $\Lambda\eta = \eta$ and, by the uniqueness of the fixed point of Λ we have $\eta = \eta^*$. Therefore, $u = u_{\eta^*}$, which concludes the proof of the uniqueness part of the theorem. □

5.3 Second-Order Hemivariational Inequalities

In this section we use the results of Sect. 5.2 to provide existence and uniqueness results of solutions to hemivariational inequalities of hyperbolic type. These results represent an evolutionary version of theorems presented in Sect. 4.3. Most of the notation and functional spaces we use in this section were already introduced on pages 109 and 121 but, for the convenience of the reader, we recall them in what follows.

Let $\Omega \subset \mathbb{R}^d$ be an open bounded subset of $\mathbb{R}^d$ with a Lipschitz continuous boundary $\partial\Omega = \Gamma$ and let Γ_C be a measurable part of Γ. Let V be a closed subspace of $H^1(\Omega;\mathbb{R}^s)$, $s \in \mathbb{N}$, $H = L^2(\Omega;\mathbb{R}^s)$, and $Z = H^\delta(\Omega;\mathbb{R}^s)$ with a fixed $\delta \in (\frac{1}{2}, 1)$. We denote by $\gamma\colon Z \to L^2(\Gamma_C;\mathbb{R}^s)$ and $\gamma_0\colon H^1(\Omega;\mathbb{R}^s) \to H^{1/2}(\Gamma_C;\mathbb{R}^s) \subset L^2(\Gamma_C;\mathbb{R}^s)$ the trace operators and, for simplicity, we write $\gamma_0 v = \gamma v$ for $v \in V$. Recall that (V, H, V^*) and (Z, H, Z^*) form evolution triples of spaces and $V \subset Z$ with compact embedding. We denote by $c_e > 0$ the embedding constant of V into Z, by $\|\gamma\|$ the norm of the trace in $\mathcal{L}(Z, L^2(\Gamma_C;\mathbb{R}^s))$, and by $\gamma^*\colon L^2(\Gamma_C;\mathbb{R}^s) \to Z^*$ the adjoint operator to γ.

Moreover, given a time interval $(0,T)$, we denote $\Sigma_C = \Gamma_C \times (0,T)$ and we introduce the spaces

$$\mathcal{V} = L^2(0,T;V), \quad \mathcal{Z} = L^2(0,T;Z), \quad \widehat{\mathcal{H}} = L^2(0,T;H),$$

$$\mathcal{Z}^* = L^2(0,T;Z^*), \quad \mathcal{V}^* = L^2(0,T;V^*), \quad \mathcal{W} = \{\, v \in \mathcal{V} \mid v' \in \mathcal{V}^* \,\}.$$

A local existence result. Let $A\colon (0,T) \times V \to V^*$ and $B\colon V \to V^*$ be given operators, let f, h, j_1, j_2 denote prescribed functions, and let u_0, v_0 represent initial conditions. With these data we consider the following Cauchy problem for hyperbolic hemivariational inequalities.

Problem 5.18. *Find $u \in \mathcal{V}$ such that $u' \in \mathcal{W}$ and*

$$\left.\begin{aligned}
&\langle u''(t) + A(t,u'(t)) + Bu(t), v\rangle_{V^*\times V}\\
&\quad + \int_{\Gamma_C} \Big(j_1^0(t,\gamma u(t);\gamma v) + h(\gamma u(t),\gamma u'(t))\, j_2^0(t,\gamma u'(t);\gamma v)\Big)\, d\Gamma\\
&\quad \geq \langle f(t), v\rangle_{V^*\times V} \text{ for all } v \in V, \text{ a.e. } t \in (0,T),\\
&u(0) = u_0, \quad u'(0) = v_0.
\end{aligned}\right\}$$

An inequality as above is called a *second-order hemivariational inequality* or a *hemivariational inequality of hyperbolic type*.

In the study of Problem 5.18 we consider the following hypotheses.

$$\left.\begin{aligned}
&h\colon \Gamma_C \times \mathbb{R}^s \times \mathbb{R}^s \to \mathbb{R} \text{ is such that}\\
&\text{(a) } h(\cdot,\eta_1,\eta_2) \text{ is measurable on } \Gamma_C \text{ for all } \eta_1,\eta_2 \in \mathbb{R}^s.\\
&\text{(b) } h(x,\cdot,\cdot) \text{ is continuous on } \mathbb{R}^s \times \mathbb{R}^s \text{ for a.e. } x \in \Gamma_C.\\
&\text{(c) } 0 \leq h(x,\eta_1,\eta_2) \leq \overline{h} \text{ for all } \eta_1,\eta_2 \in \mathbb{R}^s, \text{ a.e. } x \in \Gamma_C\\
&\qquad \text{with } \overline{h} > 0.
\end{aligned}\right\} \tag{5.89}$$

$$\left.\begin{aligned}
&j_i\colon \Sigma_C \times \mathbb{R}^s \to \mathbb{R},\ i = 1,2 \text{ is such that}\\
&\text{(a) } j_i(\cdot,\cdot,\xi) \text{ is measurable on } \Sigma_C \text{ for all } \xi \in \mathbb{R}^s \text{ and there}\\
&\qquad \text{exists } e_i \in L^2(\Gamma_C;\mathbb{R}^s) \text{ such that } j_i(\cdot,\cdot,e_i(\cdot)) \in L^1(\Sigma_C).\\
&\text{(b) } j_i(x,t,\cdot) \text{ is locally Lipschitz on } \mathbb{R}^s \text{ for a.e. } (x,t) \in \Sigma_C.\\
&\text{(c) } \|\partial j_i(x,t,\xi)\|_{\mathbb{R}^s} \leq \overline{c}_{0i}(t) + \overline{c}_{1i}\|\xi\|_{\mathbb{R}^s} \text{ for all } \xi \in \mathbb{R}^s,\\
&\qquad \text{a.e. } (x,t) \in \Sigma_C \text{ with } \overline{c}_{0i},\overline{c}_{1i} \geq 0, \overline{c}_{0i} \in L^\infty(0,T).\\
&\text{(d) Either } j_1(x,t,\cdot), j_2(x,t,\cdot) \text{ or } -j_1(x,t,\cdot), -j_2(x,t,\cdot)\\
&\qquad \text{are regular on } \mathbb{R}^s \text{ for a.e. } (x,t) \in \Sigma_C.
\end{aligned}\right\} \tag{5.90}$$

The main existence result for Problem 5.18 reads as follows.

Theorem 5.19. *Assume that* (5.1), (5.2), (5.4), (5.89), (5.90) *hold and*

$$\alpha > 6 \max\{\overline{c}_{11}, \overline{c}_{12}\overline{h}\}\, c_e^2\, \|\gamma\|^2\, T. \tag{5.91}$$

Then Problem 5.18 *has at least one solution.*

Proof. We apply Theorem 5.11. To this end, we consider the space $X = L^2(\Gamma_C;\mathbb{R}^s)$, the operator $R\colon Z \times Z \to X \times X$ given by $R(w,z) = (\gamma w, \gamma z)$ for all $w, z \in Z$ and the functional $J\colon (0,T) \times L^2(\Gamma_C;\mathbb{R}^s)^4 \to \mathbb{R}$ defined by

$$J(t,w,z,u,v) = \int_{\Gamma_C} \Big(j_1(x,t,\gamma u) + h(x,R(w,z))\, j_2(x,t,\gamma v) \Big)\, d\Gamma \tag{5.92}$$

for all w, z, u, $v \in L^2(\Gamma_C;\mathbb{R}^s)$, a.e. $t \in (0,T)$. It is clear that the operator R belongs to $\mathcal{L}(Z \times Z, X \times X)$, i.e., (5.60) holds. We prove that under the hypotheses (5.89) and (5.90) the functional J defined by (5.92) satisfies condition (5.63). To make the proof transparent, we first state the properties of the integrand of J. Let $j\colon \Sigma_C \times \mathbb{R}^s \times \mathbb{R}^s \times \mathbb{R}^s \times \mathbb{R}^s \to \mathbb{R}$ be given by

$$j(x,t,\eta_1,\eta_2,\xi_1,\xi_2) = j_1(x,t,\xi_1) + h(x,\eta_1,\eta_2)\, j_2(x,t,\xi_2)$$

for all η_1, η_2, ξ_1, $\xi_2 \in \mathbb{R}^s$, a.e. $(x,t) \in \Sigma_C$. It follows from the hypotheses (5.89) and (5.90) that j has the following properties:

(a) $j(\cdot,\cdot,\eta_1,\eta_2,\xi_1,\xi_2)$ is measurable on Σ_C, for all η_1, η_2, ξ_1, $\xi_2 \in \mathbb{R}^s$ and there exists $e \in L^2(\Gamma_C;\mathbb{R}^s \times \mathbb{R}^s)$ given by $e(\cdot) = (e_1(\cdot), e_2(\cdot))$ such that for all $\overline{w} \in L^2(\Gamma_C;\mathbb{R}^s \times \mathbb{R}^s)$, we have $j(\cdot,\cdot,\overline{w}(\cdot),e(\cdot)) \in L^1(\Sigma_C)$.

(b) $j(x,t,\cdot,\cdot,\xi_1,\xi_2)$ is continuous on $\mathbb{R}^s \times \mathbb{R}^s$ for all ξ_1, $\xi_2 \in \mathbb{R}^s$, a.e. $(x,t) \in \Sigma_C$ and $j(x,t,\eta_1,\eta_2,\cdot,\cdot)$ is locally Lipschitz on $\mathbb{R}^s \times \mathbb{R}^s$ for all η_1, $\eta_2 \in \mathbb{R}^s$, a.e. $(x,t) \in \Sigma_C$.

(c) $\|\partial j(x,t,\eta_1,\eta_2,\xi_1,\xi_2)\|_{\mathbb{R}^s \times \mathbb{R}^s} \le \overline{c}_{01}(t) + \overline{c}_{02}(t)\,\overline{h} + \overline{c}_{11}\, \|\xi_1\|_{\mathbb{R}^s} + \overline{c}_{12}\,\overline{h}\, \|\xi_2\|_{\mathbb{R}^s}$ for all η_1, η_2, ξ_1, $\xi_2 \in \mathbb{R}^s$, a.e. $(x,t) \in \Sigma_C$, where the subdifferential of j is taken with respect to the pair (ξ_1,ξ_2).

(d) Either $j(x,t,\eta_1,\eta_2,\cdot,\cdot)$ or $-j(x,t,\eta_1,\eta_2,\cdot,\cdot)$ is regular on $\mathbb{R}^s \times \mathbb{R}^s$ for all η_1, $\eta_2 \in \mathbb{R}^s$, a.e. $(x,t) \in \Sigma_C$.

(e) $j^0(x,t,\cdot,\cdot,\cdot,\cdot;\rho_1,\rho_2)$ is upper semicontinuous on $(\mathbb{R}^s)^4$ for all ρ_1, $\rho_2 \in \mathbb{R}^s$, a.e. $(x,t) \in \Sigma_C$.

The proof of the properties above follows directly from the hypotheses (5.89) and (5.90) and, for this reason, we skip them. Nevertheless, to provide an example, we restrict ourselves to present only the proof of the property (e). For all $n \in \mathbb{N}$, let $(\eta_{1n},\eta_{2n},\xi_{1n},\xi_{2n}) \in (\mathbb{R}^s)^4$ and assume that $(\eta_{1n},\eta_{2n},\xi_{1n},\xi_{2n}) \to (\eta_1,\eta_2,\xi_1,\xi_2)$ as $n \to \infty$, where $(\eta_1,\eta_2,\xi_1,\xi_2) \in (\mathbb{R}^s)^4$. Let also ρ_1, $\rho_2 \in \mathbb{R}^s$. From Lemma 3.39(3) and Proposition 3.23 (ii), for a.e. $(x,t) \in \Sigma_C$, we have

$$
\begin{aligned}
&\limsup j^0(x,t,\eta_{1n},\eta_{2n},\xi_{1n},\xi_{2n};\rho_1,\rho_2)\\
&\quad= \limsup \left(j_1^0(x,t,\xi_{1n};\rho_1) + h(x,\eta_{1n},\eta_{2n})\, j_2^0(x,t,\xi_{2n};\rho_2)\right)\\
&\quad\le \limsup j_1^0(x,t,\xi_{1n};\rho_1) + h(x,\eta_1,\eta_2) \limsup j_2^0(x,t,\xi_{2n};\rho_2)\\
&\qquad + \limsup\, (h(x,\eta_{1n},\eta_{2n}) - h(x,\eta_1,\eta_2))\;\, j_2^0(x,t,\xi_{2n};\rho_2)\\
&\quad\le j_1^0(x,t,\xi_1;\rho_1) + h(x,\eta_1,\eta_2)\, j_2^0(x,t,\xi_2;\rho_2)\\
&\qquad + \|\rho_2\|_{\mathbb{R}^s}\, (\overline{c}_{20}(t) + \overline{c}_{21}\, \|\xi_{2n}\|_{\mathbb{R}^s}) \lim\, |h(x,\eta_{1n},\eta_{2n}) - h(x,\eta_1,\eta_2)|\\
&\quad= j_1^0(x,t,\xi_1;\rho_1) + h(x,\eta_1,\eta_2)\, j_2^0(x,t,\xi_2;\rho_2) = j^0(x,\eta_1,\eta_2,\xi_1,\xi_2;\rho_1,\rho_2),
\end{aligned}
$$

which proves the upper semicontinuity of $j^0(x,t,\cdot,\cdot,\cdot,\cdot;\rho_1,\rho_2)$.

Next, using the properties (a)–(e) of the integrand j of J, from Theorem 3.47, we deduce that

(i) $J(t,\cdot,\cdot,\cdot,\cdot)$ is well defined and finite on $L^2(\Gamma_C;\mathbb{R}^s)^4$, a.e. $t \in (0,T)$.

(ii) $J(\cdot,w,z,u,v)$ is measurable on $(0,T)$ for all w, z, u, $v \in L^2(\Gamma_C;\mathbb{R}^s)$.

(iii) $J(t,w,z,\cdot,\cdot)$ is Lipschitz continuous on bounded subsets of $L^2(\Gamma_C;\mathbb{R}^s)^2$ for all $w, z \in L^2(\Gamma_C;\mathbb{R}^s)$, a.e. $t \in (0,T)$.

(iv) For all w, z, u, $v \in L^2(\Gamma_C;\mathbb{R}^s)$, a.e. $t \in (0,T)$, we have

$$\|\partial J(t,w,z,u,v)\|_{L^2(\Gamma_C;\mathbb{R}^s)^2} \le c_0(t) + c_1\, \|(u,v)\|_{L^2(\Gamma_C;\mathbb{R}^s)^2}$$

with $c_0(t) = \sqrt{3\,\text{meas}\,(\Gamma_C)}\left(\overline{c}_{10}(t) + \overline{c}_{20}\overline{h}\right)$, $c_1 = \sqrt{3}\max\{\overline{c}_{11}, \overline{c}_{21}\overline{h}\}$, where the subdifferential of J is taken with respect to (u,v).

(v) The multifunction $\partial J(t,\cdot,\cdot,\cdot,\cdot)\colon L^2(\Gamma_C;\mathbb{R}^s)^4 \to 2^{L^2(\Gamma_C;\mathbb{R}^s)^2}$ has a closed graph in the $L^2(\Gamma_C;\mathbb{R}^s)^4 \times (w\text{–}L^2(\Gamma_C;\mathbb{R}^s)^2)$ topology for a.e. $t \in (0,T)$

(vi) For all w, z, u, v, $\overline{u}$, $\overline{v} \in L^2(\Gamma_C;\mathbb{R}^s)$, a.e. $t \in (0,T)$, we have

$$J^0(t,w,z,u,v;\overline{u},\overline{v}) = \int_{\Gamma_C} j^0(x,t,w(x),z(x),u(x),v(x);\overline{u}(x),\overline{v}(x))\, d\Gamma.$$

From the properties (i)–(vi), we deduce that the functional J given by (5.92) satisfies the conditions (5.63). Moreover, since $c_2 = 0$, it follows that (5.91) implies (5.61). We use now Theorem 5.11 to obtain that Problem 5.9 has at least a solution.

Finally, we show that every solution to Problem 5.9 is also a solution to Problem 5.18. Let $u \in \mathcal{V}$ be a solution to Problem 5.9. According to Definition 5.2 this means that $u' \in \mathcal{W}$ and there is $\zeta \in \mathcal{Z}^*$ such that

$$\left.\begin{aligned}&u''(t) + A(t, u'(t)) + Bu(t) + \zeta(t) = f(t) \ \text{ a.e. } t \in (0, T),\\&\zeta(t) \in S\, R^*\, \partial J(t, R(u(t), u'(t)), R(u(t), u'(t))) \ \text{ a.e. } t \in (0, T),\\&u(0) = u_0, \ \ u'(0) = v_0.\end{aligned}\right\}$$

Hence, it follows that

$$\zeta(t) = \zeta_1(t) + \zeta_2(t), \quad (\zeta_1(t), \zeta_2(t)) = (\gamma^* z_1(t), \gamma^* z_2(t))$$

and

$$(z_1(t), z_2(t)) \in \partial J(t, \gamma u(t), \gamma u'(t), \gamma u(t), \gamma u'(t))$$

for a.e. $t \in (0, T)$. The latter inclusion is equivalent to

$$\begin{aligned}&\langle z_1(t), \overline{u}\rangle_{L^2(\Gamma_C;\mathbb{R}^s)} + \langle z_2(t), \overline{v}\rangle_{L^2(\Gamma_C;\mathbb{R}^s)}\\&\quad \le J^0(t, \gamma u(t), \gamma u'(t), \gamma u(t), \gamma u'(t); \overline{u}, \overline{v})\end{aligned}$$

for all $\overline{u}$, $\overline{v} \in L^2(\Gamma_C; \mathbb{R}^s)$, a.e. $t \in (0, T)$. Therefore, from the property (vi), we obtain that

$$\begin{aligned}&\langle f(t) - u''(t) - A(t, u'(t)) - Bu(t), v\rangle_{V^*\times V} = \langle \zeta(t), v\rangle_{Z^*\times Z}\\&= \langle \gamma^* z_1(t) + \gamma^* z_2(t), v\rangle_{Z^*\times Z} = \langle z_1(t) + z_2(t), \gamma v\rangle_{L^2(\Gamma_C;\mathbb{R}^s)}\\&\le J^0(t, \gamma u(t), \gamma u'(t), \gamma u(t), \gamma u'(t); \gamma v, \gamma v)\\&= \int_{\Gamma_C} \Big(j_1^0(t, \gamma u(t); \gamma v) + h(\gamma u(t), \gamma u'(t))\, j_2^0(t, \gamma u'(t); \gamma v)\Big)\, d\Gamma\end{aligned}$$

for all $v \in V$ and a.e. $t \in (0, T)$. It results from above that u is a solution to Problem 5.18, which completes the proof. □

We underline that the result of Theorem 5.19 represents a local existence result, since the length of the time interval has to satisfy the smallness condition (5.91). In what follows we consider a particular case of Problem 5.18 in which $j_1 = 0$. In this case we are able to remove this smallness assumption, i.e., we establish a global existence result, see Theorem 5.21. In the proof of this global existence result we introduce a superpotential and, in contrast to the proof of Theorem 5.19, we consider its subdifferential with respect to only one variable.

A global existence result. Let $A\colon (0,T) \times V \to V^*$, $B\colon V \to V^*$ be given operators, let f, h and j be prescribed functions and let u_0, v_0 represent given initial conditions. With these data we consider the following Cauchy problem for hyperbolic hemivariational inequalities.

Problem 5.20. *Find $u \in \mathcal{V}$ such that $u' \in \mathcal{W}$ and*

$$\left.\begin{aligned} &\langle u''(t) + A(t,u'(t)) + Bu(t), v\rangle_{V^*\times V} \\ &\qquad + \int_{\Gamma_C} h(\gamma u(t), \gamma u'(t))\, j^0(t, \gamma u'(t); \gamma v)\, d\Gamma \ge \langle f(t), v\rangle_{V^*\times V} \\ &\qquad\qquad \text{for all } v \in V, \text{ a.e. } t \in (0,T), \\ &u(0) = u_0, \ \ u'(0) = v_0. \end{aligned}\right\}$$

In the study of Problem 5.20 we need the following hypothesis on the function j.

$$\left.\begin{aligned} &j\colon \Sigma_C \times \mathbb{R}^s \to \mathbb{R} \text{ is such that} \\ &\text{(a) } j(\cdot,\cdot,\xi) \text{ is measurable on } \Sigma_C \text{ for all } \xi \in \mathbb{R}^s \text{ and there} \\ &\qquad \text{exists } e \in L^2(\Gamma_C;\mathbb{R}^s) \text{ such that } j(\cdot,\cdot,e(\cdot)) \in L^1(\Sigma_C). \\ &\text{(b) } j(x,t,\cdot) \text{ is locally Lipschitz on } \mathbb{R}^s \text{ for a.e. } (x,t) \in \Sigma_C. \\ &\text{(c) } \|\partial j(x,t,\xi)\|_{\mathbb{R}^s} \le \overline{c}_0(t) + \overline{c}_1 \|\xi\|_{\mathbb{R}^s} \text{ for all } \xi \in \mathbb{R}^s, \\ &\qquad \text{a.e. } (x,t) \in \Sigma_C \text{ with } \overline{c}_0, \overline{c}_1 \ge 0, \overline{c}_0 \in L^\infty(0,T). \\ &\text{(d) Either } j(x,t,\cdot) \text{ or } -j(x,t,\cdot) \text{ is regular on } \mathbb{R}^s \text{ for} \\ &\qquad \text{a.e. } (x,t) \in \Sigma_C. \end{aligned}\right\} \tag{5.93}$$

The main existence result for Problem 5.20 reads as follows.

Theorem 5.21. *Assume that* (5.1), (5.2), (5.4), (5.89), (5.93) *hold and*

$$\alpha > 6\,\overline{c}_1\, c_e^2\, \overline{h}\, \|\gamma\|^2. \tag{5.94}$$

Then Problem 5.20 *has at least one solution.*

Proof. We apply Theorem 5.13. We consider the space $X = L^2(\Gamma_C;\mathbb{R}^s)$, the operator $M = \gamma\colon Z \to X$ and the functional $J\colon (0,T) \times L^2(\Gamma_C;\mathbb{R}^s)^3 \to \mathbb{R}$ defined by

$$J(t,w,u,v) = \int_{\Gamma_C} h(x, w(x), u(x))\, j(x,t,v(x))\, d\Gamma \tag{5.95}$$

for all w, u, $v \in L^2(\Gamma_C;\mathbb{R}^s)$, a.e. $t \in (0,T)$. It is clear that $\gamma \in \mathcal{L}(Z,X)$, i.e., (5.64) holds. We prove that under the hypotheses (5.89) and (5.93) the functional J defined by (5.95) satisfies condition (5.63). To make the proof transparent, we first state the properties of the integrand of J.

Let $j_1 \colon \Sigma_C \times \mathbb{R}^s \times \mathbb{R}^s \times \mathbb{R}^s \to \mathbb{R}$ be given by

$$j_1(x,t,\eta_1,\eta_2,\xi) = h(x,\eta_1,\eta_2)\, j(x,t,\xi)$$

for all $\eta_1, \eta_2, \xi \in \mathbb{R}^s$, a.e. $(x,t) \in \Sigma_C$. It follows from the hypotheses (5.89) and (5.93) that j_1 has the following properties.

(a) $j_1(\cdot,\cdot,\eta_1,\eta_2,\xi)$ is measurable on Σ_C, for all $\eta_1, \eta_2, \xi \in \mathbb{R}^s$ and there exists $e \in L^2(\Gamma_C;\mathbb{R}^s)$ such that for all $v, w \in L^2(\Gamma_C;\mathbb{R}^s)$, we have $j_1(\cdot,\cdot,v(\cdot),w(\cdot),e(\cdot)) \in L^1(\Sigma_C)$.
(b) $j_1(x,t,\cdot,\cdot,\xi)$ is continuous on $\mathbb{R}^s \times \mathbb{R}^s$ for all $\xi \in \mathbb{R}^s$, a.e. $(x,t) \in \Sigma_C$ and $j_1(x,t,\eta_1,\eta_2,\cdot)$ is locally Lipschitz on $\mathbb{R}^s$, for all $\eta_1, \eta_2 \in \mathbb{R}^s$, a.e. $(x,t) \in \Sigma_C$.
(c) $\|\partial j_1(x,t,\eta_1,\eta_2,\xi)\|_{\mathbb{R}^s} \le \overline{c}_0(t)\,\overline{h} + \overline{c}_1\,\overline{h}\,\|\xi\|_{\mathbb{R}^s}$ for all $\eta_1, \eta_2, \xi \in \mathbb{R}^s$, a.e. $(x,t) \in \Sigma_C$, where the subdifferential of j_1 is taken with respect to the last variable.
(d) Either $j_1(x,t,\eta_1,\eta_2,\cdot)$ or $-j_1(x,t,\eta_1,\eta_2,\cdot)$ is regular on $\mathbb{R}^s$, for all $\eta_1, \eta_2 \in \mathbb{R}^s$, a.e. $(x,t) \in \Sigma_C$.
(e) $j_1^0(x,t,\cdot,\cdot,\cdot;\rho)$ is upper semicontinuous on $(\mathbb{R}^s)^3$ for all $\rho \in \mathbb{R}^s$, a.e. $(x,t) \in \Sigma_C$.

The proof of properties follows directly from the hypotheses (5.89) and (5.93) and, for this reason, we provide only the proof of the property (e). Let $(\eta_{1n},\eta_{2n},\xi_n) \in (\mathbb{R}^s)^3$ be such that $(\eta_{1n},\eta_{2n},\xi_n) \to (\eta_1,\eta_2,\xi)$ as $n \to \infty$, where $(\eta_1,\eta_2,\xi) \in (\mathbb{R}^s)^3$. Also, let $\rho \in \mathbb{R}^s$. From Proposition 3.23 (ii), for a.e. $(x,t) \in \Sigma_C$, we have

$$\begin{aligned}
&\limsup j_1^0(x,t,\eta_{1n},\eta_{2n},\xi_n;\rho) \\
&\quad = \limsup h(x,\eta_{1n},\eta_{2n})\, j^0(x,t,\xi_n;\rho) \\
&\quad \le \limsup [(h(x,\eta_{1n},\eta_{2n}) - h(x,\eta_1,\eta_2))\, j^0(x,t,\xi_n;\rho) \\
&\qquad + h(x,\eta_1,\eta_2)\, j^0(x,t,\xi_n;\rho)] \\
&\quad \le \|\rho\|_{\mathbb{R}^s} (\overline{c}_0(t) + \overline{c}_1 \|\xi_n\|_{\mathbb{R}^s}) \lim |h(x,\eta_{1n},\eta_{2n}) - h(x,\eta_1,\eta_2)| \\
&\qquad + h(x,\eta_1,\eta_2) \limsup j^0(x,t,\xi_n;\rho) \\
&\quad \le h(x,\eta_1,\eta_2)\, j^0(x,t,\xi;\rho) \\
&\quad = j_1^0(x,\eta_1,\eta_2,\xi;\rho),
\end{aligned}$$

which proves the upper semicontinuity of $j_1^0(x,t,\cdot,\cdot,\cdot;\rho)$.

Next, using the properties (a)–(e) of the integrand j_1 of J, from Theorem 3.47, we deduce that

(i) $J(t,\cdot,\cdot,\cdot)$ is well defined and finite on $L^2(\Gamma_C;\mathbb{R}^s)^3$, for a.e. $t \in (0,T)$.
(ii) $J(\cdot,w,u,v)$ is measurable on $(0,T)$ for all $w, u, v \in L^2(\Gamma_C;\mathbb{R}^s)$.
(iii) $J(t,w,u,\cdot)$ is Lipschitz continuous on bounded subsets of $L^2(\Gamma_C;\mathbb{R}^s)$ for all $w, u \in L^2(\Gamma_C;\mathbb{R}^s)$, a.e. $t \in (0,T)$.

(iv) For all $w, u, v \in L^2(\Gamma_C;\mathbb{R}^s)$, a.e. $t \in (0,T)$, we have

$$\|\partial J(t,w,u,v)\|_{L^2(\Gamma_C;\mathbb{R}^s)} \le c_0(t) + c_1 \|v\|_{L^2(\Gamma_C;\mathbb{R}^s)}$$

with $c_0(t) = \sqrt{3 \operatorname{meas}(\Gamma_C)}\,\overline{c}_0(t)\,\overline{h}$ and $c_1 = \sqrt{3}\,\overline{c}_1\,\overline{h}$, where ∂J is taken with respect to the last variable.

(v) The multifunction $\partial J(t,\cdot,\cdot,\cdot)\colon L^2(\Gamma_C;\mathbb{R}^s)^3 \to 2^{L^2(\Gamma_C;\mathbb{R}^s)}$ has a closed graph in $L^2(\Gamma_C;\mathbb{R}^s)^3 \times (w\text{–}L^2(\Gamma_C;\mathbb{R}^s))$ topology for a.e. $t \in (0,T)$.

From the properties (i)–(v), we deduce that the functional J given by (5.95) satisfies the conditions (5.63). Moreover, since $c_2 = c_3 = 0$, then (5.94) implies (5.65). The conclusion of Theorem 5.21 is now a direct consequence of Theorem 5.13. □

A global existence and uniqueness result. In the following we prove that if A is strongly monotone, h is a nonnegative constant function and j satisfies an additional hypothesis, then the solution to Problem 5.20 is unique. Consider the following evolutionary hemivariational inequality which is a counterpart of the stationary hemivariational inequality in Problem 4.19.

Problem 5.22. *Find $u \in \mathcal{V}$ such that $u' \in \mathcal{W}$ and*

$$\left.\begin{aligned} &\langle u''(t) + A(t,u'(t)) + Bu(t), v\rangle_{V^*\times V} + \int_{\Gamma_C} j^0(t,\gamma u'(t);\gamma v)\,d\Gamma \\ &\qquad\qquad \ge \langle f(t), v\rangle_{V^*\times V} \ \text{ for all } v \in V, \text{ a.e. } t \in (0,T), \\ &u(0) = u_0, \ \ u'(0) = v_0. \end{aligned}\right\}$$

In the study of Problem 5.22 we need the following hypotheses involving the function j and the smallness conditions below, as well.

$$\left.\begin{aligned} &j\colon \Sigma_C \times \mathbb{R}^s \to \mathbb{R} \text{ is such that} \\ &\text{(a) } j(\cdot,\cdot,\xi) \text{ is measurable on } \Sigma_C \text{ for all } \xi \in \mathbb{R}^s \text{ and there} \\ &\quad\ \text{exists } e \in L^2(\Gamma_C;\mathbb{R}^s) \text{ such that } j(\cdot,\cdot,e(\cdot)) \in L^1(\Sigma_C). \\ &\text{(b) } j(x,t,\cdot) \text{ is locally Lipschitz on } \mathbb{R}^s \text{ for a.e. } (x,t) \in \Sigma_C. \\ &\text{(c) } \|\partial j(x,t,\xi)\|_{\mathbb{R}^s} \le \overline{c}_0(t) + \overline{c}_1\|\xi\|_{\mathbb{R}^s} \text{ for all } \xi \in \mathbb{R}^s, \\ &\quad\ \text{a.e. } (x,t) \in \Sigma_C \text{ with } \overline{c}_0, \overline{c}_1 \ge 0, \overline{c}_0 \in L^\infty(0,T). \\ &\text{(d) Either } j(x,t,\cdot) \text{ or } -j(x,t,\cdot) \text{ is regular on } \mathbb{R}^s \text{ for} \\ &\quad\ \text{a.e. } (x,t) \in \Sigma_C. \\ &\text{(e) } (\zeta_1 - \zeta_2)\cdot(\xi_1 - \xi_2) \ge -m_2\|\xi_1 - \xi_2\|^2_{\mathbb{R}^s} \text{ for all } \zeta_i,\ \xi_i \in \mathbb{R}^s, \\ &\quad\ \zeta_i \in \partial j(x,t,\xi_i),\ i = 1,2, \text{ a.e. } (x,t) \in \Sigma_C \text{ with } m_2 \ge 0. \end{aligned}\right\} \qquad (5.96)$$

$$m_1 \geq m_2 \, c_e^2 \, \|\gamma\|^2. \tag{5.97}$$

$$\alpha > 6 \, \overline{c}_1 \, c_e^2 \, \|\gamma\|^2. \tag{5.98}$$

We have the following existence and uniqueness result.

Theorem 5.23. *Assume that* (5.2), (5.4), (5.67), (5.96)–(5.98) *hold. Then Problem* 5.22 *has a unique solution. Moreover, if u_i denotes the solution to Problem* 5.22 *corresponding to $f = f_i \in \mathcal{V}^*$, $i = 1, 2$, then there exists $c > 0$ such that*

$$\|u_1(t) - u_2(t)\|_V^2 + \int_0^t \|u_1'(s) - u_2'(s)\|_V^2 \, ds \leq c \int_0^t \|f_1(s) - f_2(s)\|_{V^*}^2 \, ds \tag{5.99}$$

for all $t \in [0, T]$.

Proof. We use Theorem 5.15 with the space $X = L^2(\Gamma_C; \mathbb{R}^s)$, the operator $M = \gamma \colon Z \to X$ and the functional $J \colon (0, T) \times L^2(\Gamma_C; \mathbb{R}^s) \to \mathbb{R}$ defined by

$$J(t, u) = \int_{\Gamma_C} j(x, t, u(x)) \, d\Gamma \quad \text{for all } \; u \in L^2(\Gamma_C; \mathbb{R}^s), \text{ a.e. } t \in (0, T).$$

We note that (5.64) holds and the conditions (5.97) and (5.98) imply (5.69) and (5.70), respectively. Moreover, under the hypothesis (5.96), from Theorem 3.47 we deduce the following properties.

(i) $J(\cdot, u)$ is measurable on $(0, T)$ for all $u \in L^2(\Gamma_C; \mathbb{R}^s)$.
(ii) $J(t, \cdot)$ is Lipschitz continuous on bounded subsets of $L^2(\Gamma_C; \mathbb{R}^s)$ for a.e. $t \in (0, T)$.
(iii) For all $u \in L^2(\Gamma_C; \mathbb{R}^s)$, a.e. $t \in (0, T)$, we have

$$\|\partial J(t, u)\|_{L^2(\Gamma_C; \mathbb{R}^s)} \leq c_0(t) + c_1 \, \|u\|_{L^2(\Gamma_C; \mathbb{R}^s)}$$

with $c_0(t) = \sqrt{3 \, \mathrm{meas}\,(\Gamma_C)} \, \overline{c}_0(t)$ and $c_1 = \sqrt{3} \, \overline{c}_1$.
(iv) For all $u \in L^2(\Gamma_C; \mathbb{R}^s)$, a.e. $t \in (0, T)$, we have

$$\partial J(t, u) = \int_{\Gamma_C} \partial j(x, t, u(x)) \, d\Gamma.$$

(v) For all $u, v \in L^2(\Gamma_C; \mathbb{R}^s)$, a.e. $t \in (0, T)$, we have

$$J^0(t, u; v) = \int_{\Gamma_C} j^0(x, t, u(x); v(x)) \, d\Gamma.$$

The proof of these properties follows by using similar arguments to those already used in the proof of Theorem 5.21 and, for this reason, we skip it.

Next, from the properties (i)–(iii), it is easy to see that J satisfies conditions (5.68)(a)–(c). In what follows, we show that J satisfies condition (5.68)(d), too. Indeed, let $t \in (0,T) \setminus N$ with meas $(N) = 0$ and let $u_i, z_i \in L^2(\Gamma_C;\mathbb{R}^s)$ be such that $z_i \in \partial J(t,u_i)$, $i = 1, 2$. From the property (iv), it follows that there exists $\zeta_i \in L^2(\Gamma_C;\mathbb{R}^s)$ such that $\zeta_i(x) \in \partial j(x,t,u_i(x))$ for a.e. $x \in \Gamma_C$ and

$$\langle z_i, v\rangle_{L^2(\Gamma_C;\mathbb{R}^s)} = \int_{\Gamma_C} \zeta_i(x)\cdot v(x)\,d\Gamma \quad \text{for all } v \in L^2(\Gamma_C;\mathbb{R}^s).$$

Then, using (5.96) we obtain

$$\begin{aligned}
\langle z_1 - z_2, u_1 - u_2\rangle_{L^2(\Gamma_C;\mathbb{R}^s)} &= \int_{\Gamma_C} (\zeta_1(x) - \zeta_2(x))\cdot(u_1(x) - u_2(x))\,d\Gamma \\
&\geq -m_2 \int_{\Gamma_C} \|u_1(x) - u_2(x)\|^2_{\mathbb{R}^s}\,d\Gamma \\
&= -m_2\,\|u_1 - u_2\|^2_{L^2(\Gamma_C;\mathbb{R}^s)}
\end{aligned}$$

and, therefore, we conclude that condition (5.68)(d) is satisfied.

Further, from Theorem 5.15 we deduce that there exists a unique solution $u \in \mathcal{V}$ such that $u' \in \mathcal{W}$, which solves

$$\left.\begin{aligned}
&u''(t) + A(t,u'(t)) + Bu(t) + \gamma^*\,\partial J(t,\gamma u'(t)) \ni f(t) \\
&\qquad\qquad\qquad\qquad\qquad\qquad\text{a.e. } t \in (0,T), \\
&u(0) = u_0, \quad u'(0) = v_0.
\end{aligned}\right\} \tag{5.100}$$

In what follows we show that $u \in \mathcal{V}$ with $u' \in \mathcal{W}$ is a solution to (5.100) if and only if it solves Problem 5.22. To this end, assume that $u \in \mathcal{V}$ is such that $u' \in \mathcal{W}$ and u is solution to (5.100). This means that there exists $\zeta \in L^2(\Gamma_C;\mathbb{R}^s)$ such that

$$\left.\begin{aligned}
&u''(t) + A(t,u'(t)) + Bu(t) + \gamma^*\,\zeta(t) = f(t) \ \text{ a.e. } t \in (0,T), \\
&\zeta(t) \in \partial J(t,\gamma u'(t)) \ \text{ a.e. } t \in (0,T), \\
&u(0) = u_0, \quad u'(0) = v_0.
\end{aligned}\right\} \tag{5.101}$$

Let $v \in V$. Then

$$\begin{aligned}
&\langle u''(t) + A(t,u'(t)) + Bu(t), v\rangle_{V^*\times V} + \langle \zeta, \gamma v\rangle_{L^2(\Gamma_C;\mathbb{R}^s)} \\
&\quad = \langle f(t), v\rangle_{V^*\times V} \ \text{ a.e. } t \in (0,T),
\end{aligned} \tag{5.102}$$

and the definition of the subdifferential together with the property (v) gives

$$\langle \zeta, \gamma v \rangle_{L^2(\Gamma_C;\mathbb{R}^s)} \le J^0(t, \gamma u'(t); \gamma v)$$

$$= \int_{\Gamma_C} j^0(x, t, \gamma u'(x,t); \gamma v(x))\, d\Gamma \ \text{ a.e. } t \in (0,T). \tag{5.103}$$

Relations (5.101)–(5.103) imply that u is a solution to Problem 5.22.

Conversely, let u be a solution to Problem 5.22. Using the property (v), we know that

$$\langle u''(t) + A(t, u'(t)) + Bu(t), v \rangle_{V^* \times V} + J^0(t, \gamma u'(t); \gamma v) \ge \langle f(t), v \rangle_{V^* \times V}$$

for all $v \in V$ and a.e. $t \in (0,T)$. We use Proposition 3.37 to see that

$$J^0(t, \gamma u'(t); \gamma v) = (J \circ \gamma)^0(t, u'(t); v) \ \text{ and } \ \partial(J \circ \gamma)(t, u'(t)) = \gamma^* \partial J(t, \gamma u'(t)),$$

and, therefore, we have

$$\langle f(t) - u''(t) - A(t, u'(t)) - Bu(t), v \rangle_{V^* \times V}$$

$$\le J^0(t, \gamma u'(t); \gamma v) = (J \circ \gamma)^0(t, u'(t); v)$$

for all $v \in V$, a.e. $t \in (0,T)$, and

$$f(t) - u''(t) - A(t, u'(t)) - Bu(t) \in \partial(J \circ \gamma)(t, u'(t)) = \gamma^* \partial J(t, \gamma u'(t))$$

for a.e. $t \in (0,T)$. We conclude from here that $u \in \mathcal{V}$ with $u' \in \mathcal{W}$ is a solution to the evolutionary inclusion (5.100).

The arguments above prove the existence and uniqueness part of the theorem. To end the proof, let $f_i \in \mathcal{V}^*$ and denote by u_i the unique solution to Problem 5.22 corresponding to $f = f_i$, $i = 1, 2$. We have $u_i \in \mathcal{V}$, $u_i' \in \mathcal{W}$ and

$$u_1''(s) + A(s, u_1'(s)) + Bu_1(s) + \gamma^* \zeta_1(s) = f_1(s) \text{ a.e. } s \in (0,T), \tag{5.104}$$

$$u_2''(s) + A(s, u_2'(t)) + Bu_2(s) + \gamma^* \zeta_2(s) = f_2(s) \text{ a.e. } s \in (0,T), \tag{5.105}$$

$$\zeta_i(s) \in \partial J(s, \gamma u_i'(s)) \ \text{ a.e. } \ s \in (0,T),\ i = 1, 2, \tag{5.106}$$

$$u_1(0) = u_2(0) = u_0,\ u_1'(0) = u_2'(0) = v_0. \tag{5.107}$$

Subtracting (5.105) from (5.104), multiplying the result by $u_1'(s)-u_2'(s)$, integrating by parts and using the initial conditions (5.107), we obtain

$$\frac{1}{2}\,\|u_1'(t)-u_2'(t)\|_H^2+\int_0^t \langle A(s,u_1'(s))-A(s,u_2'(s)),u_1'(s)-u_2'(s)\rangle_{V^*\times V}\,ds$$

$$+\int_0^t \langle Bu_1(s)-Bu_2(s),u_1'(s)-u_2'(s)\rangle_{V^*\times V}\,ds$$

$$+\int_0^t \langle \gamma^*\,\zeta_1(s)-\gamma^*\,\zeta_2(s),u_1'(s)-u_2'(s)\rangle_{Z^*\times Z}\,ds$$

$$=\int_0^t \langle f_1(s)-f_2(s),u_1'(s)-u_2'(s)\rangle_{V^*\times V}\,ds \qquad (5.108)$$

for all $t\in[0,T]$. Also, since J satisfies (5.68)(d), from (5.106), we find

$$\int_0^t \langle \gamma^*(\zeta_1(s)-\zeta_2(s)),u_1'(s)-u_2'(s)\rangle_{Z^*\times Z}\,ds$$

$$=\int_0^t \langle \zeta_1(s)-\zeta_2(s),\gamma u_1'(s)-\gamma u_2'(s)\rangle_{L^2(\Gamma_C;\mathbb{R}^s)}\,ds$$

$$\geq -m_2\int_0^t \|\gamma u_1'(s)-\gamma u_2'(s)\|_{L^2(\Gamma_C;\mathbb{R}^s)}^2\,ds$$

$$\geq -m_2\,c_e^2\,\|\gamma\|^2\int_0^t \|u_1'(s)-u_2'(s)\|_V^2\,ds. \qquad (5.109)$$

Finally, conditions (5.2) and (5.107) imply

$$\int_0^t \langle Bu_1(s)-Bu_2(s),u_1'(s)-u_2'(s)\rangle_{V^*\times V}\,ds$$

$$=\frac{1}{2}\int_0^t \frac{d}{ds}\langle B(u_1(s)-u_2(s)),u_1(s)-u_2(s)\rangle_{V^*\times V}\,ds$$

$$=\frac{1}{2}\,\langle B(u_1(t)-u_2(t)),u_1(t)-u_2(t)\rangle_{V^*\times V}\geq 0 \qquad (5.110)$$

for all $t \in [0, T]$. We combine (5.108)–(5.110) and use the properties (5.67) of the operator A to obtain

$$\frac{1}{2} \|u_1'(t) - u_2'(t)\|_H^2 + \widetilde{c} \int_0^t \|u_1'(s) - u_2'(s)\|_V^2 \, ds$$

$$\leq \int_0^t \|f_1(s) - f_2(s)\|_{V^*} \|u_1'(s) - u_2'(s)\|_V \, ds \quad (5.111)$$

for all $t \in [0, T]$ with $\widetilde{c} = m_1 - m_2 c_e^2 \|\gamma\|^2$. Note that by the smallness assumption (5.97) we get $\widetilde{c} > 0$. Moreover, from (5.111) we have

$$\widetilde{c} \|u_1' - u_2'\|_{L^2(0,t;V)}^2 \leq \|f_1 - f_2\|_{L^2(0,t;V^*)} \|u_1' - u_2'\|_{L^2(0,t;V)}$$

for all $t \in [0, T]$, and, therefore,

$$\|u_1' - u_2'\|_{L^2(0,t;V)} \leq \frac{1}{\widetilde{c}} \|f_1 - f_2\|_{L^2(0,t;V^*)} \quad (5.112)$$

for all $t \in [0, T]$. On the other hand, using the initial conditions (5.107), we have

$$u_i(t) = u_0 + \int_0^t u_i'(s) \, ds$$

for all $t \in [0, T]$, $i = 1, 2$, which implies that

$$\|u_1(t) - u_2(t)\|_V \leq \int_0^t \|u_1'(s) - u_2'(s)\|_V \, ds \leq \sqrt{T} \|u_1' - u_2'\|_{L^2(0,t;V)}$$

for all $t \in [0, T]$. Using now (5.112) yields

$$\|u_1(t) - u_2(t)\|_V \leq \frac{\sqrt{T}}{\widetilde{c}} \|f_1 - f_2\|_{L^2(0,t;V^*)} \quad (5.113)$$

for all $t \in [0, T]$. Finally, combining (5.112) and (5.113), we obtain the estimate (5.99) which completes the proof of the theorem. □

Second-order hemivariational inequalities with Volterra integral term. We conclude this section with a result on the unique weak solvability of evolutionary hemivariational inequality involving the Volterra integral term. The problem we are interested in is formulated as follows.

Problem 5.24. *Find $u \in \mathcal{V}$ such that $u' \in \mathcal{W}$ and*

$$\left.\begin{aligned} &\langle u''(t) + A(t,u'(t)) + Bu(t) + \int_0^t C(t-s)u(s)\,ds, v\rangle_{V^*\times V} \\ &\quad + \int_{\Gamma_C} j^0(t,\gamma u'(t);\gamma v)\,d\Gamma \geq \langle f(t), v\rangle_{V^*\times V} \\ &\qquad\qquad\qquad \text{for all } v \in V, \text{ a.e. } t \in (0,T), \\ &u(0) = u_0, \ \ u'(0) = v_0. \end{aligned}\right\}$$

Theorem 5.25. *Assume that* (5.2), (5.4), (5.67), (5.74), (5.96)–(5.98) *hold. Then Problem* 5.24 *admits a unique solution.*

Proof. We provide two proofs of this theorem. In the first proof the unique solvability of Problem 5.24 follows from Theorem 5.17. Indeed, taking the space $X = L^2(\Gamma_C;\mathbb{R}^s)$ and the operator $M = \gamma\colon Z \to X$, it is clear that the condition (5.64) holds, while the assumptions (5.97) and (5.98) entail (5.69) and (5.70), respectively. We consider the functional $J\colon (0,T) \times L^2(\Gamma_C;\mathbb{R}^s) \to \mathbb{R}$ defined by

$$J(t,u) = \int_{\Gamma_C} j(x,t,u(x))\,d\Gamma \ \text{ for all } \ u \in L^2(\Gamma_C;\mathbb{R}^s), \text{ a.e. } t \in (0,T).$$

Arguments similar to those used in the proof of Theorem 5.23 show that the hypothesis (5.96) on the integrand j implies condition (5.68) on J. Therefore, using Theorem 5.17 we obtain the existence of a unique solution $u \in \mathcal{V}$ with $u' \in \mathcal{W}$ to the inclusion

$$\left.\begin{aligned} &u''(t) + A(t,u'(t)) + Bu(t) + \int_0^t C(t-s)u(s)\,ds + \gamma^*\,\partial J(t,\gamma u'(t)) \ni f(t) \\ &\qquad\qquad\qquad \text{a.e. } t \in (0,T), \\ &u(0) = u_0, \ \ u'(0) = v_0. \end{aligned}\right\}$$

And, as in the final part of the proof of Theorem 5.23, we deduce that u is a solution of the previous Cauchy problem if and only if u solves Problem 5.24. This completes the first proof of the theorem.

For the second proof, let $\eta \in \mathcal{V}^*$ be fixed. We consider the following intermediate problem.

Problem P_η. *Find $u \in \mathcal{V}$ such that $u' \in \mathcal{W}$ and*

$$\left.\begin{aligned}&\langle u''(t) + A(t, u'(t)) + Bu(t), v\rangle_{V^*\times V} + \int_{\Gamma_C} j^0(t, \gamma u'(t); \gamma v)\, d\Gamma \\ &\qquad \geq \langle f(t) - \eta(t), v\rangle_{V^*\times V} \quad \text{for all } v \in V, \text{ a.e. } t \in (0, T), \\ &u(0) = u_0, \ \ u'(0) = v_0.\end{aligned}\right\}$$

From Theorem 5.23, we know that Problem P_η admits a unique solution and

$$\|u_1(t) - u_2(t)\|_V^2 \leq c \int_0^t \|\eta_1(s) - \eta_2(s)\|_{V^*}^2\, ds \quad \text{for all } t \in [0, T],$$

where c is a positive constant and u_i denotes the solution to Problem P_η corresponding to $\eta = \eta_i$, $i = 1, 2$. Next, we consider the operator $\Lambda\colon \mathcal{V}^* \to \mathcal{V}^*$ defined by

$$(\Lambda\eta)(t) = \int_0^t C(t - s)u_\eta(s)\, ds \ \text{ for all } \ \eta \in \mathcal{V}^*, \ \text{a.e. } t \in (0, T),$$

where $u_\eta \in \mathcal{V}$ is the unique solution to Problem P_η. Analogously as in the proof of Theorem 5.17, we deduce that the operator Λ is well defined, has a unique fixed point $\eta^* \in \mathcal{V}^*$ and $u = u_{\eta^*}$ is the unique solution to Problem 5.24. This completes the second proof of the theorem. □

Bibliographical Notes

The interest in hemivariational inequalities has originated in mechanical problems. The inequality problems in mechanics can be divided into two main classes, that of variational inequalities, which is concerned with convex energy functions (potentials), and that of hemivariational inequalities which is concerned with nonconvex energy functions (superpotentials). In this sense hemivariational inequalities are generalizations of variational inequalities.

The theory of variational inequalities was developed in early sixties. Basic references concerning this theory are [17, 30, 31, 82, 128, 131, 149, 202]. The numerical analysis of various classes of variational inequalities was treated in [86, 87] and also in [103, 108]. An excellent reference in the study of numerical analysis of plasticity problems via the theory of variational inequalities is [99], and results on optimal control of variational inequalities can be found in [23].

The notion of hemivariational inequality was introduced and studied in the early eighties in [202–204]. By means of hemivariational inequality, problems involving nonmonotone and multivalued constitutive laws and boundary conditions can be treated mathematically. These multivalued relations are derived from nonsmooth and nonconvex superpotentials by using the notion of generalized gradient introduced and studied in [46–48]. The Clarke theory of subdifferentiation for locally Lipschitz functions has been motivated by the fact that a convex function is locally Lipschitz in the interior of its effective domain and a locally Lipschitz function is differentiable almost everywhere (Rademacher's theorem). For a description of various problems arising in mechanics and engineering sciences which lead to hemivariational inequalities we refer to [202–205]. There, various results on the analysis of hemivariational inequalities are also discussed. This includes results on existence and uniqueness of the solutions, eigenvalues, optimal control, numerical approximation. Additional results on the mathematical theory of stationary hemivariational inequalities can be found in [195]. For the numerical analysis of hemivariational inequalities, including their finite element approach, we refer to [104].

Most of the results presented in Chaps. 4 and 5 of this book are based on our research and were obtained in the recent papers [181–188].

Some of the stationary inclusions considered in Sects. 4.1 and 4.2 and the hemivariational inequalities of Sect. 4.3 were studied earlier in [164] and [173] in the context of inequality problems for the stationary Navier–Stokes-type operators related to the flow of a viscous incompressible fluid. Additional results can be found in [167], in the study of an inequality problem for static frictional contact for a piezoelastic body, and in [152] in the study of a mathematical model which describes the static elastic contact with subdifferential boundary conditions. A result on history-dependent stationary subdifferential inclusions has been recently proved in [189].

Other issues for stationary inclusions and hemivariational inequalities which are not discussed in Chap. 4, like homogenization and inverse problems, were considered in [152, 163, 171].

The Cauchy problem for the second-order evolutionary inclusion introduced in Problem 5.1 has been studied in [67] with $F\colon (0,T)\times H\times H \to 2^H$, [175] in the case when B is linear, continuous, symmetric, and coercive operator, and in [165,212], in the case when B is linear, continuous, symmetric, and nonnegative. Other existence results for this evolutionary inclusion with a multifunction F independent of u have been provided in [165, 168]. Details on the proof of Lemma 5.5 can be found in Lemma 1.9 of [199], see also Lemma 11 of [165].

Existence results for second-order nonlinear evolutionary inclusions similar to those presented in Sects. 5.1 and 5.2 can be found also in [3, 28, 160, 161, 176, 209–211]. Nevertheless, we underline that none of the results in these papers can be applied in the study of the nonlinear evolutionary inclusions we consider in Chap. 5. The reason arises in the restrictive hypotheses on the multivalued term which, in the papers mentioned above in this paragraph, was supposed to have values in the pivot space H. And, as it is shown in Chaps. 7 and 8 of this book, the variational formulation of various contact problems leads to hemivariational inequalities for which the associated multivalued mapping has values in a dual space, which is larger than the space H. To provide an appropriate mathematical tool in solving the corresponding inequalities arising in Contact Mechanics, we had to develop the new results in the study of nonlinear inclusions, presented in Sects. 5.1–5.3. Moreover, we underline that for some of the existence results presented in Chap. 5 we have used a method which is different to that used in [67]. It combines a surjectivity result for pseudomonotone operators with the Banach contraction principle.

The first existence result for the evolutionary inclusion of subdifferential type in Problem 5.9 where the Clarke subdifferential is calculated with respect to two variables has been obtained in [174]. There, a general method for the study of dynamic viscoelastic contact problems involving subdifferential boundary conditions was presented. Within the framework of evolutionary hemivariational inequalities, this method represents a new approach which unifies several other methods used in the study of viscoelastic contact and allows to obtain new existence and uniqueness results.

The result on the unique solvability of the subdifferential inclusion with Volterra integral term (Theorem 5.17 in Sect. 5.2) was written following [181].

The research on hyperbolic hemivariational inequalities arising in the study of nonlinear boundary value problems has been initiated in [205–207] (where models characterized by one-dimensional reaction-velocity laws have been considered), in [208] (where the Galerkin method has been used) as well as in [199] (where a surjectivity result for multivalued operators has been employed). Hyperbolic hemivariational inequality problems with a subdifferential relation depending on the first-order derivative were considered in [88, 104, 166] while evolutionary inequalities with a multivalued term depending on the unknown function were treated in [84, 89, 94, 104, 208] (where one-dimensional wave equation was considered), and in [162, 166, 212].

The results of Sect. 5.3 are based on our research papers [168, 172, 174, 181, 182]. A local existence result for Problem 5.18 has been firstly obtained in [174] in a particular case when the function h is constant and B is a coercive operator, by exploiting Theorem 4 in [175]. We address as an open problem the uniqueness of solution to Problem 5.18 when h is not the constant function. A result on existence of a global solution to Problem 5.20 without the smallness assumption of Theorem 5.21 and under stronger hypothesis on the superpotential j can be found in [168]. For particular cases of Problem 5.20 see [172] (where the function h is independent of the last variable) and [182] (where the operators A and B are assumed to be linear, continuous, and coercive). Theorem 5.23 on the unique solvability of Problem 5.22 has been obtained in [165] under an additional hypothesis on the superpotential. For further investigation of the evolutionary hemivariational inequality in Problem 5.22 see [88, 206, 208]. The question of the unique solvability of Problem 5.24 has been firstly studied in [181] and the related variational inequalities with a Volterra-type operator have been studied in [223, 240].

Existence results for the solutions to evolutionary hemivariational inequalities of second order similar to those considered in Sect. 5.3 can be found in [25, 26, 93, 137, 147, 148, 162, 165–168, 174, 178, 181–188, 212], see [180] for a survey. Recent books and monographs on mathematical theory of hemivariational inequalities include [38, 90, 91, 104, 192, 195]. We refer the reader there for a wealth of additional information about these and related topics. Other results on evolutionary second-order hemivariational inequalities concern the asymptotic behavior of solutions and can be found in [126, 177, 179, 213]. The regularization of such inequalities as well as the study of the noncoercive case were provided in [151, 256]. References on problems described by a system of evolutionary hemivariational inequalities include [61, 62, 65] and, finally, we recall that a result on the sensitivity of solutions of evolutionary hemivariational inequalities was obtained in [63].

We also remark that the literature concerning first-order evolutionary hemivariational inequalities, which are not touched in this book, is extensive. For various results, details, and comments we refer to the survey [170] and the references therein.

Part III
Modeling and Analysis of Contact Problems

Chapter 6
Modeling of Contact Problems

In this chapter we deal with the mathematical modeling of the processes of contact between a deformable body and a foundation. We present the physical setting, the variables which determine the state of the system, the balance equations, the material's behavior which is reflected in the constitutive law, and the boundary conditions for the system variables. In particular, we provide a description of the frictional contact conditions, including versions of the Coulomb law of dry friction and its regularizations. Then we extend our description to the case of piezoelectric materials, i.e., materials which present a coupling between mechanical and electrical properties. In this chapter, all the variables are assumed to have sufficient degree of smoothness consistent with developments they are involved in. Moreover, as usual in the literature devoted to Contact Mechanics, everywhere in the rest of the book we denote vectors and tensors by bold-face letters.

6.1 Physical Setting

We consider the general physical setting shown in Fig. 6.1 that we describe in what follows. A deformable body occupies, in the reference configuration, an open bounded connected set $\Omega \subset \mathbb{R}^d$ $(d = 1, 2, 3)$ with boundary $\partial\Omega = \Gamma$, assumed to be Lipschitz. We denote by $\boldsymbol{\nu} = (\nu_i)$ the unit outward normal vector and by $\boldsymbol{x} = (x_i) \in \overline{\Omega} = \Omega \cup \Gamma$ the position vector. Here and below, the indices i, j, k, l run from 1 to d; an index that follows a comma indicates a derivative with respect to the corresponding component of the spatial variable $\boldsymbol{x}$ and the summation convention over repeated indices is adopted. We denote by $\mathbb{S}^d$ the space of second-order symmetric tensors on $\mathbb{R}^d$ or, equivalently, the space of symmetric matrices of order d. We recall that the canonical inner products and the corresponding norms on $\mathbb{R}^d$ and $\mathbb{S}^d$ are given by

S. Migórski et al., *Nonlinear Inclusions and Hemivariational Inequalities*, Advances in Mechanics and Mathematics 26, DOI 10.1007/978-1-4614-4232-5_6,

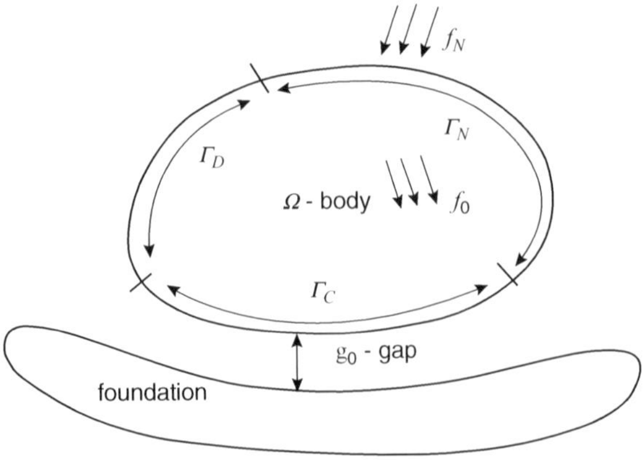

Fig. 6.1 The physical setting; Γ_C is the contact surface

$$\boldsymbol{u}\cdot\boldsymbol{v}=u_i v_i,\quad \|\boldsymbol{v}\|_{\mathbb{R}^d}=(\boldsymbol{v}\cdot\boldsymbol{v})^{1/2}\qquad \text{for all } \boldsymbol{u}=(u_i),\ \boldsymbol{v}=(v_i)\in\mathbb{R}^d,$$

$$\boldsymbol{\sigma}:\boldsymbol{\tau}=\sigma_{ij}\tau_{ij},\quad \|\boldsymbol{\tau}\|_{\mathbb{S}^d}=(\boldsymbol{\tau}:\boldsymbol{\tau})^{1/2}\quad \text{for all}\quad \boldsymbol{\sigma}=(\sigma_{ij}),\ \boldsymbol{\tau}=(\tau_{ij})\in\mathbb{S}^d,$$

respectively.

We also assume that the boundary Γ is composed of three sets $\overline{\Gamma}_D$, $\overline{\Gamma}_N$, and $\overline{\Gamma}_C$, with mutually disjoint relatively open sets Γ_D, Γ_N, and Γ_C, such that meas $(\Gamma_D)>0$. The body is clamped on Γ_D and time-dependent surface tractions of density $\boldsymbol{f}_N$ act on Γ_N. The body is, or can arrive, in contact on Γ_C with an obstacle, the so-called foundation. At each time instant Γ_C is divided into two parts: one part where the body and the foundation are in contact and the other part where they are separated. The boundary of the contact part is a free boundary, determined by the solution of the problem. We assume that in the reference configuration there exists a gap, denoted by g_0, between Γ_C and the foundation, which is measured along the outer normal $\boldsymbol{\nu}$.

We are interested in mathematical models which describe the evolution of the mechanical state of the body during the time interval $[0,T]$, with $T>0$. To this end, we denote by $\boldsymbol{\sigma}=\boldsymbol{\sigma}(\boldsymbol{x},t)=(\sigma_{ij}(\boldsymbol{x},t))$ the *stress field* and by $\boldsymbol{u}=\boldsymbol{u}(\boldsymbol{x},t)=(u_i(\boldsymbol{x},t))$ the *displacement field* where, here and below, t denotes the time variable. The functions $\boldsymbol{u}:\Omega\times[0,T]\to\mathbb{R}^d$ and $\boldsymbol{\sigma}:\Omega\times[0,T]\to\mathbb{S}^d$ will play the role of the unknowns of the contact problem. From time to time, we suppress the explicit dependence of various quantities on the spatial variable $\boldsymbol{x}$, or both $\boldsymbol{x}$ and t; i.e., when it is convenient to do so, we write $\boldsymbol{\sigma}(t)$ and $\boldsymbol{u}(t)$, or even $\boldsymbol{\sigma}$ and $\boldsymbol{u}$.

The equation of motion that governs the evolution of the mechanical state of the body is

$$\rho\boldsymbol{u}''=\operatorname{Div}\boldsymbol{\sigma}+\boldsymbol{f}_0\quad\text{in}\quad\Omega\times(0,T),\tag{6.1}$$

where ρ is the mass density and $\boldsymbol{f}_0$ is the density of applied forces, such as gravity or electromagnetic forces. Here "Div" is the divergence operator, i.e., $\operatorname{Div}\boldsymbol{\sigma}=(\sigma_{ij,j})$, and recall that $\sigma_{ij,j} = \frac{\partial \sigma_{ij}}{\partial x_j}$. Also, as usual, the prime denotes the time derivative; thus $\boldsymbol{u}' = \frac{\partial \boldsymbol{u}}{\partial t}$ and $\boldsymbol{u}'' = \frac{\partial^2 \boldsymbol{u}}{\partial t^2}$ represent the velocity field and the acceleration field, respectively.

When the external forces and tractions vary slowly with time, and the accelerations in the system are rather small and can be neglected, we omit the inertial term in the equation of motion and obtain the equation of equilibrium

$$\operatorname{Div} \boldsymbol{\sigma} + \boldsymbol{f}_0 = \boldsymbol{0} \quad \text{in} \quad \Omega \times (0, T). \tag{6.2}$$

Processes modeled by the equation of motion (6.1) are called *dynamic processes*. Processes modeled by the equation of equilibrium (6.2) are called *quasistatic*, if at least one derivative of the unknowns $\boldsymbol{u}$ and $\boldsymbol{\sigma}$ appears in the rest of equations or boundary conditions, and *static*, in the opposite case. Also, note that among the static processes we distinguish the *time-dependent processes*, in which the data and the unknowns depend on time and the *time-independent processes*, in which the time variable does not appear. And, note that in this last case, the equation of equilibrium is valid in Ω, i.e.,

$$\operatorname{Div} \boldsymbol{\sigma} + \boldsymbol{f}_0 = \boldsymbol{0} \quad \text{in} \quad \Omega. \tag{6.3}$$

Note also that in this book we deal only with static and dynamic contact processes. Models and variational analysis of various quasistatic contact processes can be found in [102, 233], for instance.

Since the body is clamped on Γ_D, we impose the displacement boundary condition

$$\boldsymbol{u} = \boldsymbol{0} \quad \text{on} \quad \Gamma_D \times (0, T). \tag{6.4}$$

The traction boundary condition is

$$\boldsymbol{\sigma \nu} = \boldsymbol{f}_N \quad \text{on} \quad \Gamma_N \times (0, T). \tag{6.5}$$

It states that the stress vector $\boldsymbol{\sigma \nu}$ is given on part Γ_N of the boundary, during the contact process. Note that in the case of time-independent processes the boundary conditions (6.4) and (6.5) hold on Γ_D and Γ_N, respectively, i.e.,

$$\boldsymbol{u} = \boldsymbol{0} \quad \text{on} \quad \Gamma_D, \tag{6.6}$$

$$\boldsymbol{\sigma \nu} = \boldsymbol{f}_N \quad \text{on} \quad \Gamma_N. \tag{6.7}$$

On Γ_C we will specify contact conditions. The contact may be frictionless or frictional, with a rigid or with a deformable foundation and, in the case of frictional contact, a number of different friction conditions may be employed. A survey of the various contact and frictional conditions used in this book will be provided in Sect. 6.3.

We also recall the strain-displacement relation

$$\boldsymbol{\varepsilon}(\boldsymbol{u}) = (\varepsilon_{ij}(\boldsymbol{u})), \quad \varepsilon_{ij}(\boldsymbol{u}) = \frac{1}{2}(u_{i,j} + u_{j,i}) \quad \text{in} \quad \Omega \times (0,T) \tag{6.8}$$

which defines the linearized (or the small) strain tensor. Sometimes we omit the explicit dependence of $\boldsymbol{\varepsilon}$ on $\boldsymbol{u}$ by writing $\boldsymbol{\varepsilon}$ instead of $\boldsymbol{\varepsilon}(\boldsymbol{u})$ and we note that in the case of time-independent processes (6.8) holds in Ω. Finally, note that all the problems studied in this book are formulated in the framework of small strain theory.

At this stage the description of our model is not complete yet, since we have more unknown functions than equations. Indeed, in the case $d = 3$ we have three equations in (6.1) or (6.2) and six relations in (6.8) (taking into account the symmetry of $\boldsymbol{\varepsilon}$) for the 15 unknowns $(\boldsymbol{u}, \boldsymbol{\sigma}, \boldsymbol{\varepsilon})$ (taking into account the symmetry of $\boldsymbol{\sigma}$, as well). When $d = 2$, there are eight unknown functions and we only have two equations and three relations.

Physical considerations also indicate that the description of the problem so far is incomplete. The equation of motion (6.1) as well as the equation of equilibrium (6.2) or (6.3) are valid for all materials, since they are derived from the principle of momentum conservation. In addition to the kinematics and the balance laws that apply to all materials, we need a description of the particular behavior of the material the body is made of. This information is the content of the so-called *constitutive equation*, or *constitutive law*, or *constitutive relation* of the material, and it provides the remaining equations for the model.

6.2 Constitutive Laws

The relationship between the stresses $\boldsymbol{\sigma}$ and the strains $\boldsymbol{\varepsilon}$ which cause them is given by the constitutive law, which characterizes a specific material. It describes the deformations of the body resulting from the action of forces and tractions. Although they must satisfy some basic axioms and invariance principles, constitutive laws originate mostly from experience. A general description of several diagnostic experiments which provide information needed in constructing constitutive relations for specific materials, made with "standard" universal testing machines, can be found in [57, 71, 117]. Rheological considerations used to derive constitutive laws can be found in [71, 102, 239], for instance. In this book we consider constitutive laws for nonlinearly elastic and viscoelastic materials and, for the viscoelastic materials, we consider both the case of short and long memory.

Elastic constitutive laws. A general elastic constitutive law is given by

$$\boldsymbol{\sigma} = \mathcal{F}\boldsymbol{\varepsilon}(\boldsymbol{u}) \tag{6.9}$$

where $\mathcal{F}$ is the *elasticity operator*, assumed to be nonlinear. We allow $\mathcal{F}$ to depend on the location of the point; consequently, all that follows is valid for nonhomogeneous materials. We use the shorthand notation $\mathcal{F}\boldsymbol{\varepsilon}(\boldsymbol{u})$ for $\mathcal{F}(\boldsymbol{x}, \boldsymbol{\varepsilon}(\boldsymbol{u}))$.

In particular, if $\mathcal{F}$ is a linear operator, (6.9) leads to the constitutive law of linearly elastic materials,

$$\sigma_{ij} = f_{ijkl}\varepsilon_{kl}, \tag{6.10}$$

where σ_{ij} are the components of the stress tensor $\boldsymbol{\sigma}$ and f_{ijkl} are the components of the elasticity tensor $\mathcal{F}$. Usually, the components f_{ijkl} belong to $L^\infty(\Omega)$ and satisfy the usual properties of symmetry and ellipticity, i.e.,

$$f_{ijkl} = f_{jikl} = f_{klij},$$

and there exists $m_{\mathcal{F}} > 0$ such that

$$f_{ijkl}\varepsilon_{ij}\varepsilon_{kl} \geq m_{\mathcal{F}}\|\boldsymbol{\varepsilon}\|^2_{\mathbb{S}^d} \quad \text{for all} \quad \boldsymbol{\varepsilon} = (\varepsilon_{ij}) \in \mathbb{S}^d.$$

Due to the symmetry, when $d = 3$ there are only 21 independent coefficients, when $d = 2$ there are only four independent coefficients, and when $d = 1$ there is only one component in the elasticity tensor.

Let $d = 3$. When the material is linear and isotropic, the elasticity tensor is characterized by only two coefficients. Thus, the constitutive law of a linearly elastic isotropic material is given by

$$\boldsymbol{\sigma} = 2\,\mu\,\boldsymbol{\varepsilon}(\boldsymbol{u}) + \lambda\,\mathrm{tr}(\boldsymbol{\varepsilon}(\boldsymbol{u}))\,\boldsymbol{I}_3 \tag{6.11}$$

so that the elasticity operator is

$$\mathcal{F}(\boldsymbol{\varepsilon}) = 2\,\mu\,\boldsymbol{\varepsilon} + \lambda\,\mathrm{tr}(\boldsymbol{\varepsilon})\,\boldsymbol{I}_3. \tag{6.12}$$

Here λ and μ are the *Lamé coefficients* and satisfy $\lambda > 0$, $\mu > 0$, $\mathrm{tr}(\boldsymbol{\varepsilon}(\boldsymbol{u}))$ denotes the trace of the tensor $\boldsymbol{\varepsilon}(\boldsymbol{u})$,

$$\mathrm{tr}(\boldsymbol{\varepsilon}(\boldsymbol{u})) = \varepsilon_{kk}(\boldsymbol{u}),$$

and $\boldsymbol{I}_3$ denotes the identity tensor of the second order on $\mathbb{R}^3$. In components, we have

$$\sigma_{ij} = 2\,\mu\,\boldsymbol{\varepsilon}_{ij}(\boldsymbol{u}) + \lambda\,\varepsilon_{kk}(\boldsymbol{u})\,\delta_{ij} \tag{6.13}$$

where δ_{ij} is the Kronecker symbol, i.e., δ_{ij} are the components of the unit matrix 3×3.

Besides the linear case described above, a second example of elastic constitutive law of the form (6.9) is provided by

$$\boldsymbol{\sigma} = \mathcal{E}\boldsymbol{\varepsilon}(\boldsymbol{u}) + \beta\,(\boldsymbol{\varepsilon}(\boldsymbol{u}) - \mathcal{P}_K\boldsymbol{\varepsilon}(\boldsymbol{u})). \tag{6.14}$$

Here $\mathcal{E}$ is a linear or a nonlinear operator, $\beta > 0$, K is a closed convex subset of $\mathbb{S}^d$ such that $\mathbf{0} \in K$ and $\mathcal{P}_K : \mathbb{S}^d \to K$ denotes the projection operator. The corresponding elasticity operator is nonlinear and is given by

$$\mathcal{F}(\boldsymbol{\varepsilon}) = \mathcal{E}\boldsymbol{\varepsilon} + \beta\,(\boldsymbol{\varepsilon} - \mathcal{P}_K\boldsymbol{\varepsilon}). \tag{6.15}$$

Usually the operator $\mathcal{E}$ is chosen to be linear, i.e., is a fourth-order tensor, and the set K is defined by

$$K = \{\boldsymbol{\varepsilon} \in \mathbb{S}^d \mid \mathcal{G}(\boldsymbol{\varepsilon}) \le 0\}, \tag{6.16}$$

where $\mathcal{G}: \mathbb{S}^d \to \mathbb{R}$ is a convex continuous function such that $\mathcal{G}(\mathbf{0}) < 0$. A well-known example is the *von Mises function*

$$\mathcal{G}(\boldsymbol{\varepsilon}) = \frac{1}{2}\,\|\boldsymbol{\varepsilon}^D\|_{\mathbb{S}^d}^2 - g^2. \tag{6.17}$$

Here $\boldsymbol{\varepsilon}^D$ represents the deviatoric part of $\boldsymbol{\varepsilon}$, defined by

$$\boldsymbol{\varepsilon}^D = \boldsymbol{\varepsilon} - \frac{1}{d}\,\mathrm{tr}(\boldsymbol{\varepsilon})\,\boldsymbol{I}_d, \tag{6.18}$$

$\boldsymbol{I}_d$ is the identity tensor of the second order on $\mathbb{R}^d$ and g is a positive constant. The corresponding set (6.16) is given by

$$K = \{\boldsymbol{\varepsilon} \in \mathbb{S}^d \mid \|\boldsymbol{\varepsilon}^D\|_{\mathbb{S}^d} \le g\sqrt{2}\}. \tag{6.19}$$

Using the convexity of the norm it is easy to see that the set K defined by (6.19) is a convex subset of $\mathbb{S}^d$. Moreover, since $g > 0$ it follows that $\mathbf{0} \in K$ and, therefore, K is not empty. Finally, since $\boldsymbol{\varepsilon} \mapsto \boldsymbol{\varepsilon}^D$ is a continuous mapping, it follows that K is a closed subset of $\mathbb{S}^d$. The convex set defined by (6.19) is called the *von Mises convex*.

A third family of elasticity operators is provided by nonlinear *Hencky materials* (for detail, cf. e.g. [262]). For a Hencky material, the stress–strain relation is

$$\boldsymbol{\sigma} = k_0\,\mathrm{tr}(\boldsymbol{\varepsilon}(\boldsymbol{u}))\,\boldsymbol{I}_d + \psi(\|\boldsymbol{\varepsilon}^D(\boldsymbol{u})\|_{\mathbb{S}^d}^2)\,\boldsymbol{\varepsilon}^D(\boldsymbol{u}), \tag{6.20}$$

so that the elasticity operator is

$$\mathcal{F}(\boldsymbol{\varepsilon}) = k_0\,\mathrm{tr}(\boldsymbol{\varepsilon})\,\boldsymbol{I}_d + \psi(\|\boldsymbol{\varepsilon}^D\|_{\mathbb{S}^d}^2)\,\boldsymbol{\varepsilon}^D. \tag{6.21}$$

Here, $k_0 > 0$ is a material coefficient, $\psi: \mathbb{R} \to \mathbb{R}$ is a constitutive function and, again, $\boldsymbol{I}_d$ and $\boldsymbol{\varepsilon}^D = \boldsymbol{\varepsilon}^D(\boldsymbol{u})$ denote the identity tensor of the second order on $\mathbb{R}^d$, and the deviatoric part of $\boldsymbol{\varepsilon} = \boldsymbol{\varepsilon}(\boldsymbol{u})$ respectively. The function ψ is assumed to be piecewise continuously differentiable, and there exist positive constants c_1, c_2, d_1, and d_2, such that for $\xi \ge 0$,

$$\psi(\xi) \le d_1, \quad -c_1 \le \psi'(\xi) \le 0, \quad c_2 \le \psi(\xi) + 2\,\psi'(\xi)\,\xi \le d_2. \tag{6.22}$$

Viscoelastic constitutive laws with short memory. A general *viscoelastic* constitutive law with short memory is given by

$$\boldsymbol{\sigma}(t) = \mathcal{A}(t, \boldsymbol{\varepsilon}(\boldsymbol{u}'(t))) + \mathcal{B}\boldsymbol{\varepsilon}(\boldsymbol{u}(t)). \tag{6.23}$$

We allow the *viscosity operator* $\mathcal{A}$ to depend on time as well as on the location of the point. So, we use the shorthand notation $\mathcal{A}(t, \boldsymbol{\varepsilon}(\boldsymbol{u}'))$ for $\mathcal{A}(\boldsymbol{x}, t, \boldsymbol{\varepsilon}(\boldsymbol{u}'(t)))$. The explicit dependence of the operator $\mathcal{A}$ with respect to the time variable makes the model more general and allows to describe situations when the viscous properties of the material depend on the temperature, which plays the role of a parameter, i.e., its evolution in time is prescribed. We also allow the *elasticity operator* $\mathcal{B}$ to depend on the location of the point, i.e., we use the shorthand notation $\mathcal{B}\boldsymbol{\varepsilon}(\boldsymbol{u})$ for $\mathcal{B}(\boldsymbol{x}, \boldsymbol{\varepsilon}(\boldsymbol{u}(t)))$. Consequently, all that follows is valid for nonhomogeneous materials.

In linearized viscoelasticity the stress tensor $\boldsymbol{\sigma} = (\sigma_{ij})$ is given by the Kelvin–Voigt relation

$$\sigma_{ij} = a_{ijkl}\varepsilon_{kl}(\boldsymbol{u}') + b_{ijkl}\varepsilon_{kl}(\boldsymbol{u}), \tag{6.24}$$

where $\mathcal{A} = (a_{ijkl})$ is the viscosity tensor and $\mathcal{B} = (b_{ijkl})$ is the elasticity tensor. Here, for simplicity, we assume that the coefficients a_{ijkl} are time-independent. Usually, the components a_{ijkl} belong to $L^\infty(\Omega)$ and satisfy the properties of symmetry and ellipticity, i.e.,

$$a_{ijkl} = a_{jikl} = a_{klij},$$

and there exists $m_{\mathcal{A}} > 0$ such that

$$a_{ijkl}\varepsilon_{ij}\varepsilon_{kl} \ge m_{\mathcal{A}}\|\boldsymbol{\varepsilon}\|^2_{\mathbb{S}^d} \quad \text{for all} \quad \boldsymbol{\varepsilon} = (\varepsilon_{ij}) \in \mathbb{S}^d.$$

Also, we assume that the components b_{ijkl} belong to $L^\infty(\Omega)$ and satisfy the same symmetry properties. Due to the symmetry, when $d = 3$ there are only 21 independent coefficients, when $d = 2$ there are only four independent coefficients, and when $d = 1$ there is only one component in each tensor.

Let $d = 3$. The constitutive law for a linearly viscoelastic isotropic material with short memory is given by

$$\boldsymbol{\sigma} = 2\,\theta\,\boldsymbol{\varepsilon}(\boldsymbol{u}') + \zeta\,\mathrm{tr}(\boldsymbol{\varepsilon}(\boldsymbol{u}'))\,\boldsymbol{I}_3 + 2\,\mu\,\boldsymbol{\varepsilon}(\boldsymbol{u}) + \lambda\,\mathrm{tr}(\boldsymbol{\varepsilon}(\boldsymbol{u}))\,\boldsymbol{I}_3 \tag{6.25}$$

or, in components,

$$\sigma_{ij} = 2\,\theta\,\varepsilon_{ij}(\boldsymbol{u}') + \zeta\,\varepsilon_{kk}(\boldsymbol{u}')\,\delta_{ij} + 2\,\mu\,\varepsilon_{ij}(\boldsymbol{u}) + \lambda\,\varepsilon_{kk}(\boldsymbol{u})\,\delta_{ij}. \tag{6.26}$$

Here, again, λ and μ are the Lamé coefficients whereas θ and ζ represent the *viscosity coefficients* which satisfy $\theta > 0$ and $\zeta \ge 0$.

A second example of viscoelastic constitutive law of the form (6.23) is provided by the nonlinear viscoelastic constitutive law

$$\boldsymbol{\sigma} = \mathcal{A}\boldsymbol{\varepsilon}(\boldsymbol{u}') + \beta\,(\boldsymbol{\varepsilon}(\boldsymbol{u}) - \mathcal{P}_K\boldsymbol{\varepsilon}(\boldsymbol{u})). \tag{6.27}$$

Here $\mathcal{A}$ is a fourth-order viscosity tensor, β is a positive coefficient, K is a closed convex subset of $\mathbb{S}^d$ such that $\boldsymbol{0} \in K$ and, again, $\mathcal{P}_K : \mathbb{S}^d \to K$ denotes the projection operator. And, as a concrete example we can choose the von Mises convex (6.19).

Viscoelastic constitutive laws with long memory. A general *viscoelastic* constitutive law with long memory is given by

$$\boldsymbol{\sigma}(t) = \mathcal{B}(t, \boldsymbol{\varepsilon}(\boldsymbol{u}(t)) + \int_0^t \mathcal{C}(t-s, \boldsymbol{\varepsilon}(\boldsymbol{u}(s)))\,ds. \tag{6.28}$$

We allow the *elasticity operator* $\mathcal{B}$ and the *relaxation operator* $\mathcal{C}$ to depend on the location of the point. Also, as shown in (6.28), the operators $\mathcal{B}$ and $\mathcal{C}$ are supposed to be time-dependent.

In the linear case, the stress tensor $\boldsymbol{\sigma} = (\sigma_{ij})$ which satisfies (6.28) is given by

$$\sigma_{ij}(t) = b_{ijkl}\,\varepsilon_{kl}(\boldsymbol{u}(t)) + \int_0^t c_{ijkl}(t-s)\,\varepsilon_{kl}(\boldsymbol{u}(s))\,ds,$$

where $\mathcal{B} = (b_{ijkl})$ is the elasticity tensor and $\mathcal{C} = (c_{ijkl})$ is the relaxation tensor. Here, for simplicity, we assume that the coefficients b_{ijkl} are time-independent. Nevertheless, we allow the coefficients b_{ijkl} and c_{ijkl} to depend on the location of the point in the body.

Let $d = 3$. The constitutive law for a linearly viscoelastic isotropic material with long memory is given by

$$\boldsymbol{\sigma}(t) = 2\,\mu\,\boldsymbol{\varepsilon}(\boldsymbol{u}(t)) \; + \lambda\,\mathrm{tr}(\boldsymbol{\varepsilon}(\boldsymbol{u}(t)))\,\boldsymbol{I}_3 + 2\int_0^t \theta(t-s)\,\boldsymbol{\varepsilon}(\boldsymbol{u}(s))\,ds$$
$$+ \int_0^t \zeta(t-s)\,\mathrm{tr}(\boldsymbol{\varepsilon}(\boldsymbol{u}(s)))\,\boldsymbol{I}_3\,ds$$

or, in components,

$$\sigma_{ij}(t) = 2\,\mu\,\varepsilon_{ij}(\boldsymbol{u}(t)) \; + \lambda\,\varepsilon_{kk}(\boldsymbol{u}(t))\,\delta_{ij} + 2\int_0^t \theta(t-s)\,\varepsilon_{ij}(\boldsymbol{u}(s))\,ds$$
$$+ \int_0^t \zeta(t-s)\,\varepsilon_{kk}(\boldsymbol{u}(s))\,\delta_{ij}\,ds.$$

Here θ and ζ represent *relaxation coefficients* which are time-dependent.

Finally, we combine (6.23) and (6.28) to obtain the more general viscoelastic constitutive law

$$\boldsymbol{\sigma}(t) = \mathcal{A}(t, \boldsymbol{\varepsilon}(\boldsymbol{u}'(t))) + \mathcal{B}(\boldsymbol{\varepsilon}(\boldsymbol{u}(t))) + \int_0^t \mathcal{C}(t-s, \boldsymbol{\varepsilon}(\boldsymbol{u}(s)))\, ds. \tag{6.29}$$

One-dimensional examples of constitutive laws of the form (6.29) can be constructed by using rheological arguments, see for instance [123, 181]. We note that when $\mathcal{C} = \mathbf{0}$ the constitutive law (6.29) reduces to the viscoelastic constitutive law with short memory, (6.23) and, in the case $\mathcal{A} = \mathbf{0}$, it reduces to the viscoelastic constitutive law with long memory, (6.28). We conclude from above that one of the features of the viscoelastic model (6.29) consists in gathering short and long memory effects in the same constitutive law.

6.3 Contact Conditions and Friction Laws

We proceed with the description of various conditions on the contact surface Γ_C. These are divided, naturally, into the conditions in the normal direction and those in the tangential directions. To describe these conditions we denote by u_ν and $\boldsymbol{u}_\tau$ the *normal* and *tangential* components of the displacement field $\boldsymbol{u}$ on the boundary, given by

$$u_\nu = \boldsymbol{u} \cdot \boldsymbol{\nu}, \quad \boldsymbol{u}_\tau = \boldsymbol{u} - u_\nu \boldsymbol{\nu}. \tag{6.30}$$

We use similar notations for the *normal* and *tangential* components of the velocity field $\boldsymbol{u}'$ on the boundary, defined by

$$u'_\nu = \boldsymbol{u}' \cdot \boldsymbol{\nu}, \quad \boldsymbol{u}'_\tau = \boldsymbol{u}' - u'_\nu \boldsymbol{\nu}. \tag{6.31}$$

Below we refer to the tangential components $\boldsymbol{u}_\tau$ and $\boldsymbol{u}'_\tau$ as the *slip* and the *slip rate*, respectively. Sometimes, for simplicity, we extend this terminology to the magnitude of these vectors, i.e., we refer to $\|\boldsymbol{u}_\tau\|_{\mathbb{R}^d}$ and $\|\boldsymbol{u}'_\tau\|_{\mathbb{R}^d}$ as the slip and the slip rate, respectively. We also denote by σ_ν and $\boldsymbol{\sigma}_\tau$ the *normal* and *tangential* components of the stress field $\boldsymbol{\sigma}$ on the boundary, i.e.,

$$\sigma_\nu = (\boldsymbol{\sigma}\boldsymbol{\nu}) \cdot \boldsymbol{\nu}, \quad \boldsymbol{\sigma}_\tau = \boldsymbol{\sigma}\boldsymbol{\nu} - \sigma_\nu \boldsymbol{\nu}. \tag{6.32}$$

The component $\boldsymbol{\sigma}_\tau$ represents the *tangential shear* or the *friction force* on the contact surface Γ_C.

Obviously, we have the orthogonality relations $\boldsymbol{v}_\tau \cdot \boldsymbol{\nu} = 0$, $\boldsymbol{\sigma}_\tau \cdot \boldsymbol{\nu} = 0$ and, moreover, the following decomposition formula holds:

$$\boldsymbol{\sigma}\boldsymbol{\nu} \cdot \boldsymbol{v} = (\sigma_\nu \boldsymbol{\nu} + \boldsymbol{\sigma}_\tau) \cdot (v_\nu \boldsymbol{\nu} + \boldsymbol{v}_\tau) = \sigma_\nu v_\nu + \boldsymbol{\sigma}_\tau \cdot \boldsymbol{v}_\tau. \tag{6.33}$$

This formula will be used in various places in the next chapters of the book, in order to derive the variational formulation of various contact problems.

Contact conditions. We start with the presentation of the conditions in the normal direction, called also *contact conditions* or *contact laws*. In this book we consider contact conditions of the form

$$-\sigma_\nu(t) \in \partial j_\nu(t, u_\nu(t) - g_0) \quad \text{on} \quad \Gamma_C \times (0,T), \tag{6.34}$$

in which j_ν is a given function, the symbol ∂j_ν denotes the Clarke subdifferential of j_ν with respect to the last variable and, recall, g_0 represents the gap function. Here and below, as usual, we do not indicate the explicit dependence of various functions on the spatial variable $\boldsymbol{x}$ and, sometimes, on both $\boldsymbol{x}$ and t. Note that the explicit dependence of the function j_ν with respect to the time variable makes the model more general and allows to describe situations when the contact condition depends on the temperature, which plays the role of a parameter. Note also that in the study of static problems for elastic materials we remove the dependence of the variables with respect the time and, therefore, we replace the contact condition (6.34) with the condition

$$-\sigma_\nu \in \partial j_\nu(u_\nu - g_0) \quad \text{on} \quad \Gamma_C. \tag{6.35}$$

The so-called *normal compliance* contact condition represents an example of contact condition which can be cast in the subdifferential form (6.34). It describes a deformable foundation and assigns a reactive normal pressure which depends on the interpenetration of the asperities on the body surface and those on the foundation. A general form of this condition is

$$-\sigma_\nu = k_\nu\, p_\nu(u_\nu - g_0) \quad \text{on} \quad \Gamma_C \times (0,T), \tag{6.36}$$

where p_ν is a prescribed nonnegative function which vanishes when its argument is negative and k_ν is a nonnegative function, the *stiffness coefficient*. Equality (6.36) shows that when there is no contact (i.e., when $u_\nu < g_0$) then $\sigma_\nu = 0$ and, therefore, the normal pressure vanishes. When there is contact (i.e., when $u_\nu \geq g_0$) then $\sigma_\nu \leq 0$ and, therefore, the reaction of the foundation is towards the body; in this case $u_\nu - g_0$ represents a measure of the interpenetration of the surface asperities. The contact condition (6.36) was first introduced in [156, 200] and since then used in many publications, see e.g. the references in [233]. The term normal compliance was first used in [133, 134].

Now, assume that $k_\nu \colon \Gamma_C \times (0,T) \to \mathbb{R}_+$ is a prescribed function and $p_\nu \colon \mathbb{R} \to \mathbb{R}_+$ is continuous. Let $g_\nu \colon \mathbb{R} \to \mathbb{R}$ and $j_\nu \colon \Gamma_C \times (0,T) \times \mathbb{R} \to \mathbb{R}$ be the functions defined by

$$g_\nu(r) = \int_0^r p_\nu(s)\, ds \quad \text{for all} \ \ r \in \mathbb{R}, \tag{6.37}$$

$$j_\nu(\boldsymbol{x}, t, r) = k_\nu(\boldsymbol{x}, t)\, g_\nu(r) \quad \text{for all} \ \ r \in \mathbb{R}, \quad \text{a.e.}\ (\boldsymbol{x}, t) \in \Gamma_C \times (0,T). \tag{6.38}$$

Then, following Lemma 3.50(iii) on page 78, we have

$$\partial g_\nu(r) = p_\nu(r), \quad \partial j_\nu(\boldsymbol{x},t,r) = k_\nu(\boldsymbol{x},t)\,\partial g_\nu(r) = k_\nu(\boldsymbol{x},t)\,p_\nu(r)$$

for all $r \in \mathbb{R}$, a.e. $(\boldsymbol{x},t) \in \Gamma_C \times (0,T)$. Therefore, it is easy to see that the contact condition (6.36) is of the form (6.34), as stated above. Note that here the function p_ν is not assumed to be increasing and, therefore, the potential function $j_\nu(\boldsymbol{x},t,\cdot)$ is not necessarily a convex function.

An example of the normal compliance function p_ν is

$$p_\nu(r) = r_+, \tag{6.39}$$

or, more general,

$$p_\nu(r) = (r_+)^m.$$

Here $m > 0$ is the normal exponent, and $r_+ = \max\{0, r\}$ is the positive part of r. And, finally, in literature one can find condition (6.36) with the choice

$$p_\nu(r) = \begin{cases} r_+ & \text{if } r \le \delta, \\ \delta & \text{if } r > \delta, \end{cases} \tag{6.40}$$

where δ is a positive coefficient related to the wear and hardness of the surface.

Clearly, the normal compliance contact condition (6.36) recovers the case when the normal stress is prescribed on the contact surface, i.e.,

$$-\sigma_\nu = F \quad \text{on} \quad \Gamma_C \times (0,T), \tag{6.41}$$

where F is a given positive function. Such type of contact conditions arises in the study of some mechanisms and was considered by a number of authors (see, e.g., [72, 202]). It also arises in geophysics in the study of earthquakes models, see for instance [37, 111–113] and the references therein.

Note that the normal compliance contact condition (6.36) is characterized by a univalued relation between the normal displacement u_ν and the normal stress σ_ν. In Sect. 7.4 we shall present examples of contact laws expressed in terms of multivalued relations between u_ν and σ_ν, which lead to subdifferential conditions of the form (6.34) or (6.35).

In Chap. 8 of this book we also consider contact conditions of the form

$$-\sigma_\nu(t) \in \partial j_\nu(t, u_\nu'(t)) \quad \text{on} \quad \Gamma_C \times (0,T), \tag{6.42}$$

in which, again, j_ν is a given function and the symbol ∂j_ν denotes the Clarke subdifferential of j_ν with respect the last variable.

The so-called *normal damped response* contact condition represents an example of contact condition which can be cast in the subdifferential form (6.42). It describes the contact with a lubricated foundation and assigns a reactive normal pressure which depends on normal velocity on the contact surface. A general form of this condition is

$$-\sigma_\nu = p_\nu(u'_\nu) \quad \text{on} \quad \Gamma_C \times (0, T), \tag{6.43}$$

where p_ν is a prescribed function.

Consider now, for simplicity, the case when the p_ν does not depend explicitly on the variables $\boldsymbol{x}$ and t. Thus, we assume in what follows that $p_\nu\colon \mathbb{R} \to \mathbb{R}$ is continuous and we denote by $j_\nu\colon \mathbb{R} \to \mathbb{R}$ the function defined by

$$j_\nu(r) = \int_0^r p_\nu(s)\,ds \quad \text{for all } r \in \mathbb{R}. \tag{6.44}$$

Then, we have $\partial j_\nu(r) = p_\nu(r)$ for all $r \in \mathbb{R}$ and, therefore, it follows that condition (6.43) is of the form (6.42), as stated above. Note that since p_ν is not assumed to be increasing the potential function j_ν is not necessarily a convex function.

An example of normal damped function p_ν is given by

$$p_\nu(r) = d_\nu r_+,$$

in which d_ν is a positive function, the *damping resistance coefficient*. In this case condition (6.43) shows that the lubricant layer presents resistance, or damping, only when the surface moves towards the foundation, but does nothing when it recedes. A second example of normal damped function is given by

$$p_\nu(r) = d_\nu r_+ + p_0, \tag{6.45}$$

in which, again, d_ν is the damping resistance coefficient and p_0 is the pressure of the lubricant, which is given and is nonnegative. Condition (6.43) with the choice (6.45) is used in the case when the lubricant fils the gap between the body and the foundation during the contact process. Another choice if p_ν is

$$p_\nu(r) = d_\nu |r|^{q-1} r. \tag{6.46}$$

Here $d_\nu > 0$, $q > 0$ and the normal contact stress depends on a power of the normal velocity, which mimics the behavior of a nonlinear viscous dashpot. Note that in this case the normal damped function could be negative and, therefore, the normal stress could be positive. More details on the normal damped response condition can be found in [102, 233].

Remark that the normal damped response contact condition (6.43) is characterized by a univalued relation between the normal velocity u'_ν and the normal stress σ_ν. Nevertheless, following the arguments in Sect. 7.4 it results that various examples of contact laws expressed in terms of multivalued relations between u'_ν and σ_ν, which lead to subdifferential conditions of the form (6.42), can be considered.

Based on the examples and comments above, in the next chapters of this book we consider contact problems involving subdifferential conditions of the forms (6.34) and (6.42). Also, when the displacement field and the stress field do not depend on time, we also consider contact conditions of the form (6.35).

Friction laws. We turn now to the conditions in the tangential directions, called also *friction conditions* or *friction laws*. In this book we consider friction laws of the form

$$-\boldsymbol{\sigma}_\tau(t) \in h_\tau(u_\nu(t) - g_0)\, \partial j_\tau(t, \boldsymbol{u}'_\tau(t)) \quad \text{on} \quad \Gamma_C \times (0, T), \tag{6.47}$$

in which h_τ and j_τ are given functions and the symbol ∂j_τ denotes the Clarke subdifferential of j_τ with respect to the last variable. Note that the explicit dependence of the function j_τ with respect to the time variable makes the model more general and allows to describes situations when the friction condition depends on the temperature, which plays the role of a parameter. In the particular case when h_τ is a constant, say $h_\tau \equiv 1$, the friction law (6.47) becomes

$$-\boldsymbol{\sigma}_\tau(t) \in \partial j_\tau(t, \boldsymbol{u}'_\tau(t)) \quad \text{on} \quad \Gamma_C \times (0, T). \tag{6.48}$$

Examples of friction laws which can be cast in the subdifferential form (6.47) or (6.48) abound in the literature. The simplest one is the so-called *frictionless* condition in which the friction force vanishes during the process, i.e.,

$$\boldsymbol{\sigma}_\tau = \boldsymbol{0} \quad \text{on} \quad \Gamma_C \times (0, T). \tag{6.49}$$

This is an idealization of the process, since even completely lubricated surfaces generate shear resistance to tangential motion. However, (6.49) is a sufficiently good approximation of the reality in some situations. Clearly, the frictionless condition (6.49) is of the form (6.48) with $j_\tau \equiv 0$.

In the case when the friction force $\boldsymbol{\sigma}_\tau$ does not vanish on the contact surface, the contact is *frictional*. Frictional contact is usually modeled with the *Coulomb law of dry friction* or its variants. According to this law, the magnitude of the tangential traction $\boldsymbol{\sigma}_\tau$ is bounded by a positive function, the so-called *friction bound*, which is the maximal frictional resistance that the surface can generate; also, once slip starts, the frictional resistance opposes the direction of the motion and its magnitude reaches the friction bound. Thus,

$$\|\boldsymbol{\sigma}_\tau\|_{\mathbb{R}^d} \le F_b, \quad \boldsymbol{\sigma}_\tau = -F_b \frac{\boldsymbol{u}'_\tau}{\|\boldsymbol{u}'_\tau\|_{\mathbb{R}^d}} \quad \text{if} \quad \boldsymbol{u}'_\tau \ne \boldsymbol{0} \quad \text{on} \quad \Gamma_C \times (0, T), \tag{6.50}$$

where $\boldsymbol{u}'_\tau$ is the tangential velocity or slip rate and F_b represents the friction bound. On a nonhomogeneous surface F_b depends explicitly on the position $\boldsymbol{x}$ on the surface; it also could depend on time as well as on the process variables and we describe this dependence below in this section.

Note that the Coulomb law (6.50) is characterized by the existence of stick-slip zones on the contact boundary, at each time moment $t \in [0,T]$. Indeed, it follows from (6.50) that if in a point $\boldsymbol{x} \in \Gamma_C$ the inequality $\|\boldsymbol{\sigma}_\tau(\boldsymbol{x},t)\|_{\mathbb{R}^d} < F_b(\boldsymbol{x},t)$ holds, then $\boldsymbol{u}'_\tau(\boldsymbol{x},t) = \boldsymbol{0}$ and the material point $\boldsymbol{x}$ is in the so-called *stick zone*; if $\|\boldsymbol{\sigma}_\tau(\boldsymbol{x},t)\|_{\mathbb{R}^d} = F_b(\boldsymbol{x},t)$ then the point $\boldsymbol{x}$ is in the so-called *slip zone*. We conclude that Coulomb's friction law (6.50) models the phenomenon that slip may occur only when the magnitude of the friction force reaches a critical value, the friction bound F_b.

In what follows we show that the Coulomb law (6.50) leads to boundary conditions of the form (6.47) or (6.48). To this end, we first note that if $\boldsymbol{u}'_\tau$ and $\boldsymbol{\sigma}_\tau$ satisfy (6.50) then

$$F_b \|\boldsymbol{\xi}\|_{\mathbb{R}^d} - F_b \|\boldsymbol{u}'_\tau\|_{\mathbb{R}^d} \geq -\boldsymbol{\sigma}_\tau \cdot (\boldsymbol{\xi} - \boldsymbol{u}'_\tau)$$
$$\text{for all } \boldsymbol{\xi} \in \mathbb{R}^d, \text{ a.e. on } \Gamma_C \times (0,T). \tag{6.51}$$

Indeed, let $\boldsymbol{\xi} \in \mathbb{R}^d$; in the points of $\Gamma_C \times (0,T)$ where $\boldsymbol{u}'_\tau \neq \boldsymbol{0}$ we have

$$\begin{aligned}\boldsymbol{\sigma}_\tau \cdot (\boldsymbol{\xi} - \boldsymbol{u}'_\tau) &= -F_b \frac{\boldsymbol{u}'_\tau}{\|\boldsymbol{u}'_\tau\|_{\mathbb{R}^d}} \cdot (\boldsymbol{\xi} - \boldsymbol{u}'_\tau)\\ &\geq F_b \|\boldsymbol{u}'_\tau\|_{\mathbb{R}^d} - F_b \|\boldsymbol{\xi}\|_{\mathbb{R}^d},\end{aligned}$$

since $\boldsymbol{u}'_\tau \cdot \boldsymbol{u}'_\tau = \|\boldsymbol{u}'_\tau\|^2_{\mathbb{R}^d}$ and $\boldsymbol{u}'_\tau \cdot \boldsymbol{\xi} \leq \|\boldsymbol{u}'_\tau\|_{\mathbb{R}^d} \|\boldsymbol{\xi}\|_{\mathbb{R}^d}$; in the points of $\Gamma_C \times (0,T)$ where $\boldsymbol{u}'_\tau = \boldsymbol{0}$, we have

$$\begin{aligned}\boldsymbol{\sigma}_\tau \cdot (\boldsymbol{\xi} - \boldsymbol{u}'_\tau) &= \boldsymbol{\sigma}_\tau \cdot \boldsymbol{\xi} \geq -\|\boldsymbol{\sigma}_\tau\|_{\mathbb{R}^d} \|\boldsymbol{\xi}\|_{\mathbb{R}^d}\\ &\geq -F_b \|\boldsymbol{\xi}\|_{\mathbb{R}^d} = F_b \|\boldsymbol{u}'_\tau\|_{\mathbb{R}^d} - F_b \|\boldsymbol{\xi}\|_{\mathbb{R}^d},\end{aligned}$$

since $\|\boldsymbol{\sigma}_\tau\|_{\mathbb{R}^d} \leq F_b$ and $\|\boldsymbol{u}'_\tau\|_{\mathbb{R}^d} = 0$. We conclude from above that, in all cases, (6.51) holds.

In certain applications, especially where the bodies are light or the friction is very large, the function F_b in (6.50) does not depend on the process variables and behaves like a function which depends only on the position $\boldsymbol{x}$ on the contact surface and, eventually, on time. Considering

$$F_b = F_b(\boldsymbol{x},t), \tag{6.52}$$

in (6.50) leads to the *Tresca friction law*, and it simplifies considerably the analysis of the corresponding contact problem, see for instance [102, 233]. Then, using (6.51), it is easy to check that (6.50) and (6.52) lead to the subdifferential condition (6.48) with

$$j_\tau(\boldsymbol{x},t,\boldsymbol{\xi}) = F_b(\boldsymbol{x},t)\, \|\boldsymbol{\xi}\|_{\mathbb{R}^d} \quad \text{for all } \boldsymbol{\xi} \in \mathbb{R}^d, \text{ a.e. } (\boldsymbol{x},t) \in \Gamma_C \times (0,T). \tag{6.53}$$

Moreover, note that in this case j_τ is a convex function with respect to its last variable.

Often, especially in engineering literature, the friction bound F_b is chosen as

$$F_b = F_b(\sigma_\nu) = \mu\,|\sigma_\nu| \tag{6.54}$$

where $\mu \geq 0$ is the *coefficient of friction*. The choice (6.54) in (6.50) leads to the *classical version of Coulomb's law* which was intensively studied in the literature, see for instance the references in [233]. When the wear of the contacting surface is taken into account, a *modified version of Coulomb's law* is more appropriate. This law has been derived in [242–244] from thermodynamic considerations, and is given by choosing

$$F_b = \mu\,|\sigma_\nu|\,(1-\delta|\sigma_\nu|)_+ \tag{6.55}$$

in (6.50), where δ is a very small positive parameter related to the wear constant of the surface and, again, μ is the coefficient of friction.

The choice (6.36) in (6.54) leads to the friction bound

$$F_b = \mu\,k_\nu\,p_\nu(u_\nu - g_0). \tag{6.56}$$

Then, using again (6.51), it is easy to check that (6.50) leads to the subdifferential condition (6.47) with

$$\begin{aligned} &h_\tau(\boldsymbol{x},t,r) = \mu(\boldsymbol{x},t)\,k_\nu(\boldsymbol{x},t)\,p_\nu(r) \\ &\qquad \text{for all } r \in \mathbb{R},\ \text{a.e. } (\boldsymbol{x},t) \in \Gamma_C \times (0,T), \end{aligned} \tag{6.57}$$

$$\begin{aligned} &j_\tau(\boldsymbol{x},t,\boldsymbol{\xi}) = \|\boldsymbol{\xi}\|_{\mathbb{R}^d} \\ &\qquad \text{for all } \boldsymbol{\xi} \in \mathbb{R}^d,\ \text{a.e. } (\boldsymbol{x},t) \in \Gamma_C \times (0,T). \end{aligned} \tag{6.58}$$

In a similar way, the choice (6.36) in (6.55) leads to the friction bound

$$F_b = \mu\,k_\nu\,p_\nu(u_\nu - g_0)\big(1 - \delta k_\nu\,p_\nu(u_\nu - g_0)\big)_+. \tag{6.59}$$

Therefore, by (6.51) it is easy to check that (6.50) leads to the subdifferential condition (6.47) with

$$\begin{aligned} &h_\tau(\boldsymbol{x},t,r) = \mu(\boldsymbol{x},t)\,k_\nu(\boldsymbol{x},t)\,p_\nu(r)\,\big(1 - \delta(\boldsymbol{x},t)k_\nu(\boldsymbol{x},t)p_\nu(r)\big)_+ \\ &\qquad \text{for all } r \in \mathbb{R},\ \text{a.e. } (\boldsymbol{x},t) \in \Gamma_C \times (0,T), \end{aligned} \tag{6.60}$$

and j_τ given by (6.58).

The choice (6.41) in (6.54) leads to the friction bound

$$F_b = \mu\,F \tag{6.61}$$

and the choice (6.41) in (6.55) leads to the friction bound

$$F_b = \mu F(1 - \delta F)_+. \tag{6.62}$$

It follows from (6.61) and (6.62) that, in the case when μ and δ are given, the friction bound F_b is a given function defined on $\Gamma_C \times (0, T)$ and, therefore, we recover the Tresca friction law which is of the form (6.48), as shown above.

We observe that the friction coefficient μ is not an intrinsic property of a material, a body or its surface, since it depends on the contact process and the operating conditions. For instance, it could depend on the surface characteristics, on the surface geometry and structure, on the relative velocity between the contacting surfaces, on the surface temperature, on the wear or rearrangement of the surface and, therefore, on its history. A thorough description of these issues can be found in [217] and in the survey [246].

In many geophysical publications the motion of tectonic plates is modeled with the Coulomb law (6.50) in which the friction bound is assumed to depend on the magnitude of the tangential displacement, that is

$$\|\boldsymbol{\sigma}_\tau\|_{\mathbb{R}^d} \le F_b(\|\boldsymbol{u}_\tau\|_{\mathbb{R}^d}), \quad \boldsymbol{\sigma}_\tau = -F_b(\|\boldsymbol{u}_\tau\|_{\mathbb{R}^d})\frac{\boldsymbol{u}'_\tau}{\|\boldsymbol{u}'_\tau\|_{\mathbb{R}^d}} \quad \text{if} \quad \boldsymbol{u}'_\tau \neq \boldsymbol{0} \tag{6.63}$$

on $\Gamma_C \times (0, T)$. Alternatively, the friction bound could be assumed to depend on the magnitude of the tangential velocity, i.e.,

$$\|\boldsymbol{\sigma}_\tau\|_{\mathbb{R}^d} \le F_b(\|\boldsymbol{u}'_\tau\|_{\mathbb{R}^d}), \quad \boldsymbol{\sigma}_\tau = -F_b(\|\boldsymbol{u}'_\tau\|_{\mathbb{R}^d})\frac{\boldsymbol{u}'_\tau}{\|\boldsymbol{u}'_\tau\|_{\mathbb{R}^d}} \quad \text{if} \quad \boldsymbol{u}'_\tau \neq \boldsymbol{0} \tag{6.64}$$

on $\Gamma_C \times (0, T)$. Details can be found in [37, 217, 228] and the references therein. For this reason (6.63) is also called a *slip-dependent friction law* and (6.64) is also called a *slip rate-dependent friction law*. A concrete example of friction law of the form (6.64) which can be cast in the subdifferentiable form (6.48) will be provided on page 238.

Since the functional j_τ in (6.53) is nondifferentiable with respect to its last argument, several regularizations of the Coulomb law are used in the literature, mainly for numerical reasons, see e.g. [233]. A first example is given by

$$-\boldsymbol{\sigma}_\tau = F_b\,\frac{\boldsymbol{u}'_\tau}{\sqrt{\|\boldsymbol{u}'_\tau\|^2_{\mathbb{R}^d} + \rho^2}} \quad \text{on} \quad \Gamma_C \times (0, T), \tag{6.65}$$

where $\rho > 0$ represents a regularization parameter and, again, F_b is a positive function defined on $\Gamma_C \times (0, T)$. Note that the function $p_\tau\colon \mathbb{R}^d \to \mathbb{R}^d$ defined by

$$p_\tau(\boldsymbol{\xi}) = \frac{\boldsymbol{\xi}}{\sqrt{\|\boldsymbol{\xi}\|^2_{\mathbb{R}^d} + \rho^2}} \quad \text{for all} \quad \boldsymbol{\xi} \in \mathbb{R}^d$$

is the gradient of the convex Gâteaux differentiable function

$$\boldsymbol{\xi} \mapsto \sqrt{\|\boldsymbol{\xi}\|_{\mathbb{R}^d}^2 + \rho^2} - \rho.$$

Then, it is easy to see that the frictional condition (6.65) can be written in the equivalent form (6.48) with the choice

$$j_\tau(\boldsymbol{x}, t, \boldsymbol{\xi}) = F_b(\boldsymbol{x}, t)\left(\sqrt{\|\boldsymbol{\xi}\|_{\mathbb{R}^d}^2 + \rho^2} - \rho\right)$$

$$\text{for all } \boldsymbol{\xi} \in \mathbb{R}^d, \text{ a.e. } (\boldsymbol{x}, t) \in \Gamma_C \times (0, T).$$

A second example is given by

$$-\boldsymbol{\sigma}_\tau = \begin{cases} F_b\,\|\boldsymbol{u}'_\tau\|_{\mathbb{R}^d}^{\rho-1}\,\boldsymbol{u}'_\tau & \text{if } \boldsymbol{u}'_\tau \neq \boldsymbol{0}, \\ \boldsymbol{0} & \text{if } \boldsymbol{u}'_\tau = \boldsymbol{0}, \end{cases} \quad \text{on} \quad \Gamma_C \times (0, T), \tag{6.66}$$

where, again, ρ represents a positive regularization parameter. Then, it is easy to see that the frictional condition (6.66) can be written in the equivalent form (6.48) with the choice

$$j_\tau(\boldsymbol{x}, t, \boldsymbol{\xi}) = F_b(\boldsymbol{x}, t)\frac{1}{\rho + 1}\,\|\boldsymbol{\xi}\|_{\mathbb{R}^d}^{\rho+1}$$

$$\text{for all } \boldsymbol{\xi} \in \mathbb{R}^d, \text{ a.e. } (\boldsymbol{x}, t) \in \Gamma_C \times (0, T).$$

Note that the friction laws (6.65) and (6.66) describe situation when slip appears even for small tangential shears, which is the case when the surfaces are lubricated by a thin layer of non-Newtonian fluid. Relation (6.66) is called in the literature the *power-law friction*; indeed, in this case the tangential shear is proportional to a power of the tangential velocity and, in the particular case $\rho = 1$, (6.66) implies that the tangential shear is proportional to the tangential velocity. Also, note that the Coulomb law (6.50) is obtained, formally, from the friction laws (6.65) and (6.66) in the limit as $\rho \to 0$.

Based on the examples and comments above, in the next chapters of this book we consider contact problems involving subdifferential frictional conditions of the forms (6.47) and (6.48). Note that all these laws model dynamic or quasistatic frictional processes. However, sometimes we shall consider *static versions* of these laws which are obtained by replacing the tangential velocity with the tangential displacement. Thus, the static version of the friction law (6.47) is given by

$$-\boldsymbol{\sigma}_\tau(t) \in h_\tau(u_\nu(t) - g_0)\,\partial j_\tau(t, \boldsymbol{u}_\tau(t)) \quad \text{on} \quad \Gamma_C \times (0, T), \tag{6.67}$$

and the static version of the friction law (6.48) is given by

$$-\boldsymbol{\sigma}_\tau(t) \in \partial j_\tau(t, \boldsymbol{u}_\tau(t)) \quad \text{on} \quad \Gamma_C \times (0, T). \tag{6.68}$$

Also, when the displacement field and the stress field do not depend on time, the friction laws (6.67) and (6.68) read

$$-\boldsymbol{\sigma}_\tau \in h_\tau(u_\nu - g_0)\,\partial j_\tau(\boldsymbol{u}_\tau) \quad \text{on} \quad \Gamma_C, \tag{6.69}$$

and

$$-\boldsymbol{\sigma}_\tau \in \partial j_\tau(\boldsymbol{u}_\tau) \quad \text{on} \quad \Gamma_C, \tag{6.70}$$

respectively.

Concrete examples of static friction laws that can be cast in one of the subdifferential from (6.67)–(6.70) can be obtained as above and, in order to avoid repetitions, we do not mention them in detail. Nevertheless, we restrict ourselves to recall that the static version of Coulomb's law of dry friction (6.50) is given by

$$\|\boldsymbol{\sigma}_\tau\|_{\mathbb{R}^d} \le F_b, \quad \boldsymbol{\sigma}_\tau = -F_b \frac{\boldsymbol{u}_\tau}{\|\boldsymbol{u}_\tau\|_{\mathbb{R}^d}} \quad \text{if} \quad \boldsymbol{u}_\tau \neq \boldsymbol{0}. \tag{6.71}$$

Equality (6.71) holds on Γ_C in the time-independent case and on $\Gamma_C \times (0,T)$ in the time-dependent case. The choice (6.36) in (6.54) leads to the friction bound (6.56). Therefore, in the time-dependent case it is easy to check that (6.71) leads to the subdifferential condition (6.67) with the functions h_τ and j_τ given by (6.57) and (6.58), respectively. And, in the time-independent case it is easy to check that (6.71) leads to the subdifferential condition (6.69) in which

$$h_\tau(\boldsymbol{x}, r) = \mu(\boldsymbol{x})\, k_\nu(\boldsymbol{x})\, p_\nu(r) \;\; \text{for all } r \in \mathbb{R}, \text{ a.e. } \boldsymbol{x} \in \Gamma_C,$$

$$j_\tau(\boldsymbol{x}, \boldsymbol{\xi}) = \|\boldsymbol{\xi}\|_{\mathbb{R}^d} \;\; \text{for all } \boldsymbol{\xi} \in \mathbb{R}^d, \text{ a.e. } \boldsymbol{x} \in \Gamma_C.$$

The static version of the slip-dependent friction law (6.63) is given by

$$\left.\begin{aligned} &\|\boldsymbol{\sigma}_\tau\|_{\mathbb{R}^d} \le F_b(\|\boldsymbol{u}_\tau\|_{\mathbb{R}^d}), \\ &\boldsymbol{\sigma}_\tau = -F_b(\|\boldsymbol{u}_\tau\|_{\mathbb{R}^d}) \frac{\boldsymbol{u}_\tau}{\|\boldsymbol{u}_\tau\|_{\mathbb{R}^d}} \quad \text{if} \quad \boldsymbol{u}_\tau \neq \boldsymbol{0} \end{aligned}\right\} \quad \text{on } \Gamma_C. \tag{6.72}$$

A concrete example of such a friction law, which can be cast in the subdifferentiable form (6.70), will presented in Sect. 7.4 of the book.

The static friction laws are suitable for proportional loadings and can be considered as a first approximation of the evolutionary friction laws. Considering the static versions (6.67), (6.68) or (6.69), (6.70) of the friction laws (6.47), (6.48), respectively, simplifies the mathematical analysis of the corresponding mechanical problems.

6.4 Contact of Piezoelectric Materials

The piezoelectric effect is characterized by the coupling between the mechanical and electrical properties of the materials. This coupling leads to the appearance of electric potential when mechanical stress is present and, conversely, mechanical stress is generated when electric potential is applied. A deformable material which exhibits such a behavior is called a *piezoelectric* material. Piezoelectric materials are used as switches and actuators in many engineering systems, in radioelectronics, electroacoustics, and measuring equipments. Piezoelectric materials for which the mechanical properties are elastic are also called *electro-elastic* materials and piezoelectric materials for which the mechanical properties are viscoelastic are also called *electro-viscoelastic* materials.

Physical setting. To model contact problems with piezoelectric materials we refer to the physical setting shown in Fig. 6.2 that we describe in what follows. Here Ω represents the reference configuration of a piezoelectric body and, besides the partition of Γ into three sets $\overline{\Gamma}_D$, $\overline{\Gamma}_N$, and $\overline{\Gamma}_C$, which corresponds to the mechanical boundary conditions described in Sects. 6.1 and 6.3, we consider a partition of $\overline{\Gamma}_D \cup \overline{\Gamma}_N$ into two sets $\overline{\Gamma}_a$ and $\overline{\Gamma}_b$ with mutually disjoint relatively open sets Γ_a and Γ_b, which corresponds to the electrical boundary conditions.

Everywhere below we assume that meas $(\Gamma_D) > 0$ and meas $(\Gamma_a) > 0$. The body is clamped on Γ_D and time-dependent surface tractions of density $\boldsymbol{f}_N$ act on Γ_N. We also assume that the electrical potential vanishes on Γ_a and a surface electric charge of density q_b is prescribed on Γ_b. The body is, or can arrive, in contact on Γ_C with an obstacle, the so-called foundation. As in Sect. 6.1 we assume that in the reference configuration there exists a gap, denoted by g_0, between Γ_C and the foundation, which is measured along the outer normal $\boldsymbol{\nu}$. Also, we assume that the foundation is electrically conductive and its potential is maintained at φ_0. The contact is frictional and there may be also electric charges on the part of the body surface which is in contact with the foundation and which vanish when contact is lost.

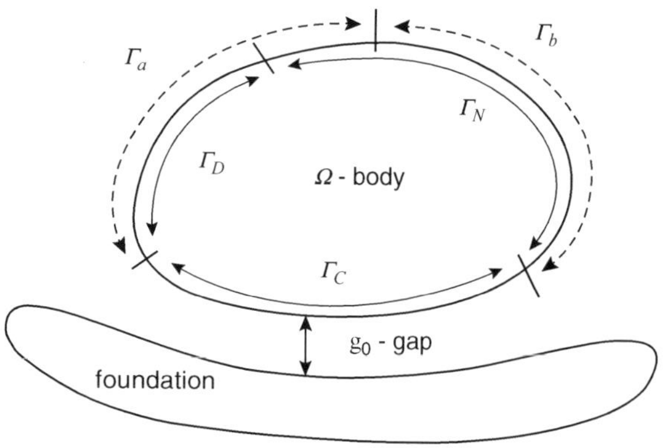

Fig. 6.2 The physical setting for piezoelectric material; Γ_C is the contact surface

We are interested in mathematical models which describe the evolution of the mechanical and electrical state of the piezoelectric body during the time interval $[0, T]$, with $T > 0$. To this end, besides the stress field $\boldsymbol{\sigma} = \boldsymbol{\sigma}(\boldsymbol{x}, t) = (\sigma_{ij}(\boldsymbol{x}, t))$ and the displacement field $\boldsymbol{u} = \boldsymbol{u}(\boldsymbol{x}, t) = (u_i(\boldsymbol{x}, t))$, we introduce the *electric displacement field* $\boldsymbol{D} = \boldsymbol{D}(\boldsymbol{x}, t) = (D_i(\boldsymbol{x}, t))$ and the *electric potential* $\varphi = \varphi(\boldsymbol{x}, t)$. The functions $\boldsymbol{u}: \Omega \times [0, T] \to \mathbb{R}^d$, $\boldsymbol{\sigma}: \Omega \times [0, T] \to \mathbb{S}^d$, $\varphi: \Omega \times [0, T] \to \mathbb{R}$ and $\boldsymbol{D}: \Omega \times [0, T] \to \mathbb{R}^d$ will play the role of the unknowns of the piezoelectric contact problem. From time to time, we suppress the explicit dependence of various quantities on the spatial variable $\boldsymbol{x}$, or both $\boldsymbol{x}$ and t.

The balance equation for the stress field is (6.1) if the mechanical process is dynamic and (6.2) or (6.3) if the mechanical process is static and time-dependent or time-independent, respectively. The balance equation for the electric displacement field is

$$\operatorname{div} \boldsymbol{D} - q_0 = 0 \quad \text{in} \quad \Omega \times (0, T) \tag{6.73}$$

if the process is time-dependent and

$$\operatorname{div} \boldsymbol{D} - q_0 = 0 \quad \text{in} \quad \Omega \tag{6.74}$$

if the process is time-independent. Here and below q_0 is the density of the volume electric charges and "div" denotes the divergence operator, i.e., $\operatorname{div} \boldsymbol{D} = D_{i,i}$.

We turn now to the boundary conditions on Γ_D, Γ_N, Γ_a, and Γ_b. First, since the piezoelectric body is clamped on Γ_D, we impose the displacement boundary condition (6.4). Moreover, the traction boundary condition on the boundary Γ_N is given by (6.5). Next, since the electric potential vanishes on Γ_a during the process, we impose the boundary condition

$$\varphi = 0 \quad \text{on} \quad \Gamma_a \times (0, T). \tag{6.75}$$

We also assume that a surface electric charge of density q_b is prescribed on Γ_b and, therefore,

$$\boldsymbol{D} \cdot \boldsymbol{\nu} = q_b \quad \text{on} \quad \Gamma_b \times (0, T). \tag{6.76}$$

Note that in the case of static time-independent processes we use (6.6) and (6.7) instead of (6.4) and (6.5), respectively. Moreover, the boundary conditions (6.75) and (6.76) hold on Γ_a and Γ_b, respectively.

Constitutive laws. We define the electric field $\boldsymbol{E}$ by relation

$$\boldsymbol{E}(\varphi) = -\nabla \varphi = -(\varphi_{,i}) \quad \text{in} \quad \Omega \times (0, T). \tag{6.77}$$

As in the case of the deformable bodies, we need a constitutive law to describe the particular behavior of the material the body is made of.

A general electro-elastic constitutive law is given by

$$\boldsymbol{\sigma} = \mathcal{F}\boldsymbol{\varepsilon}(\boldsymbol{u}) - \mathcal{P}^{\top} \boldsymbol{E}(\varphi), \tag{6.78}$$

where $\mathcal{F}$ is the *elasticity operator*, assumed to be nonlinear, $\mathcal{P} = (p_{ijk})$ represents the third-order *piezoelectric tensor* and $\mathcal{P}^\top$ denotes its transpose. Recall that the tensors $\mathcal{P}$ and $\mathcal{P}^\top$ satisfy the equality

$$\mathcal{P}\boldsymbol{\sigma} \cdot \boldsymbol{v} = \boldsymbol{\sigma} : \mathcal{P}^\top \boldsymbol{v} \quad \text{for all } \boldsymbol{\sigma} \in \mathbb{S}^d,\ \boldsymbol{v} \in \mathbb{R}^d \tag{6.79}$$

and the components of the tensor $\mathcal{P}^\top$ are given by $p^\top_{ijk} = p_{kij}$. We complete (6.78) with a constitutive equation for the electric displacement field. This equation is of the form

$$\boldsymbol{D} = \mathcal{P}\boldsymbol{\varepsilon}(\boldsymbol{u}) + \boldsymbol{\beta}\boldsymbol{E}(\varphi), \tag{6.80}$$

where $\boldsymbol{\beta} = (\beta_{ij})$ denotes the *electric permittivity tensor*. Usually, the components β_{ij} belong to $L^\infty(\Omega)$ and satisfy the usual properties of symmetry and ellipticity, i.e.,

$$\beta_{ij} = \beta_{ji},$$

and there exists $m_\beta > 0$ such that

$$\beta_{ij}\xi_i\xi_j \geq m_\beta \|\boldsymbol{\xi}\|^2_{\mathbb{R}^d} \quad \text{for all } \boldsymbol{\xi} = (\xi_i) \in \mathbb{R}^d.$$

Equation (6.78) indicates that the mechanical properties of the material are described by a nonlinear elastic constitutive relation which takes into account the dependence of the stress field on the electric field. Note that in the case when $\mathcal{P} = \boldsymbol{0}$, (6.78) reduces to the purely elastic constitutive law (6.9). Equation (6.80) describes a linear dependence of the electric displacement field $\boldsymbol{D}$ on the strain and electric fields. Constitutive laws of the form (6.78) and (6.80) have been frequently employed in the literature in order to model the behavior of piezoelectric materials, see, e.g., [27, 29, 215] and the references therein. In the linear case, the constitutive laws (6.78) and (6.80) read as follows:

$$\sigma_{ij} = f_{ijkl}\,\varepsilon_{kl}(\boldsymbol{u}) + p_{kij}\varphi_{,k}\,,$$

$$D_i = p_{ijk}\,\varepsilon_{jk}(\boldsymbol{u}) - \beta_{ij}\varphi_{,j}\,,$$

where f_{ijkl} are the components of the elasticity tensor $\mathcal{F}$.

A general electro-viscoelastic constitutive law is given by

$$\boldsymbol{\sigma}(t) = \mathcal{A}(t, \boldsymbol{\varepsilon}(\boldsymbol{u}'(t))) + \mathcal{B}\boldsymbol{\varepsilon}(\boldsymbol{u}(t)) - \mathcal{P}^\top \boldsymbol{E}(\varphi(t)) \tag{6.81}$$

in which $\mathcal{A}$ is the *viscosity operator*, assumed to be nonlinear, $\mathcal{B}$ is the *elasticity operator* and, again, $\mathcal{P}^\top$ is the transpose of the piezoelectric tensor $\mathcal{P}$. We complete (6.81) with the linear constitutive relation (6.80) for the electric displacement field. Note that the operator $\mathcal{A}$ may depend explicitly on the time variable and this is the case when the viscosity properties of the material depend on the temperature field, which plays the role of a parameter, i.e., its evolution in time is prescribed.

Equation (6.81) indicates that the mechanical properties of the material are described by a nonlinear viscoelastic constitutive relation which takes into account the dependence of the stress field on the electric field. It already has been employed in the literature, see for instance, [20, 21, 146, 185] and the references therein. Note that in the case when $\mathcal{P} = \mathbf{0}$, equation (6.81) reduces to the purely viscoelastic constitutive law (6.23).

Frictional contact conditions. In the study of static time-independent problems for electro-elastic materials we assume that the normal stress satisfies the condition

$$-\sigma_\nu \in h_\nu(\varphi - \varphi_0)\, \partial j_\nu(u_\nu - g_0) \quad \text{on} \quad \Gamma_C \tag{6.82}$$

in which both h_ν and j_ν are prescribed functions. This condition represents an extension of the contact condition (6.35) and takes into account the influence of the electric variables on the contact.

An example of contact condition of the form (6.82) can be obtained as follows. Assume that the normal stress satisfies the normal compliance contact condition (6.36) on Γ_C in which the stiffness coefficient depends on the difference between the potential on the foundation and the body surface, i.e.,

$$-\sigma_\nu = k_\nu(\varphi - \varphi_0)\, p_\nu(u_\nu - g_0) \qquad \text{on} \qquad \Gamma_C. \tag{6.83}$$

Here, both $k_\nu \colon \Gamma_C \times \mathbb{R} \to \mathbb{R}_+$ and $p_\nu \colon \mathbb{R} \to \mathbb{R}_+$ are prescribed nonnegative functions. The dependence of the stiffness coefficient on $\varphi - \varphi_0$ arises from the fact that the foundation is electrically conductive and models the influence of the electric variables on the contact. Next, assume that p_ν is a continuous function and let $j_\nu \colon \mathbb{R} \to \mathbb{R}$ be the function defined by

$$j_\nu(r) = \int_0^r p_\nu(s)\, ds \quad \text{for all } r \in \mathbb{R}. \tag{6.84}$$

Then, following Lemma 3.50(iii) on page 78, it is easy to see that the contact condition (6.83) is of the form (6.82) with $h_\nu(\boldsymbol{x}, r) = k_\nu(\boldsymbol{x}, r)$ for all $r \in \mathbb{R}$, a.e. $\boldsymbol{x} \in \Omega$. Note that here the function p_ν is not assumed to be increasing and, therefore, the superpotential function j_ν is not necessarily a convex function.

In a similar way, we shall use a friction law of the form

$$-\boldsymbol{\sigma}_\tau \in h_\tau(\varphi - \varphi_0, u_\nu - g_0)\, \partial j_\tau(\boldsymbol{u}_\tau) \qquad \text{on } \Gamma_C. \tag{6.85}$$

This law represents an extension of the static friction law (6.69) which takes into account the influence of the electric variables on the frictional contact.

Concrete examples of such kind of condition can be easily obtained from the corresponding examples in Sect. 6.2 by taking into consideration the dependence of the friction bound or, alternatively, of the coefficient of friction, on the difference $\varphi - \varphi_0$. For instance, assuming the dependence of the friction bound on the difference between the potentials of the body surface and the foundation, the friction law (6.71) leads to the following static version of Coulomb's law of dry friction:

$$\left.\begin{aligned} &\|\boldsymbol{\sigma}_\tau\|_{\mathbb{R}^d} \le F_b(\varphi - \varphi_0), \\ &\boldsymbol{\sigma}_\tau = -\,F_b(\varphi - \varphi_0)\,\frac{\boldsymbol{u}_\tau}{\|\boldsymbol{u}_\tau\|_{\mathbb{R}^d}} \quad \text{if} \quad \boldsymbol{u}_\tau \ne \boldsymbol{0} \end{aligned}\right\} \quad \text{on } \Gamma_C. \tag{6.86}$$

Again, we note that the dependence of F_b on $\varphi - \varphi_0$ models the influence of the electric variables on the friction and it is used here since the foundation is supposed to be electrically conductive.

In addition, assume that the normal stress satisfies the normal compliance contact condition (6.83) and the friction bound is given by (6.54). Then it follows that

$$F_b = \mu\, k_\nu(\varphi - \varphi_0)\, p_\nu(u_\nu - g_0).$$

With this choice for F_b, using (6.51) it is easy to check that (6.86) leads to the subdifferential condition (6.85) with

$$h_\tau(\boldsymbol{x}, r_1, r_2) = \mu(\boldsymbol{x})\, k_\nu(\boldsymbol{x}, r_1)\, p_\nu(r_2), \qquad \text{for all } r_1,\ r_2 \in \mathbb{R},\ \text{a.e. } \boldsymbol{x} \in \Gamma_C, \tag{6.87}$$

$$j_\tau(\boldsymbol{x}, \boldsymbol{\xi}) = \|\boldsymbol{\xi}\|_{\mathbb{R}^d} \quad \text{for all } \boldsymbol{\xi} \in \mathbb{R}^d,\ \text{a.e. } \boldsymbol{x} \in \Gamma_C. \tag{6.88}$$

In particular, if the function F_b does not depend on $\varphi - \varphi_0$, from (6.86) we obtain the Tresca friction law. Finally, if F_b vanishes, (6.86) reduces to the frictionless condition $\boldsymbol{\sigma}_\tau = \boldsymbol{0}$. And, obviously, these last two friction laws are of the form (6.85).

Note that in Sect. 8.3 we shall study a dynamic frictional contact problems for electro-viscoelastic materials. In the modeling of this problem we assume that the electric variables do not influence the frictional contact and, therefore, we use the contact condition (6.42) associated to the friction law (6.48). Considering dynamic models in which the frictional contact conditions depend on the difference $\varphi - \varphi_0$ leads to severe mathematical difficulties and, at the best of our knowledge, the analysis of these models represents an open problem.

Electrical contact conditions. We turn to the electrical condition on the contact surface. In the study of static process for electro-elastic materials we assume that

$$\boldsymbol{D} \cdot \boldsymbol{\nu} \in h_e(u_\nu - g_0)\, \partial j_e(\varphi - \varphi_0) \quad \text{on} \quad \Gamma_C, \tag{6.89}$$

where h_e and j_e are given real-valued functions. Condition (6.89) represents a regularized condition which may be obtained as follows.

First, recall that the foundation is assumed to be electrically conductive and its potential is maintained at φ_0. When there is no contact at a point on the surface (i.e., when $u_\nu < g_0$), the gap is assumed to be an insulator (say, it is filled with air) and, therefore, the normal component of the electric displacement field vanishes, so that there are no free electrical charges on the surface. Thus,

$$u_\nu < g_0 \implies \boldsymbol{D} \cdot \boldsymbol{\nu} = 0. \tag{6.90}$$

During the process of contact (i.e., when $u_\nu \geq g_0$) the normal component of the electric displacement field or the free charge is assumed to depend on the difference between the potential on the foundation and the body surface. Thus,

$$u_\nu \geq g_0 \quad \Longrightarrow \quad \boldsymbol{D} \cdot \boldsymbol{\nu} = k_e \, p_e(\varphi - \varphi_0), \tag{6.91}$$

where p_e is a prescribed real-valued function and k_e is a nonnegative function, the *electric conductivity coefficient*. A possible choice of the function p_e is $p_e(r) = r$. We combine (6.90), (6.91) to obtain

$$\boldsymbol{D} \cdot \boldsymbol{\nu} = k_e \, \chi_{[0,\infty)}(u_\nu - g_0) \, p_e(\varphi - \varphi_0), \tag{6.92}$$

where $\chi_{[0,\infty)}$ is the characteristic function of the interval $[0, \infty)$, i.e.,

$$\chi_{[0,\infty)}(r) = \begin{cases} 0 & \text{if } r < 0, \\ 1 & \text{if } r \geq 0. \end{cases}$$

Condition (6.92) describes perfect electrical contact and is somewhat similar to the well-known Signorini contact condition, see [102, 233] for details. Both conditions may be overidealizations in many applications. To make it more realistic, we regularize condition (6.92) with condition

$$\boldsymbol{D} \cdot \boldsymbol{\nu} = h_e(u_\nu - g_0) \, p_e(\varphi - \varphi_0) \quad \text{on } \Gamma_C, \tag{6.93}$$

in which h_e is a nonnegative function which describes the electric conductivity of the foundation. The reason for this regularization is mathematical, since we need to avoid the discontinuity in the free electric charge when contact is established. Nevertheless, we note that this regularization seems to be reasonable from physical point of view as shown in the two examples below, which provide possible choices for the function h_e.

A first choice of h_e is given by

$$h_e(r) = \begin{cases} 0 & \text{if } r < 0, \\ k_e \, \delta^{-1} \, r & \text{if } 0 \leq r \leq \delta, \\ k_e & \text{if } r > \delta, \end{cases} \tag{6.94}$$

where $\delta > 0$ is a small parameter. This choice means that during the process of contact the electric conductivity increases up to k_e as the contact among the surface asperities improves, and stabilizes when the penetration $u_\nu - g_0$ reaches the value δ. A second choice is given by

$$h_e(r) = \begin{cases} 0 & \text{if } r < -\delta, \\ k_e \dfrac{r+\delta}{\delta} & \text{if } -\delta \le r \le 0, \\ k_e & \text{if } r > 0, \end{cases} \tag{6.95}$$

where, again, $\delta > 0$ is a small parameter. This choice means that the air is electrically conductive under the critical thickness δ and behaves like an insulator only above a critical thickness, which justifies the use of the electric conductivity coefficient $h_e(u_\nu - g_0)$ instead of $k_e\, \chi_{[0,\infty)}(u_\nu - g_0)$.

Now, assume that $h_e\colon \Gamma_C \times \mathbb{R} \to \mathbb{R}_+$ is a given nonnegative function and $p_e\colon \mathbb{R} \to \mathbb{R}_+$ is continuous. Let $j_e\colon \mathbb{R} \to \mathbb{R}$ be the function defined by

$$j_e(r) = \int_0^r p_e(s)\, ds \quad \text{for all } r \in \mathbb{R}. \tag{6.96}$$

Then, using Lemma 3.50(iii) it is easy to see that the boundary condition (6.93) is of the form (6.89). Note that here the function p_e is not assumed to be increasing and, therefore, the superpotential function j_e is not necessarily a convex function. Note also that when $h_e \equiv 0$ then (6.93) leads to

$$\boldsymbol{D} \cdot \boldsymbol{\nu} = 0 \quad \text{on } \Gamma_C. \tag{6.97}$$

Condition (6.97) models the case when the obstacle is a perfect insulator and was used in [29, 153, 167, 237].

Remark that the electrical contact condition (6.93) is characterized by a univalued relation between the electric potential φ and the normal component of the electric displacement field $\boldsymbol{D}\cdot\boldsymbol{\nu}$. Nevertheless, following the arguments in Sect. 7.4, it results that various examples of contact laws expressed in terms of multivalued relations between φ and $\boldsymbol{D} \cdot \boldsymbol{\nu}$, which lead to subdifferential conditions of the form (6.89), can be considered.

Finally, we note that in the study of dynamic contact problems with electro-viscoelastic materials we assume that the electric charges do not depend on penetration. Therefore, instead of condition (6.89) we shall use the simplified boundary condition

$$\boldsymbol{D} \cdot \boldsymbol{\nu} \in \partial j_e(\varphi - \varphi_0) \quad \text{on } \Gamma_C \times (0, T). \tag{6.98}$$

Considering dynamic piezoelectric contact models in which the electrical contact condition depends on the penetration $u_\nu - g_0$ leads to severe mathematical difficulties and, at the best of our knowledge, the analysis of such models represents an open problem.

Based on the equations and boundary conditions presented in this chapter, in Chaps. 7 and 8 of this book we construct and analyze various mathematical models which describe frictional contact processes with elastic, viscoelastic, and piezoelectric materials.

Chapter 7
Analysis of Static Contact Problems

In this chapter we illustrate the use of the abstract results obtained in Chap. 4, in the study of three representative static frictional contact problems for deformable bodies. In the first two problems we model the material's behavior with a nonlinear elastic constitutive law and with a viscoelastic constitutive law with long memory, respectively, and we describe the frictional contact with subdifferential boundary conditions. In the third problem the deformable body is assumed to be piezoelectric and, therefore, we model its behavior with an electro-elastic constitutive law. And, again, the contact conditions, including the electrical conditions on the contact surface, are of subdifferential type. For each problem we provide a variational formulation. For the first two problems it is in a form of a hemivariational inequality for the displacement field and, for the third one, it is in a form of a system of hemivariational inequalities in which the unknowns are the displacement and electric potential fields. Then, we use the abstract existence and uniqueness results presented in Chap. 4 to prove the weak solvability of the corresponding contact problems and, under additional assumptions, their unique weak solvability. Finally, we present concrete examples of constitutive laws and frictional contact conditions for which our results work and provide the related mechanical interpretation. Everywhere in this chapter we use the notation introduced in Chap. 6.

7.1 An Elastic Frictional Problem

We refer to the physical setting described in Sect. 6.1. We assume that the process is static, the material is elastic, and the external forces and tractions do not depend on time. We recall that, in the reference configuration, the elastic body occupies an open bounded connected set $\Omega \subset \mathbb{R}^d$ with boundary $\Gamma = \partial\Omega$, assumed to be Lipschitz continuous. It is also assumed that Γ consists of three sets $\overline{\Gamma}_D$, $\overline{\Gamma}_N$, and $\overline{\Gamma}_C$, with mutually disjoint relatively open sets Γ_D, Γ_N, and Γ_C, such that meas$\,(\Gamma_D) > 0$. The classical model for the process is as follows.

S. Migórski et al., *Nonlinear Inclusions and Hemivariational Inequalities*, Advances in Mechanics and Mathematics 26, DOI 10.1007/978-1-4614-4232-5_7,

Problem 7.1. *Find a displacement field* $\boldsymbol{u}\colon \Omega \to \mathbb{R}^d$ *and a stress field* $\boldsymbol{\sigma}\colon \Omega \to \mathbb{S}^d$ *such that*

$$\operatorname{Div}\boldsymbol{\sigma} + \boldsymbol{f}_0 = \boldsymbol{0} \quad \text{in } \Omega, \tag{7.1}$$

$$\boldsymbol{\sigma} = \mathcal{F}\boldsymbol{\varepsilon}(\boldsymbol{u}) \quad \text{in } \Omega, \tag{7.2}$$

$$\boldsymbol{u} = \boldsymbol{0} \quad \text{on } \Gamma_D, \tag{7.3}$$

$$\boldsymbol{\sigma}\boldsymbol{\nu} = \boldsymbol{f}_N \quad \text{on } \Gamma_N, \tag{7.4}$$

$$-\sigma_\nu \in \partial j_\nu(u_\nu - g_0) \quad \text{on } \Gamma_C, \tag{7.5}$$

$$-\boldsymbol{\sigma}_\tau \in h_\tau(u_\nu - g_0)\,\partial j_\tau(\boldsymbol{u}_\tau) \quad \text{on } \Gamma_C. \tag{7.6}$$

We describe now problem (7.1)–(7.6) and provide explanation of the equations and the boundary conditions. First, (7.1) is the equilibrium equation, (6.3), and we use it here since the process is time-independent. Next, (7.2) represents the elastic constitutive law (6.9). Recall that $\boldsymbol{\varepsilon}(\boldsymbol{u})$ denotes the linearized strain tensor and $\mathcal{F}$ is the elasticity operator, assumed to be nonlinear. Conditions (7.3) and (7.4) are the displacement and traction boundary conditions, see (6.6) and (6.7), respectively. Condition (7.5) represents the contact condition (6.35) and, finally, condition (7.6) is the friction law (6.69). Recall that, here, h_τ, j_ν, and j_τ are prescribed functions and ∂g denotes the Clarke subdifferential of the function g.

In the study of Problem 7.1 we assume that the elasticity operator $\mathcal{F}$ satisfies

$$\left.\begin{array}{l}
\mathcal{F}\colon \Omega \times \mathbb{S}^d \to \mathbb{S}^d \text{ is such that}\\[1ex]
\quad \text{(a) } \mathcal{F}(\cdot,\boldsymbol{\varepsilon}) \text{ is measurable on } \Omega \text{ for all } \boldsymbol{\varepsilon} \in \mathbb{S}^d.\\[1ex]
\quad \text{(b) } \mathcal{F}(\boldsymbol{x},\cdot) \text{ is continuous on } \mathbb{S}^d \text{ for a.e. } \boldsymbol{x} \in \Omega.\\[1ex]
\quad \text{(c) } \|\mathcal{F}(\boldsymbol{x},\boldsymbol{\varepsilon})\|_{\mathbb{S}^d} \le a_0(\boldsymbol{x}) + a_1\|\boldsymbol{\varepsilon}\|_{\mathbb{S}^d} \text{ for all } \boldsymbol{\varepsilon} \in \mathbb{S}^d,\\
\qquad\quad \text{a.e. } \boldsymbol{x} \in \Omega \text{ with } a_0 \in L^2(\Omega), a_0 \ge 0, a_1 > 0.\\[1ex]
\quad \text{(d) } \mathcal{F}(\boldsymbol{x},\boldsymbol{\varepsilon}) : \boldsymbol{\varepsilon} \ge \widehat{\alpha}\|\boldsymbol{\varepsilon}\|^2_{\mathbb{S}^d} \text{ for all } \boldsymbol{\varepsilon} \in \mathbb{S}^d, \text{ a.e. } \boldsymbol{x} \in \Omega\\
\qquad\quad \text{with } \widehat{\alpha} > 0.\\[1ex]
\quad \text{(e) } (\mathcal{F}(\boldsymbol{x},\boldsymbol{\varepsilon}_1) - \mathcal{F}(\boldsymbol{x},\boldsymbol{\varepsilon}_2)) : (\boldsymbol{\varepsilon}_1 - \boldsymbol{\varepsilon}_2) \ge 0 \text{ for all } \boldsymbol{\varepsilon}_1, \boldsymbol{\varepsilon}_2 \in \mathbb{S}^d,\\
\qquad\quad \text{a.e. } \boldsymbol{x} \in \Omega.
\end{array}\right\} \tag{7.7}$$

The functions j_ν, j_τ, and h_τ satisfy

$$\left.\begin{array}{l}
j_\nu\colon \Gamma_C \times \mathbb{R} \to \mathbb{R} \text{ is such that} \\
\text{(a) } j_\nu(\cdot, r) \text{ is measurable on } \Gamma_C \text{ for all } r \in \mathbb{R} \text{ and there exists} \\
\qquad e_1 \in L^2(\Gamma_C) \text{ such that } j_\nu(\cdot, e_1(\cdot)) \in L^1(\Gamma_C). \\
\text{(b) } j_\nu(\boldsymbol{x}, \cdot) \text{ is locally Lipschitz on } \mathbb{R} \text{ for a.e. } \boldsymbol{x} \in \Gamma_C. \\
\text{(c) } |\partial j_\nu(\boldsymbol{x}, r)| \le c_{0\nu} + c_{1\nu}|r| \text{ for all } r \in \mathbb{R}, \text{ a.e. } \boldsymbol{x} \in \Gamma_C \\
\qquad \text{with } c_{0\nu}, c_{1\nu} \ge 0. \\
\text{(d) } j_\nu^0(\boldsymbol{x}, r; -r) \le d_\nu\,(1 + |r|) \text{ for all } r \in \mathbb{R}, \text{ a.e. } \boldsymbol{x} \in \Gamma_C \\
\qquad \text{with } d_\nu \ge 0.
\end{array}\right\} \tag{7.8}$$

$$\left.\begin{array}{l}
j_\tau\colon \Gamma_C \times \mathbb{R}^d \to \mathbb{R} \text{ is such that} \\
\text{(a) } j_\tau(\cdot, \boldsymbol{\xi}) \text{ is measurable on } \Gamma_C \text{ for all } \boldsymbol{\xi} \in \mathbb{R}^d \text{ and there exists} \\
\qquad \mathbf{e}_2 \in L^2(\Gamma_C; \mathbb{R}^d) \text{ such that } j_\tau(\cdot, \mathbf{e}_2(\cdot)) \in L^1(\Gamma_C). \\
\text{(b) } j_\tau(\boldsymbol{x}, \cdot) \text{ is locally Lipschitz on } \mathbb{R}^d \text{ for a.e. } \boldsymbol{x} \in \Gamma_C. \\
\text{(c) } \|\partial j_\tau(\boldsymbol{x}, \boldsymbol{\xi})\|_{\mathbb{R}^d} \le c_{0\tau} + c_{1\tau}\|\boldsymbol{\xi}\|_{\mathbb{R}^d} \text{ for all } \boldsymbol{\xi} \in \mathbb{R}^d, \text{ a.e. } \boldsymbol{x} \in \Gamma_C \\
\qquad \text{with } c_{0\tau}, c_{1\tau} \ge 0. \\
\text{(d) } j_\tau^0(\boldsymbol{x}, \boldsymbol{\xi}; -\boldsymbol{\xi}) \le d_\tau\,(1 + \|\boldsymbol{\xi}\|_{\mathbb{R}^d}) \text{ for all } \boldsymbol{\xi} \in \mathbb{R}^d, \text{ a.e. } \boldsymbol{x} \in \Gamma_C \\
\qquad \text{with } d_\tau \ge 0.
\end{array}\right\} \tag{7.9}$$

$$\left.\begin{array}{l}
h_\tau\colon \Gamma_C \times \mathbb{R} \to \mathbb{R} \text{ is such that} \\
\text{(a) } h_\tau(\cdot, r) \text{ is measurable on } \Gamma_C \text{ for all } r \in \mathbb{R}. \\
\text{(b) } h_\tau(\boldsymbol{x}, \cdot) \text{ is continuous on } \mathbb{R} \text{ for a.e. } \boldsymbol{x} \in \Gamma_C. \\
\text{(c) } 0 \le h_\tau(\boldsymbol{x}, r) \le \overline{h}_\tau \text{ for all } r \in \mathbb{R}, \text{ a.e. } \boldsymbol{x} \in \Gamma_C \text{ with } \overline{h}_\tau > 0.
\end{array}\right\} \tag{7.10}$$

The forces and traction densities satisfy

$$\boldsymbol{f}_0 \in L^2(\Omega; \mathbb{R}^d), \quad \boldsymbol{f}_N \in L^2(\Gamma_N; \mathbb{R}^d), \tag{7.11}$$

whereas the gap satisfies

$$g_0 \in L^\infty(\Gamma_C), \quad g_0 \ge 0 \text{ a.e. on } \Gamma_C. \tag{7.12}$$

To present the variational formulation of Problem 7.1 we use the spaces

$$H = L^2(\Omega; \mathbb{R}^d), \qquad \mathcal{H} = \{\, \boldsymbol{\tau} = (\tau_{ij}) \mid \tau_{ij} = \tau_{ji} \in L^2(\Omega) \,\} = L^2(\Omega; \mathbb{S}^d),$$

$$\mathcal{H}_1 = \{\, \boldsymbol{\tau} \in \mathcal{H} \mid \operatorname{Div} \boldsymbol{\tau} \in H \,\},$$

introduced in Sect. 2.3. Recall that these are real Hilbert spaces with the inner products defined on page 35. For the displacement field we also use the space

$$V = \{\, \boldsymbol{v} \in H^1(\Omega;\mathbb{R}^d) \mid \boldsymbol{v} = \boldsymbol{0} \text{ on } \Gamma_D \,\},$$

which is a closed subspace of $H^1(\Omega;\mathbb{R}^d)$. On V we consider the inner product and the corresponding norm given by

$$\langle \boldsymbol{u}, \boldsymbol{v}\rangle_V = \langle \boldsymbol{\varepsilon}(\boldsymbol{u}), \boldsymbol{\varepsilon}(\boldsymbol{v})\rangle_{\mathcal{H}}, \quad \|\boldsymbol{v}\|_V = \|\boldsymbol{\varepsilon}(\boldsymbol{v})\|_{\mathcal{H}} \text{ for all } \boldsymbol{u}, \boldsymbol{v} \in V.$$

Since $\text{meas}(\Gamma_D) > 0$, it follows from the Korn inequality that $(V, \langle\cdot,\cdot\rangle_V)$ is a Hilbert space. Moreover, it is well known that the inclusions $V \subset H \subset V^*$ are continuous and compact where, here and below, V^* denotes the dual space of V.

We turn now to the variational formulation of the contact problem (7.1)–(7.6). To this end, we assume in what follows that $\boldsymbol{u}$ and $\boldsymbol{\sigma}$ are sufficiently smooth functions which solve (7.1)–(7.6). Let $\boldsymbol{v} \in V$. We use the equilibrium (7.1) and the Green formula (2.7) to find that

$$\langle \boldsymbol{\sigma}, \boldsymbol{\varepsilon}(\boldsymbol{v})\rangle_{\mathcal{H}} = \int_\Omega \boldsymbol{f}_0 \cdot \boldsymbol{v}\, dx + \int_\Gamma \boldsymbol{\sigma\nu} \cdot \boldsymbol{v}\, d\Gamma. \tag{7.13}$$

Next, we take into account the fact that $\boldsymbol{v} = \boldsymbol{0}$ on Γ_D, the traction boundary condition (7.4) and identity (6.33) to see that

$$\int_\Gamma \boldsymbol{\sigma\nu} \cdot \boldsymbol{v}\, d\Gamma = \int_{\Gamma_N} \boldsymbol{f}_N \cdot \boldsymbol{v}\, d\Gamma + \int_{\Gamma_C} (\sigma_\nu v_\nu + \boldsymbol{\sigma}_\tau \cdot \boldsymbol{v}_\tau)\, d\Gamma. \tag{7.14}$$

On the other hand, from Definition 3.22 of the Clarke subdifferential combined with the inclusions (7.5) and (7.6), we have

$$-\sigma_\nu v_\nu \le j_\nu^0(u_\nu - g_0; v_\nu), \quad -\boldsymbol{\sigma}_\tau \cdot \boldsymbol{v}_\tau \le h_\tau(u_\nu - g_0)\, j_\tau^0(\boldsymbol{u}_\tau; \boldsymbol{v}_\tau) \text{ on } \Gamma_C,$$

which imply that

$$\int_{\Gamma_C} (\sigma_\nu v_\nu + \boldsymbol{\sigma}_\tau \cdot \boldsymbol{v}_\tau)\, d\Gamma \ge -\int_{\Gamma_C} \Big(j_\nu^0(u_\nu - g_0; v_\nu) + h_\tau(u_\nu - g_0)\, j_\tau^0(\boldsymbol{u}_\tau; \boldsymbol{v}_\tau) \Big)\, d\Gamma. \tag{7.15}$$

Consider the element $\boldsymbol{f} \in V^*$ given by

$$\langle \boldsymbol{f}, \boldsymbol{v}\rangle_{V^*\times V} = \langle \boldsymbol{f}_0, \boldsymbol{v}\rangle_H + \langle \boldsymbol{f}_N, \boldsymbol{v}\rangle_{L^2(\Gamma_N;\mathbb{R}^d)} \quad \text{for all } \boldsymbol{v} \in V. \tag{7.16}$$

We combine (7.13)–(7.16) to see that

$$\langle \boldsymbol{\sigma}, \boldsymbol{\varepsilon}(\boldsymbol{v})\rangle_{\mathcal{H}} + \int_{\Gamma_C} \Big(j_\nu^0(u_\nu - g_0; v_\nu) + h_\tau(u_\nu - g_0)\, j_\tau^0(\boldsymbol{u}_\tau; \boldsymbol{v}_\tau)\Big) d\Gamma$$

$$\geq \langle \boldsymbol{f}, \boldsymbol{v}\rangle_{V^*\times V} \quad \text{for all } \boldsymbol{v} \in V. \tag{7.17}$$

Finally, we substitute (7.2) in (7.17) to derive the following variational formulation of Problem 7.1, in terms of displacement field.

Problem 7.2. *Find a displacement field $\boldsymbol{u} \in V$ such that*

$$\langle \mathcal{F}\boldsymbol{\varepsilon}(\boldsymbol{u}), \boldsymbol{\varepsilon}(\boldsymbol{v})\rangle_{\mathcal{H}} + \int_{\Gamma_C} \Big(j_\nu^0(u_\nu - g_0; v_\nu) + h_\tau(u_\nu - g_0)\, j_\tau^0(\boldsymbol{u}_\tau; \boldsymbol{v}_\tau)\Big) d\Gamma$$

$$\geq \langle \boldsymbol{f}, \boldsymbol{v}\rangle_{V^*\times V} \quad \text{for all } \boldsymbol{v} \in V. \tag{7.18}$$

Our main results in the study of Problem 7.2 concern the existence and uniqueness of the solution and are presented in Theorems 7.3 and 7.5, respectively. In order to state these results, as in Chap. 4, we consider the space $Z = H^\delta(\Omega;\mathbb{R}^d)$ with $\delta \in (1/2, 1)$ and we denote by $\gamma\colon Z \to L^2(\Gamma;\mathbb{R}^d)$ the trace operator, by $\|\gamma\|$ its norm in $\mathcal{L}(Z, L^2(\Gamma;\mathbb{R}^d))$, and by $c_e > 0$ the embedding constant of V into Z.

Theorem 7.3. *Assume that* (7.7) *and* (7.10)–(7.12) *are satisfied. Assume, in addition, that one of the following hypotheses:*

(i) (7.8)(a)–(c), (7.9)(a)–(c), *and* $\widehat{\alpha} > \sqrt{3}\left(c_{1\nu} + c_{1\tau}\overline{h}_\tau\right) c_e^2 \|\gamma\|^2$
(ii) (7.8) *and* (7.9)

holds, and

$$\left.\begin{array}{l} \text{either } j_\nu(\boldsymbol{x},\cdot) \text{ and } j_\tau(\boldsymbol{x},\cdot) \text{ are regular} \\ \text{or } -j_\nu(\boldsymbol{x},\cdot) \text{ and } -j_\tau(\boldsymbol{x},\cdot) \text{ are regular} \end{array}\right\} \tag{7.19}$$

for a.e. $\boldsymbol{x} \in \Gamma_C$. *Then Problem 7.2 has at least one solution.*

Proof. The existence of a solution to Problem 7.2 follows from Corollary 4.18 on page 113. To provide it we introduce the operator $A\colon V \to V^*$ defined by

$$\langle A\boldsymbol{u}, \boldsymbol{v}\rangle_{V^*\times V} = \langle \mathcal{F}\boldsymbol{\varepsilon}(\boldsymbol{u}), \boldsymbol{\varepsilon}(\boldsymbol{v})\rangle_{\mathcal{H}} \text{ for } \boldsymbol{u}, \boldsymbol{v} \in V. \tag{7.20}$$

We claim that under hypothesis (7.7), A is pseudomonotone and coercive. In fact, by (7.7)(c) and the Hölder inequality, we have

$$\begin{aligned} |\langle A\boldsymbol{u}, \boldsymbol{v}\rangle_{V^*\times V}| &\leq \int_\Omega \|\mathcal{F}\boldsymbol{\varepsilon}(\boldsymbol{u})\|_{\mathbb{S}^d} \|\boldsymbol{\varepsilon}(\boldsymbol{v})\|_{\mathbb{S}^d}\, dx \\ &\leq \left(\int_\Omega \left(a_0(\boldsymbol{x}) + a_1\|\boldsymbol{\varepsilon}(\boldsymbol{u})\|_{\mathbb{S}^d}\right)^2 dx\right)^{1/2} \|\boldsymbol{v}\|_V \\ &\leq \sqrt{2}\left(\|a_0\|_{L^2(\Omega)} + a_1\|\boldsymbol{u}\|_V\right)\|\boldsymbol{v}\|_V \end{aligned}$$

for all $\boldsymbol{u}, \boldsymbol{v} \in V$. This gives $\|A\boldsymbol{u}\|_{V^*} \le \sqrt{2}\left(\|a_0\|_{L^2(\Omega)} + a_1\|\boldsymbol{u}\|_V\right)$ for all $\boldsymbol{u} \in V$ and implies the boundedness of A.

We show that A is monotone and continuous. To this end we use (7.7)(e) to see that

$$\langle A\boldsymbol{u}_1 - A\boldsymbol{u}_2, \boldsymbol{u}_1 - \boldsymbol{u}_2\rangle_{V^*\times V} = \int_\Omega (\mathcal{F}\boldsymbol{\varepsilon}(\boldsymbol{u}_1) - \mathcal{F}\boldsymbol{\varepsilon}(\boldsymbol{u}_2)) : (\boldsymbol{\varepsilon}(\boldsymbol{u}_1) - \boldsymbol{\varepsilon}(\boldsymbol{u}_2))\, dx \ge 0$$

for all $\boldsymbol{u}_1, \boldsymbol{u}_2 \in V$, which shows that A is monotone. Next, let $\boldsymbol{u}_n \to \boldsymbol{u}$ in V which implies that $\boldsymbol{\varepsilon}(\boldsymbol{u}_n) \to \boldsymbol{\varepsilon}(\boldsymbol{u})$ in $L^2(\Omega;\mathbb{S}^d)$. By Theorem 2.39 there exist a subsequence $\{\boldsymbol{u}_{n_k}\}$ and a function $w \in L^2(\Omega)$ such that $\boldsymbol{\varepsilon}(\boldsymbol{u}_{n_k})(\boldsymbol{x}) \to \boldsymbol{\varepsilon}(\boldsymbol{u})(\boldsymbol{x})$ in $\mathbb{S}^d$ for a.e. $\boldsymbol{x} \in \Omega$, as $n_k \to \infty$, and $\|\boldsymbol{\varepsilon}(\boldsymbol{u}_{n_k})(\boldsymbol{x})\|_{\mathbb{S}^d} \le w(\boldsymbol{x})$ for a.e. $\boldsymbol{x} \in \Omega$ and $k \in \mathbb{N}$. Since $\mathcal{F}(\boldsymbol{x},\cdot)$ is continuous on $\mathbb{S}^d$, we have

$$\mathcal{F}(\boldsymbol{x}, \boldsymbol{\varepsilon}(\boldsymbol{u}_{n_k})(\boldsymbol{x})) \to \mathcal{F}(\boldsymbol{x}, \boldsymbol{\varepsilon}(\boldsymbol{u})(\boldsymbol{x})) \quad \text{in } \mathbb{S}^d$$

for a.e. $\boldsymbol{x} \in \Gamma_C$ and, consequently,

$$\|\mathcal{F}(\boldsymbol{x}, \boldsymbol{\varepsilon}(\boldsymbol{u}_{n_k})(\boldsymbol{x})) - \mathcal{F}(\boldsymbol{x}, \boldsymbol{\varepsilon}(\boldsymbol{u})(\boldsymbol{x}))\|^2_{\mathbb{S}^d} \to 0$$

a.e. $\boldsymbol{x} \in \Omega$, as $n_k \to \infty$. By hypothesis (7.7)(c), we get

$$\begin{aligned}
&\|\mathcal{F}(\boldsymbol{x}, \boldsymbol{\varepsilon}(\boldsymbol{u}_{n_k})(\boldsymbol{x})) - \mathcal{F}(\boldsymbol{x}, \boldsymbol{\varepsilon}(\boldsymbol{u})(\boldsymbol{x}))\|^2_{\mathbb{S}^d} \\
&\quad \le 2\left(a_0(\boldsymbol{x}) + a_1\|\boldsymbol{\varepsilon}(\boldsymbol{u}_{n_k})(\boldsymbol{x})\|_{\mathbb{S}^d}\right)^2 + 2\left(a_0(\boldsymbol{x}) + a_1\|\boldsymbol{\varepsilon}(\boldsymbol{u})(\boldsymbol{x})\|_{\mathbb{S}^d}\right)^2 \\
&\quad \le 8a_0^2(\boldsymbol{x}) + 4a_1^2\left(w^2(\boldsymbol{x}) + \|\boldsymbol{\varepsilon}(\boldsymbol{u})(\boldsymbol{x})\|^2_{\mathbb{S}^d}\right)
\end{aligned}$$

a.e. $\boldsymbol{x} \in \Omega$. Hence, by the Lebesgue-dominated convergence theorem (Theorem 2.38 on page 42) we obtain

$$\|\mathcal{F}\boldsymbol{\varepsilon}(\boldsymbol{u}_{n_k}) - \mathcal{F}\boldsymbol{\varepsilon}(\boldsymbol{u})\|^2_{\mathcal{H}} = \int_\Omega \|\mathcal{F}\boldsymbol{\varepsilon}(\boldsymbol{u}_{n_k}) - \mathcal{F}\boldsymbol{\varepsilon}(\boldsymbol{u})\|^2_{\mathbb{S}^d}\, dx \to 0, \quad \text{as } n_k \to \infty.$$

On the other hand, by the Hölder inequality, we have

$$\begin{aligned}
\langle A\boldsymbol{u}_{n_k} - A\boldsymbol{u}, \boldsymbol{v}\rangle_{V^*\times V} &= \int_\Omega \left(\mathcal{F}\boldsymbol{\varepsilon}(\boldsymbol{u}_{n_k}) - \mathcal{F}\boldsymbol{\varepsilon}(\boldsymbol{u})\right) : \boldsymbol{\varepsilon}(\boldsymbol{v})\, dx \\
&\le \|\mathcal{F}\boldsymbol{\varepsilon}(\boldsymbol{u}_{n_k}) - \mathcal{F}\boldsymbol{\varepsilon}(\boldsymbol{u})\|_{\mathcal{H}}\|\boldsymbol{\varepsilon}(\boldsymbol{v})\|_{\mathcal{H}} \quad \text{for all } \boldsymbol{v} \in V.
\end{aligned}$$

We conclude from here that $A\boldsymbol{u}_{n_k} \to A\boldsymbol{u}$ in V^*. Next, Proposition 1.14 implies that $A\boldsymbol{u}_n \to A\boldsymbol{u}$ in V^*, which shows that A is continuous.

It follows from above that the operator A is bounded, monotone, and hemicontinuous. From Theorem 3.69(i), we deduce that A is pseudomonotone. The coercivity of A immediately follows from (7.7)(d), i.e.

$$\langle A\boldsymbol{u}, \boldsymbol{u}\rangle_{V^*\times V} = \int_\Omega \mathcal{F}\boldsymbol{\varepsilon}(\boldsymbol{u}) : \boldsymbol{\varepsilon}(\boldsymbol{u})\, dx \ge \widehat{\alpha}\int_\Omega \|\boldsymbol{\varepsilon}(\boldsymbol{u})\|^2_{\mathbb{S}^d}\, dx = \widehat{\alpha}\,\|\boldsymbol{u}\|^2_V.$$

Hence, the operator A satisfies the hypotheses (4.1).

Now we consider the functions $j_1(\boldsymbol{x},\boldsymbol{\xi}) = j_\nu(\boldsymbol{x},\xi_\nu - g_0(\boldsymbol{x}))$, $h_1 \equiv 1$, $j_2(\boldsymbol{x},\boldsymbol{\xi}) = j_\tau(\boldsymbol{x},\boldsymbol{\xi}_\tau)$ and $h_2(\boldsymbol{x},\boldsymbol{\eta}) = h_\tau(\boldsymbol{x},\eta_\nu - g_0(\boldsymbol{x}))$ for $\boldsymbol{x} \in \Gamma_C$, $\boldsymbol{\xi}$, $\boldsymbol{\eta} \in \mathbb{R}^d$. It is clear that h_1 and h_2 satisfy (4.40). We verify the hypothesis (4.41) with $s = d$. To this end, we observe that

$$j_1(\boldsymbol{x},\boldsymbol{\xi}) = j_\nu(\boldsymbol{x}, N_\nu\boldsymbol{\xi} - g_0(\boldsymbol{x})), \quad j_2(\boldsymbol{x},\boldsymbol{\xi}) = j_\tau(\boldsymbol{x}, N_\tau\boldsymbol{\xi}),$$

where the operators $N_\nu \in L^\infty(\Gamma;\mathcal{L}(\mathbb{R}^d,\mathbb{R}))$, $N_\tau \in L^\infty(\Gamma;\mathcal{L}(\mathbb{R}^d,\mathbb{R}^d))$ are given by $N_\nu\boldsymbol{\xi} = \xi_\nu = \boldsymbol{\xi}\cdot\boldsymbol{\nu}(\boldsymbol{x})$ and $N_\tau\boldsymbol{\xi} = \boldsymbol{\xi}_\tau = \boldsymbol{\xi} - \xi_\nu\boldsymbol{\nu}(\boldsymbol{x})$ for $\boldsymbol{\xi} \in \mathbb{R}^d$, respectively. The operators N_ν, N_τ as well as the vector $\boldsymbol{\nu}$ depend on $\boldsymbol{x} \in \Gamma$ but, for simplicity of notation, in what follows we skip their dependence on $\boldsymbol{x}$. We note that j_1 and j_2 not only depend on the variable $\boldsymbol{x} \in \Gamma_C$ explicitly but also implicitly, via their dependence on N_ν and N_τ. We recall that the adjoint operators $N_\nu^* \in L^\infty(\Gamma;\mathcal{L}(\mathbb{R},\mathbb{R}^d))$ and $N_\tau^* \in L^\infty(\Gamma;\mathcal{L}(\mathbb{R}^d,\mathbb{R}^d))$ are given by $N_\nu^* r = r\,\boldsymbol{\nu}$ and $N_\tau^*\boldsymbol{\xi} = \boldsymbol{\xi}_\tau$ for $r \in \mathbb{R}$ and $\boldsymbol{\xi} \in \mathbb{R}^d$, respectively. By the hypotheses on j_ν and j_τ, it is clear that $j_i(\cdot,\boldsymbol{\xi})$, $i = 1, 2$ are measurable on Γ_C for all $\boldsymbol{\xi} \in \mathbb{R}^d$ and $j_i(\boldsymbol{x},\cdot)$, $i = 1, 2$ are locally Lipschitz on $\mathbb{R}^d$ for a.e. $\boldsymbol{x} \in \Gamma_C$. Moreover, by Corollary 3.48 applied to j_ν and j_τ, we have $j_1(\cdot,\mathbf{e}(\cdot))$, $j_2(\cdot,\mathbf{e}(\cdot)) \in L^1(\Gamma_C)$ for all $\mathbf{e} \in L^2(\Gamma_C;\mathbb{R}^d)$, which implies (4.41)(a). From Proposition 3.37, we obtain

$$j_1^0(\boldsymbol{x},\boldsymbol{\xi};\boldsymbol{\varrho}) \le j_\nu^0(\boldsymbol{x},\xi_\nu - g_0(\boldsymbol{x});\varrho_\nu), \quad j_2^0(\boldsymbol{x},\boldsymbol{\xi};\boldsymbol{\varrho}) \le j_\tau^0(\boldsymbol{x},\boldsymbol{\xi}_\tau;\boldsymbol{\varrho}_\tau),$$

$$\partial j_1(\boldsymbol{x},\boldsymbol{\xi}) \subseteq \partial j_\nu(\boldsymbol{x},\xi_\nu - g_0(\boldsymbol{x}))\,\boldsymbol{\nu}, \quad \partial j_2(\boldsymbol{x},\boldsymbol{\xi}) \subseteq [\partial j_\tau(\boldsymbol{x},\boldsymbol{\xi}_\tau)]_\tau$$

for all $\boldsymbol{\xi}$, $\boldsymbol{\varrho} \in \mathbb{R}^d$, a.e. $\boldsymbol{x} \in \Gamma_C$. Moreover, using the properties of the generalized directional derivative in Proposition 3.23(i) and (iii), we get

$$\begin{aligned} j_1^0(\boldsymbol{x},\boldsymbol{\xi};-\boldsymbol{\xi}) &\le j_\nu^0(\boldsymbol{x},\xi_\nu - g_0(\boldsymbol{x});-\xi_\nu) \\ &\le j_\nu^0(\boldsymbol{x},\xi_\nu - g_0(\boldsymbol{x});-(\xi_\nu - g_0(\boldsymbol{x}))) + j_\nu^0(\boldsymbol{x},\xi_\nu - g_0(\boldsymbol{x});-g_0(\boldsymbol{x})) \\ &\le d_\nu(1 + |\xi_\nu - g_0(\boldsymbol{x})|) + \max\{\zeta(-g_0(\boldsymbol{x})) \mid \zeta \in \partial j_\nu(\boldsymbol{x},\xi_\nu - g_0(\boldsymbol{x}))\} \\ &\le d_\nu(1 + |g_0(\boldsymbol{x})| + \|\boldsymbol{\xi}\|_{\mathbb{R}^d}) + |g_0(\boldsymbol{x})|\,(c_{0\nu} + c_{1\nu}|\xi_\nu - g_0(\boldsymbol{x})|) \\ &\le d_{01}\,(1 + \|\boldsymbol{\xi}\|_{\mathbb{R}^d}) \end{aligned}$$

for all $\boldsymbol{\xi} \in \mathbb{R}^d$, a.e. $\boldsymbol{x} \in \Gamma_C$ with $d_{01} \ge 0$. We also have

$$j_2^0(\boldsymbol{x},\boldsymbol{\xi};-\boldsymbol{\xi}) \le j_\tau^0(\boldsymbol{x},\boldsymbol{\xi}_\tau;-\boldsymbol{\xi}_\tau) \le d_\tau\,(1 + \|\boldsymbol{\xi}_\tau\|_{\mathbb{R}^d}) \le d_\tau\,(1 + \|\boldsymbol{\xi}\|_{\mathbb{R}^d})$$

and, in addition,

$$\|\partial j_1(\boldsymbol{x},\boldsymbol{\xi})\|_{\mathbb{R}^d} = |\partial j_\nu(\boldsymbol{x},\xi_\nu - g_0(\boldsymbol{x}))| \le c_{0\nu} + c_{1\nu}|g_0(\boldsymbol{x})| + c_{1\nu}\|\boldsymbol{\xi}\|_{\mathbb{R}^d},$$

$$\|\partial j_2(\boldsymbol{x},\boldsymbol{\xi})\|_{\mathbb{R}^d} \le \|\partial j_\tau(\boldsymbol{x},\boldsymbol{\xi}_\tau)\|_{\mathbb{R}^d} \le c_{0\tau} + c_{1\tau}\|\boldsymbol{\xi}\|_{\mathbb{R}^d}$$

for all $\boldsymbol{\xi} \in \mathbb{R}^d$, a.e. $\boldsymbol{x} \in \Gamma_C$. Note that from (7.19) we infer that either $j_1(\boldsymbol{x},\cdot)$ and $j_2(\boldsymbol{x},\cdot)$ or $-j_1(\boldsymbol{x},\cdot)$ and $-j_2(\boldsymbol{x},\cdot)$ are regular, respectively, for a.e. $\boldsymbol{x} \in \Gamma_C$. Hence, both functions j_1 and j_2 satisfy the hypothesis (4.41). Now we apply Corollary 4.18 with $\boldsymbol{f} \in V^*$ given by (7.16) and deduce that Problem 7.2 has at least one solution, which ends the proof. □

The uniqueness of a solution to Problem 7.2 under the general assumption (7.10) represents an open problem which, clearly, deserves more investigation. However, it can be obtained in particular cases, for instance if h_τ is a nonnegative constant function. To present this uniqueness result we consider a version of Problem 7.2 in the case when $h_\tau(\cdot) = k_\tau$ where k_τ represents a given constant.

Problem 7.4. *Find a displacement field $\boldsymbol{u} \in V$ such that*

$$\langle \mathcal{F}\boldsymbol{\varepsilon}(\boldsymbol{u}), \boldsymbol{\varepsilon}(\boldsymbol{v})\rangle_{\mathcal{H}} + \int_{\Gamma_C} \Big(j_\nu^0(u_\nu - g_0; v_\nu) + k_\tau\, j_\tau^0(\boldsymbol{u}_\tau; \boldsymbol{v}_\tau)\Big)\, d\Gamma$$
$$\geq \langle \boldsymbol{f}, \boldsymbol{v}\rangle_{V^*\times V} \quad \text{for all } \boldsymbol{v} \in V.$$

We have the following existence and uniqueness result.

Theorem 7.5. *Assume that the conditions* (7.7)(a)–(d), (7.11), (7.12) *and* (7.19) *hold, $k_\tau \geq 0$ and*

$$\left.\begin{array}{l}
\text{(a) } (\mathcal{F}(\boldsymbol{x},\boldsymbol{\varepsilon}_1) - \mathcal{F}(\boldsymbol{x},\boldsymbol{\varepsilon}_2)) : (\boldsymbol{\varepsilon}_1 - \boldsymbol{\varepsilon}_2) \geq m_{\mathcal{F}} \|\boldsymbol{\varepsilon}_1 - \boldsymbol{\varepsilon}_2\|^2_{\mathbb{S}^d} \\
\quad \textit{for all } \boldsymbol{\varepsilon}_1, \boldsymbol{\varepsilon}_2 \in \mathbb{S}^d, \textit{ a.e. } \boldsymbol{x} \in \Omega \textit{ with } m_{\mathcal{F}} > 0. \\
\text{(b) } (\zeta_1 - \zeta_2)(r_1 - r_2) \geq -m_\nu |r_1 - r_2|^2 \textit{ for all } \zeta_i \in \partial j_\nu(\boldsymbol{x}, r_i), \\
\quad r_i \in \mathbb{R},\ i = 1,2, \textit{ a.e. } \boldsymbol{x} \in \Gamma_C \textit{ with } m_\nu \geq 0. \\
\text{(c) } (\boldsymbol{\zeta}_1 - \boldsymbol{\zeta}_2)\cdot(\boldsymbol{\xi}_1 - \boldsymbol{\xi}_2) \geq -m_\tau \|\boldsymbol{\xi}_1 - \boldsymbol{\xi}_2\|^2_{\mathbb{R}^d} \textit{ for all} \\
\quad \boldsymbol{\zeta}_i \in \partial j_\tau(\boldsymbol{x}, \boldsymbol{\xi}_i), \boldsymbol{\xi}_i \in \mathbb{R}^d,\ i = 1,2, \textit{ a.e. } \boldsymbol{x} \in \Gamma_C \textit{ with } m_\tau \geq 0. \\
\text{(d) } m_{\mathcal{F}} > \max\{m_\nu, m_\tau k_\tau\}\, c_e^2\, \|\gamma\|^2.
\end{array}\right\} \qquad (7.21)$$

Assume, in addition, that one of the following hypotheses:

(i) (7.8)(a)–(c), (7.9)(a)–(c) *and* $\widehat{\alpha} > \sqrt{3}\,(c_{1\nu} + c_{1\tau}k_\tau)\; c_e^2\, \|\gamma\|^2$
(ii) (7.8) *and* (7.9)

is satisfied. Then Problem 7.4 has a unique solution.

Proof. We apply Theorem 4.20 with the operator $A\colon V \to V^*$ defined by (7.20), the function $j\colon \Gamma_C \times \mathbb{R}^d \to \mathbb{R}$ given by

$$j(\boldsymbol{x},\boldsymbol{\xi}) = j_\nu(\boldsymbol{x}, \xi_\nu - g_0(\boldsymbol{x})) + k_\tau\, j_\tau(\boldsymbol{x}, \boldsymbol{\xi}_\tau) \quad \text{for } (\boldsymbol{x},\boldsymbol{\xi}) \in \Gamma_C \times \mathbb{R}^d, \qquad (7.22)$$

and $\boldsymbol{f} \in V^*$ defined by (7.16). To this end, we verify below the hypotheses of this theorem.

First, using arguments similar to those used in the proof of Theorem 7.3, we obtain that the operator A is pseudomonotone and coercive with constant $\alpha = \widehat{\alpha} > 0$. Also, we note that A is a strongly monotone operator. Indeed, by (7.21)(a), we obtain

$$\begin{aligned}\langle A\boldsymbol{u}_1 - A\boldsymbol{u}_2, \boldsymbol{u}_1 - \boldsymbol{u}_2\rangle_{V^*\times V} &= \int_\Omega (\mathcal{F}\boldsymbol{\varepsilon}(\boldsymbol{u}_1) - \mathcal{F}\boldsymbol{\varepsilon}(\boldsymbol{u}_2)) : (\boldsymbol{\varepsilon}(\boldsymbol{u}_1) - \boldsymbol{\varepsilon}(\boldsymbol{u}_2))\, dx \\ &\geq m_{\mathcal{F}} \int_\Omega \|\boldsymbol{\varepsilon}(\boldsymbol{u}_1) - \boldsymbol{\varepsilon}(\boldsymbol{u}_2)\|^2_{\mathbb{S}^d}\, dx = m_{\mathcal{F}}\, \|\boldsymbol{u}_1 - \boldsymbol{u}_2\|^2_V\end{aligned}$$

for all $\boldsymbol{u}_1, \boldsymbol{u}_2 \in V$. Hence, the hypothesis (4.12) holds with $m_1 = m_{\mathcal{F}} > 0$.

Next, we study the properties of the function j defined by (7.22). Similarly as in the proof of Theorem 7.3, we observe that

$$j(\boldsymbol{x}, \boldsymbol{\xi}) = j_\nu(\boldsymbol{x}, N_\nu\boldsymbol{\xi} - g_0(\boldsymbol{x})) + k_\tau\, j_\tau(\boldsymbol{x}, N_\tau\boldsymbol{\xi}) \quad \text{for } (\boldsymbol{x}, \boldsymbol{\xi}) \in \Gamma_C \times \mathbb{R}^d$$

with the operators $N_\nu \in L^\infty(\Gamma; \mathcal{L}(\mathbb{R}^d, \mathbb{R}))$, $N_\tau \in L^\infty(\Gamma; \mathcal{L}(\mathbb{R}^d, \mathbb{R}^d))$ defined by $N_\nu\boldsymbol{\xi} = \xi_\nu = \boldsymbol{\xi} \cdot \boldsymbol{\nu}(\boldsymbol{x})$ and $N_\tau\boldsymbol{\xi} = \boldsymbol{\xi}_\tau = \boldsymbol{\xi} - \xi_\nu\boldsymbol{\nu}(\boldsymbol{x})$ for $\boldsymbol{\xi} \in \mathbb{R}^d$, respectively. The operators N_ν, N_τ as well as the vector $\boldsymbol{\nu}$ depend on $\boldsymbol{x} \in \Gamma$ but, again, for simplicity of notation, we skip their dependence on $\boldsymbol{x}$. It is obvious to see that $j(\cdot, \boldsymbol{\xi})$ is measurable on Γ_C for all $\boldsymbol{\xi} \in \mathbb{R}^d$ and $j(\boldsymbol{x}, \cdot)$ is locally Lipschitz on $\mathbb{R}^d$ for a.e. $\boldsymbol{x} \in \Gamma_C$. Furthermore, applying Corollary 3.48 to the functions j_ν and j_τ, we get $j(\cdot, \mathbf{e}(\cdot)) \in L^1(\Gamma_C)$ for all $\mathbf{e} \in L^2(\Gamma_C; \mathbb{R}^d)$, i.e., (4.42)(a) holds. Also, from Proposition 3.37, we obtain

$$j^0(\boldsymbol{x}, \boldsymbol{\xi}; \boldsymbol{\varrho}) \leq j_\nu^0(\boldsymbol{x}, \xi_\nu - g_0(\boldsymbol{x}); \varrho_\nu) + k_\tau\, j_\tau^0(\boldsymbol{x}, \boldsymbol{\xi}_\tau; \boldsymbol{\varrho}_\tau),$$

$$\partial j(\boldsymbol{x}, \boldsymbol{\xi}) \subseteq \partial j_\nu(\boldsymbol{x}, \xi_\nu - g_0(\boldsymbol{x}))\, \boldsymbol{\nu} + k_\tau\, [\partial j_\tau(\boldsymbol{x}, \boldsymbol{\xi}_\tau)]_\tau$$

for all $\boldsymbol{\xi}, \boldsymbol{\varrho} \in \mathbb{R}^d$, a.e. $\boldsymbol{x} \in \Gamma_C$. Repeating the previous calculations, we get

$$\begin{aligned}\|\partial j(\boldsymbol{x}, \boldsymbol{\xi})\|_{\mathbb{R}^d} &\leq c_{0\nu} + c_{1\nu}|\xi_\nu - g_0(\boldsymbol{x})| + c_{0\tau}k_\tau + c_{1\tau}k_\tau\, \|\boldsymbol{\xi}_\tau\|_{\mathbb{R}^d} \\ &\leq (c_{0\nu} + c_{1\nu}|g_0(\boldsymbol{x})| + c_{0\tau}k_\tau) + (c_{1\nu} + c_{1\tau}\, k_\tau)\|\boldsymbol{\xi}\|_{\mathbb{R}^d}\end{aligned}$$

for all $\boldsymbol{\xi} \in \mathbb{R}^d$, a.e. $\boldsymbol{x} \in \Gamma_C$, which shows that (4.42)(c) is satisfied with $\overline{c}_1 = c_{1\nu} + c_{1\tau}\, k_\tau \geq 0$. Analogously, we obtain

$$\begin{aligned}j^0(\boldsymbol{x}, \boldsymbol{\xi}; -\boldsymbol{\xi}) &\leq d_\nu(1 + |\xi_\nu - g_0(\boldsymbol{x})|) + |g_0(\boldsymbol{x})|\, (c_{0\nu} + c_{1\nu}|\xi_\nu - g_0(\boldsymbol{x})|) \\ &\quad + d_\tau k_\tau(1 + \|\boldsymbol{\xi}_\tau\|_{\mathbb{R}^d}) \leq \overline{d}_0\, (1 + \|\boldsymbol{\xi}\|_{\mathbb{R}^d})\end{aligned}$$

for all $\boldsymbol{\xi} \in \mathbb{R}^d$, a.e. $\boldsymbol{x} \in \Gamma_C$ with $\overline{d}_0 \geq 0$. Moreover, from the regularity conditions (7.19), we infer that either $j(\boldsymbol{x},\cdot)$ or $-j(\boldsymbol{x},\cdot)$ is regular on $\mathbb{R}^d$, for a.e. $\boldsymbol{x} \in \Gamma_C$. It remains to check the relaxed monotonicity condition (4.42)(d) on j. We consider $\boldsymbol{\zeta}_i \in \partial j(\boldsymbol{x},\boldsymbol{\xi}_i)$, $i = 1, 2$, where $\boldsymbol{x} \in \Gamma_C$, $\boldsymbol{\zeta}_i, \boldsymbol{\xi}_i \in \mathbb{R}^d$. Hence, $\boldsymbol{\zeta}_i = \overline{\zeta}_i\,\boldsymbol{\nu} + [\overline{\overline{\boldsymbol{\zeta}}}_i]_\tau$ with $\overline{\zeta}_i \in \mathbb{R}$, $\overline{\overline{\boldsymbol{\zeta}}}_i \in \mathbb{R}^d$, $\overline{\zeta}_i \in \partial j_\nu(\boldsymbol{x},\xi_{i\nu} - g_0(\boldsymbol{x}))$, $\overline{\overline{\boldsymbol{\zeta}}}_i \in k_\tau\,\partial j_\tau(\boldsymbol{x},\boldsymbol{\xi}_{i\tau})$. Using (7.21)(b), (c) and the equality $\boldsymbol{\varrho}\cdot\boldsymbol{\eta}_\tau = \boldsymbol{\varrho}_\tau\cdot\boldsymbol{\eta}$ for $\boldsymbol{\varrho}, \boldsymbol{\eta} \in \mathbb{R}^d$, we have

$$\begin{aligned}(\boldsymbol{\zeta}_1 - \boldsymbol{\zeta}_2)\cdot(\boldsymbol{\xi}_1 - \boldsymbol{\xi}_2) &= (\overline{\zeta}_1 - \overline{\zeta}_2)\,\boldsymbol{\nu}\cdot(\boldsymbol{\xi}_1 - \boldsymbol{\xi}_2) + [\overline{\overline{\boldsymbol{\zeta}}}_1 - \overline{\overline{\boldsymbol{\zeta}}}_2]_\tau\cdot(\boldsymbol{\xi}_1 - \boldsymbol{\xi}_2)\\ &= (\overline{\zeta}_1 - \overline{\zeta}_2)(\xi_{1\nu} - \xi_{2\nu}) + (\overline{\overline{\boldsymbol{\zeta}}}_1 - \overline{\overline{\boldsymbol{\zeta}}}_2)\cdot(\boldsymbol{\xi}_{1\tau} - \boldsymbol{\xi}_{2\tau})\\ &\geq -m_\nu|\xi_{1\nu} - \xi_{2\nu}|^2 - m_\tau k_\tau\|\boldsymbol{\xi}_{1\tau} - \boldsymbol{\xi}_{2\tau}\|^2_{\mathbb{R}^d}\\ &\geq -\max\{m_\nu, m_\tau k_\tau\}\,\|\boldsymbol{\xi}_1 - \boldsymbol{\xi}_2\|^2_{\mathbb{R}^d},\end{aligned} \tag{7.23}$$

which proves (4.42)(d) with $m_2 = \max\{m_\nu, m_\tau k_\tau\} \geq 0$. We conclude from here that the function j, given by (7.22), satisfies (4.42).

Finally, we note that (4.43) is satisfied and the inequality in the hypothesis (i) of Theorem 4.20 holds. Theorem 7.5 is now a consequence of Theorem 4.20. □

A couple of functions $(\boldsymbol{u},\boldsymbol{\sigma})$ which satisfies (7.2) and (7.18) is called a *weak solution* of Problem 7.1. We remark that under the assumptions of Theorem 7.3 there exists at least one weak solution of Problem 7.1. To describe precisely the regularity of the weak solution, we note that the constitutive relation (7.2), assumption (7.7), and regularity $\boldsymbol{u} \in V$ show that $\boldsymbol{\sigma} \in \mathcal{H}$. Moreover, using (7.2) and (7.18), it follows that (7.17) holds for all $\boldsymbol{v} \in V$. Next, we test (7.17) with $\boldsymbol{v} = \boldsymbol{\varphi}$ where $\boldsymbol{\varphi}$ is an arbitrary element of the space $C_0^\infty(\Omega;\mathbb{R}^d) \subset V$. Since $\boldsymbol{\varphi}$ vanishes on Γ it follows that

$$\int_{\Gamma_C}\Big(j_\nu^0(u_\nu - g_0;\varphi_\nu) + h_\tau(u_\nu - g_0)\,j_\tau^0(\boldsymbol{u}_\tau;\boldsymbol{\varphi}_\tau)\Big)\,d\Gamma = 0$$

and, therefore, we obtain that

$$\langle\boldsymbol{\sigma},\boldsymbol{\varepsilon}(\boldsymbol{\varphi})\rangle_{\mathcal{H}} \geq \langle\boldsymbol{f},\boldsymbol{\varphi}\rangle_{V^*\times V}\quad\text{for all } \boldsymbol{\varphi} \in C_0^\infty(\Omega;\mathbb{R}^d).$$

We replace now $\boldsymbol{\varphi}$ by $-\boldsymbol{\varphi}$ in the previous inequality to see that

$$\langle\boldsymbol{\sigma},\boldsymbol{\varepsilon}(\boldsymbol{\varphi})\rangle_{\mathcal{H}} = \langle\boldsymbol{f},\boldsymbol{\varphi}\rangle_{V^*\times V}\quad\text{for all } \boldsymbol{\varphi} \in C_0^\infty(\Omega;\mathbb{R}^d)$$

and, using the definition (7.16) of $\boldsymbol{f}$ we deduce

$$\langle\boldsymbol{\sigma},\boldsymbol{\varepsilon}(\boldsymbol{\varphi})\rangle_{\mathcal{H}} = \langle\boldsymbol{f}_0,\boldsymbol{\varphi}\rangle_H\quad\text{for all } \boldsymbol{\varphi} \in C_0^\infty(\Omega;\mathbb{R}^d). \tag{7.24}$$

Equation (7.24) combined with the definition (2.5) of the divergence and deformation operators implies that

$$\operatorname{Div} \boldsymbol{\sigma} + \boldsymbol{f}_0 = \mathbf{0} \quad \text{in } \Omega.$$

It follows now from (7.11) that $\operatorname{Div} \boldsymbol{\sigma} \in H$. We conclude that the weak solution of Problem 7.1 satisfies $(\boldsymbol{u}, \boldsymbol{\sigma}) \in V \times \mathcal{H}_1$. In addition, we recall that under the assumptions of Theorem 7.5 the weak solution is unique.

7.2 A Viscoelastic Frictional Problem

For the problem studied in this section the process is static and the behavior of the material's is described with a viscoelastic constitutive law with long memory. Therefore, in contrast with the problem studied in Sect. 7.1, the problem studied in this section is time-dependent. We denote by $[0, T]$ the time interval of interest with $T > 0$, and we refer to the physical setting described in Sect. 6.1. Also, recall that below we use the notation $Q = \Omega \times (0, T)$, $\Sigma_D = \Gamma_D \times (0, T)$, $\Sigma_N = \Gamma_N \times (0, T)$, and $\Sigma_C = \Gamma_C \times (0, T)$. Then, the classical model for the process is as follows.

Problem 7.6. *Find a displacement field $\boldsymbol{u}\colon Q \to \mathbb{R}^d$ and a stress field $\boldsymbol{\sigma}\colon Q \to \mathbb{S}^d$ such that*

$$\operatorname{Div} \boldsymbol{\sigma}(t) + \boldsymbol{f}_0(t) = \mathbf{0} \quad \text{in } Q, \tag{7.25}$$

$$\boldsymbol{\sigma}(t) = \mathcal{B}(t, \boldsymbol{\varepsilon}(\boldsymbol{u}(t))) + \int_0^t \mathcal{C}(t-s, \boldsymbol{\varepsilon}(\boldsymbol{u}(s)))\, ds \quad \text{in } Q, \tag{7.26}$$

$$\boldsymbol{u}(t) = \mathbf{0} \quad \text{on } \Sigma_D, \tag{7.27}$$

$$\boldsymbol{\sigma}(t)\boldsymbol{\nu} = \boldsymbol{f}_N(t) \quad \text{on } \Sigma_N, \tag{7.28}$$

$$-\sigma_\nu(t) \in \partial j_\nu(t, u_\nu(t)) \quad \text{on } \Sigma_C, \tag{7.29}$$

$$-\boldsymbol{\sigma}_\tau(t) \in \partial j_\tau(t, \boldsymbol{u}_\tau(t)) \quad \text{on } \Sigma_C. \tag{7.30}$$

We provide now some comments on equations and conditions in (7.25)–(7.30). First, (7.25) is the equation of equilibrium, (6.2), and we use it since we assume that the inertial term in the equation of motion is neglected. Equation (7.26) represents the viscoelastic constitutive law with long memory, (6.28), in which $\mathcal{B}$ and $\mathcal{C}$ denote the elasticity and the relaxation operators, respectively. Conditions (7.27) and (7.28) are the displacement and the traction boundary conditions, see (6.4) and (6.5). Finally, condition (7.29) is the contact condition (6.34) with $g_0 = 0$, and (7.30) represents the frictional condition (6.68), both introduced in Sect. 6.3. Recall that here j_ν and j_τ are given functions and ∂j_ν, ∂j_τ denote the Clarke subdifferential of j_ν and j_τ with respect to their last variables.

Note that the explicit dependence of $\mathcal{B}$, j_ν, and j_τ on the time variable makes the problem more general and allows to model situations when these functions depend on the temperature, if the evolution in time of the temperature is prescribed. Note also that, even if the data and the unknowns in Problem 7.6 depend on time, no derivatives of the unknowns with respect to the time are involved in the statement of this problem. We conclude that in Problem 7.6 the time variable plays the role of a parameter and, using the terminology on page 177 we refer to this problem as a static time-dependent problem.

In the study of Problem 7.6 we consider the following assumptions on the elasticity operator $\mathcal{B}$ and on the relaxation operator $\mathcal{C}$.

$$\left.\begin{array}{l}
\mathcal{B}\colon Q \times \mathbb{S}^d \to \mathbb{S}^d \text{ is such that}\\
\text{(a) } \mathcal{B}(\cdot,\cdot,\boldsymbol{\varepsilon}) \text{ is measurable on } Q \text{ for all } \boldsymbol{\varepsilon} \in \mathbb{S}^d.\\
\text{(b) } \mathcal{B}(\boldsymbol{x},t,\cdot) \text{ is continuous on } \mathbb{S}^d \text{ for a.e. } (\boldsymbol{x},t) \in Q.\\
\text{(c) } (\mathcal{B}(\boldsymbol{x},t,\boldsymbol{\varepsilon}_1) - \mathcal{B}(\boldsymbol{x},t,\boldsymbol{\varepsilon}_2)) : (\boldsymbol{\varepsilon}_1 - \boldsymbol{\varepsilon}_2) \geq m_{\mathcal{B}} \|\boldsymbol{\varepsilon}_1 - \boldsymbol{\varepsilon}_2\|_{\mathbb{S}^d}^2\\
\qquad \text{for all } \boldsymbol{\varepsilon}_1, \boldsymbol{\varepsilon}_2 \in \mathbb{S}^d, \text{ a.e. } (\boldsymbol{x},t) \in Q \text{ with } m_{\mathcal{B}} > 0.\\
\text{(d) } \|\mathcal{B}(\boldsymbol{x},t,\boldsymbol{\varepsilon})\|_{\mathbb{S}^d} \leq a_0(\boldsymbol{x},t) + a_1\|\boldsymbol{\varepsilon}\|_{\mathbb{S}^d} \text{ for all } \boldsymbol{\varepsilon} \in \mathbb{S}^d,\\
\qquad \text{a.e. } (\boldsymbol{x},t) \in Q \text{ with } a_0 \in L^2(Q), a_0 \geq 0 \text{ and } a_1 > 0.\\
\text{(e) } \mathcal{B}(\boldsymbol{x},t,\boldsymbol{0}) = \boldsymbol{0} \text{ for a.e. } (\boldsymbol{x},t) \in Q.
\end{array}\right\} \quad (7.31)$$

$$\left.\begin{array}{l}
\mathcal{C}\colon Q \times \mathbb{S}^d \to \mathbb{S}^d \text{ is such that}\\
\text{(a) } \mathcal{C}(\boldsymbol{x},t,\boldsymbol{\varepsilon}) = c(\boldsymbol{x},t)\boldsymbol{\varepsilon} \text{ for all } \boldsymbol{\varepsilon} \in \mathbb{S}^d, \text{a.e. } (\boldsymbol{x},t) \in Q.\\
\text{(b) } c(\boldsymbol{x},t) = \big(c_{ijkl}(\boldsymbol{x},t)\big) \text{ with } c_{ijkl} = c_{jikl} = c_{lkij},\\
\qquad c_{ijkl} \in L^2(0,T;L^\infty(\Omega)).
\end{array}\right\} \quad (7.32)$$

The contact potentials j_ν and j_τ satisfy the following hypotheses.

$$\left.\begin{array}{l}
j_\nu\colon \Sigma_C \times \mathbb{R} \to \mathbb{R} \text{ is such that}\\
\text{(a) } j_\nu(\cdot,\cdot,r) \text{ is measurable on } \Sigma_C \text{ for all } r \in \mathbb{R} \text{ and there}\\
\qquad \text{exists } e_1 \in L^2(\Gamma_C) \text{ such that } j_\nu(\cdot,\cdot,e_1(\cdot)) \in L^1(\Sigma_C).\\
\text{(b) } j_\nu(\boldsymbol{x},t,\cdot) \text{ is locally Lipschitz on } \mathbb{R} \text{ for a.e. } (\boldsymbol{x},t) \in \Sigma_C.\\
\text{(c) } |\partial j_\nu(\boldsymbol{x},t,r)| \leq c_{0\nu} + c_{1\nu}|r| \text{ for all } r \in \mathbb{R}, \text{ a.e. } (\boldsymbol{x},t) \in \Sigma_C\\
\qquad \text{with } c_{0\nu}, c_{1\nu} \geq 0.\\
\text{(d) } (\zeta_1 - \zeta_2)(r_1 - r_2) \geq -m_\nu|r_1 - r_2|^2 \text{ for all } \zeta_i \in \partial j_\nu(\boldsymbol{x},t,r_i),\\
\qquad r_i \in \mathbb{R}, i = 1,2, \text{ a.e. } (\boldsymbol{x},t) \in \Sigma_C \text{ with } m_\nu \geq 0.\\
\text{(e) } j_\nu^0(\boldsymbol{x},t,r;-r) \leq d_\nu(1 + |r|) \text{ for all } r \in \mathbb{R}, \text{ a.e. } (\boldsymbol{x},t) \in \Sigma_C\\
\qquad \text{with } d_\nu \geq 0.
\end{array}\right\} \quad (7.33)$$

$$
\left.
\begin{array}{l}
j_\tau \colon \Sigma_C \times \mathbb{R}^d \to \mathbb{R} \text{ is such that} \\
\quad \text{(a) } j_\tau(\cdot,\cdot,\boldsymbol{\xi}) \text{ is measurable on } \Sigma_C \text{ for all } \boldsymbol{\xi} \in \mathbb{R}^d \text{ and there} \\
\qquad \text{exists } \mathbf{e}_2 \in L^2(\Gamma_C;\mathbb{R}^d) \text{ such that } j_\tau(\cdot,\cdot,\mathbf{e}_2(\cdot)) \in L^1(\Sigma_C). \\
\quad \text{(b) } j_\tau(\boldsymbol{x},t,\cdot) \text{ is locally Lipschitz on } \mathbb{R}^d \text{ for a.e. } (\boldsymbol{x},t) \in \Sigma_C. \\
\quad \text{(c) } \|\partial j_\tau(\boldsymbol{x},t,\boldsymbol{\xi})\|_{\mathbb{R}^d} \le c_{0\tau} + c_{1\tau}\|\boldsymbol{\xi}\|_{\mathbb{R}^d} \text{ for all } \boldsymbol{\xi} \in \mathbb{R}^d, \\
\qquad \text{a.e. } (\boldsymbol{x},t) \in \Sigma_C \text{ with } c_{0\tau}, c_{1\tau} \ge 0. \\
\quad \text{(d) } (\boldsymbol{\zeta}_1 - \boldsymbol{\zeta}_2)\cdot(\boldsymbol{\xi}_1 - \boldsymbol{\xi}_2) \ge -m_\tau \|\boldsymbol{\xi}_1 - \boldsymbol{\xi}_2\|^2_{\mathbb{R}^d} \text{ for all} \\
\qquad \boldsymbol{\zeta}_i \in \partial j_\tau(\boldsymbol{x},t,\boldsymbol{\xi}_i), \boldsymbol{\xi}_i \in \mathbb{R}^d,\ i = 1,2, \text{ a.e. } (\boldsymbol{x},t) \in \Sigma_C \\
\qquad \text{with } m_\tau \ge 0. \\
\quad \text{(e) } j_\tau^0(\boldsymbol{x},t,\boldsymbol{\xi};-\boldsymbol{\xi}) \le d_\tau\,(1 + \|\boldsymbol{\xi}\|_{\mathbb{R}^d}) \text{ for all } \boldsymbol{\xi} \in \mathbb{R}^d, \\
\qquad \text{a.e. } (\boldsymbol{x},t) \in \Sigma_C \text{ with } d_\tau \ge 0.
\end{array}
\right\} \quad (7.34)
$$

Finally, the forces and traction densities satisfy

$$
\boldsymbol{f}_0 \in L^2(0,T;L^2(\Omega;\mathbb{R}^d)), \qquad \boldsymbol{f}_N \in L^2(0,T;L^2(\Gamma_N;\mathbb{R}^d)). \tag{7.35}
$$

We turn now to the variational formulation of Problem 7.6 and, to this end, we use the spaces H, $\mathcal{H}$, $\mathcal{H}_1$, V, and V^* introduced in Sect. 7.1. We assume that $(\boldsymbol{u},\boldsymbol{\sigma})$ is a couple of sufficiently smooth functions which solve (7.25)–(7.30). Let $\boldsymbol{v} \in V$ and $t \in (0,T)$. We use the equilibrium (7.25) and the Green formula (2.7) to find that

$$
\langle \boldsymbol{\sigma}(t), \boldsymbol{\varepsilon}(\boldsymbol{v})\rangle_{\mathcal{H}} = \langle \boldsymbol{f}_0(t), \boldsymbol{v}\rangle_H + \int_\Gamma \boldsymbol{\sigma}(t)\boldsymbol{\nu}\cdot\boldsymbol{v}\,d\Gamma. \tag{7.36}
$$

We now take into account the boundary condition (7.28) and the decomposition formula (6.33) to see that

$$
\int_\Gamma \boldsymbol{\sigma}(t)\boldsymbol{\nu}\cdot\boldsymbol{v}\,d\Gamma = \int_{\Gamma_N} \boldsymbol{f}_N(t)\cdot\boldsymbol{v}\,d\Gamma + \int_{\Gamma_C} (\sigma_\nu(t)v_\nu + \boldsymbol{\sigma}_\tau(t)\cdot\boldsymbol{v}_\tau)\,d\Gamma. \tag{7.37}
$$

On the other hand, from the definition of the Clarke subdifferential and the boundary conditions (7.29) and (7.30), we have

$$
-\sigma_\nu(t)v_\nu \le j_\nu^0(t,u_\nu(t);v_\nu), \quad -\boldsymbol{\sigma}_\tau(t)\cdot\boldsymbol{v}_\tau \le j_\tau^0(t,\boldsymbol{u}_\tau(t);\boldsymbol{v}_\tau) \ \text{ on } \Gamma_C
$$

which implies that

$$
\begin{aligned}
&\int_{\Gamma_C} (\sigma_\nu(t)v_\nu + \boldsymbol{\sigma}_\tau(t)\cdot\boldsymbol{v}_\tau)\,d\Gamma \\
&\qquad \ge -\int_{\Gamma_C} \Big(j_\nu^0(t,u_\nu(t);v_\nu) + j_\tau^0(t,\boldsymbol{u}_\tau(t);\boldsymbol{v}_\tau)\Big)\,d\Gamma.
\end{aligned} \tag{7.38}
$$

For a.e. $t \in (0, T)$ consider the element $\boldsymbol{f}(t) \in V^*$ given by

$$\langle \boldsymbol{f}(t), \boldsymbol{v} \rangle_{V^* \times V} = \langle \boldsymbol{f}_0(t), \boldsymbol{v} \rangle_H + \langle \boldsymbol{f}_N(t), \boldsymbol{v} \rangle_{L^2(\Gamma_N;\mathbb{R}^d)} \tag{7.39}$$

for all $\boldsymbol{v} \in V$. We now combine (7.36)–(7.39) to see that

$$\langle \boldsymbol{\sigma}(t), \boldsymbol{\varepsilon}(\boldsymbol{v}) \rangle_{\mathcal{H}} + \int_{\Gamma_C} \left(j_\nu^0(t, u_\nu(t); v_\nu) + j_\tau^0(t, \boldsymbol{u}_\tau(t); \boldsymbol{v}_\tau) \right) d\Gamma$$

$$\geq \langle \boldsymbol{f}(t), \boldsymbol{v} \rangle_{V^* \times V} \quad \text{for all } \boldsymbol{v} \in V, \text{ a.e. } t \in (0, T). \tag{7.40}$$

We use now (7.40) and the constitutive law (7.26) to obtain the following variational formulation of Problem 7.6.

Problem 7.7. *Find a displacement field* $\boldsymbol{u} \colon (0, T) \to V$ *such that* $\boldsymbol{u} \in L^2(0, T; V)$ *and*

$$\langle \mathcal{B}(t, \varepsilon(\boldsymbol{u}(t))) + \int_0^t \mathcal{C}(t - s, \varepsilon(\boldsymbol{u}(s)))\, ds, \varepsilon(\boldsymbol{v}) \rangle_{\mathcal{H}}$$

$$+ \int_{\Gamma_C} \left(j_\nu^0(t, u_\nu(t); v_\nu) + j_\tau^0(t, \boldsymbol{u}_\tau(t); \boldsymbol{v}_\tau) \right) d\Gamma$$

$$\geq \langle \boldsymbol{f}(t), \boldsymbol{v} \rangle_{V^* \times V} \quad \text{for all } \boldsymbol{v} \in V, \text{ a.e. } t \in (0, T). \tag{7.41}$$

As usual, in this section we consider the trace operator $\gamma \colon Z \to L^2(\Gamma; \mathbb{R}^d)$ where $Z = H^\delta(\Omega; \mathbb{R}^d)$ with $\delta \in (1/2, 1)$ and its adjoint operator $\gamma^* \colon L^2(\Gamma; \mathbb{R}^d) \to Z^*$. We also denote by $\|\gamma\|$ the norm of γ in $\mathcal{L}(Z, L^2(\Gamma; \mathbb{R}^d))$ and by $c_e > 0$ the embedding constant of V into Z. Then, our main result in the study of Problem 7.7 is the following.

Theorem 7.8. *Assume that hypotheses* (7.31), (7.32), *and* (7.35) *hold. If one of the following hypotheses:*

(i) (7.33)(a)–(d), (7.34)(a)–(d) *and* $m_{\mathcal{B}} > \sqrt{3}\,(c_{1\nu} + c_{1\tau})\, c_e^2\, \|\gamma\|^2$
(ii) (7.33), (7.34)

is satisfied and

$$m_{\mathcal{B}} > \max\{m_\nu, m_\tau\}\, c_e^2\, \|\gamma\|^2 \tag{7.42}$$

holds, then Problem 7.7 has at least one solution $\boldsymbol{u} \in L^2(0, T; V)$. *If, in addition,*

$$\left.\begin{array}{l} \textit{either } \ j_\nu(\boldsymbol{x}, t, \cdot) \textit{ and } j_\tau(\boldsymbol{x}, t, \cdot) \textit{ are regular} \\ \textit{or } \ -j_\nu(\boldsymbol{x}, t, \cdot) \textit{ and } -j_\tau(\boldsymbol{x}, t, \cdot) \textit{ are regular} \end{array}\right\} \tag{7.43}$$

for a.e. $(\boldsymbol{x}, t) \in \Sigma_C$, *then the solution of Problem 7.7 is unique.*

Proof. The proof is based on Theorem 4.22 in Sect. 4.3. First, we introduce the operators $A, C: (0,T) \times V \to V^*$ defined by

$$\langle A(t,\boldsymbol{u}), \boldsymbol{v}\rangle_{V^*\times V} = \langle \mathcal{B}(t,\boldsymbol{\varepsilon}(\boldsymbol{u})), \boldsymbol{\varepsilon}(\boldsymbol{v})\rangle_{\mathcal{H}}, \tag{7.44}$$

$$\langle C(t,\boldsymbol{u}), \boldsymbol{v}\rangle_{V^*\times V} = \langle \mathcal{C}(t,\boldsymbol{\varepsilon}(\boldsymbol{u})), \boldsymbol{\varepsilon}(\boldsymbol{v})\rangle_{\mathcal{H}} \tag{7.45}$$

for all $\boldsymbol{u}, \boldsymbol{v} \in V$, a.e. $t \in (0,T)$. We also consider the function $j: \Sigma_C \times \mathbb{R}^d \to \mathbb{R}$ given by

$$j(\boldsymbol{x}, t, \boldsymbol{\xi}) = j_\nu(\boldsymbol{x}, t, \xi_\nu) + j_\tau(\boldsymbol{x}, t, \boldsymbol{\xi}_\tau) \tag{7.46}$$

for all $\boldsymbol{\xi} \in \mathbb{R}^d$, a.e. $(\boldsymbol{x}, t) \in \Sigma_C$. Next, proceeding as in the proof of Theorem 7.3, we deduce that under the assumption (7.31), the operator A defined by (7.44) is such that $A(t,\cdot)$ is bounded, continuous, coercive with constant $\alpha = m_{\mathcal{B}} > 0$ and strongly monotone with positive constant $m_1 = m_{\mathcal{B}}$, for a.e. $t \in (0,T)$. Hence, condition (4.18) is satisfied. Also, if (7.32) holds, then the operator C given by (7.45) satisfies hypothesis (4.26).

It follows from Corollary 3.48 applied to the functions j_ν and j_τ that $j(\cdot,\cdot,\mathbf{e}(\cdot)) \in L^1(\Sigma_C)$ for all $\mathbf{e} \in L^2(\Gamma_C;\mathbb{R}^d)$. Under assumptions (7.33) and (7.34), the function j defined by (7.46) satisfies (4.49) with $\overline{c}_0 \geq 0$, $\overline{c}_1 = c_{1\nu} + c_{1\tau} \geq 0$, $m_2 = \max\{m_\nu, m_\tau\} \geq 0$ and $\overline{d}_0 \geq 0$.

It is also clear that the inequality in (i) implies the analogous inequality listed in the hypothesis (i) of Theorem 4.22 and condition (7.42) implies (4.43). Therefore, we are in a position to apply the existence part of Theorem 4.22 and we deduce that Problem 7.7 has a solution $\boldsymbol{u} \in L^2(0,T;V)$. It is easy to see that the regularity hypothesis (7.43) implies the regularity of either j or $-j$, respectively. In this case, by the uniqueness part of Theorem 4.22, we deduce the uniqueness of the solution to Problem 7.7, which concludes the proof. □

A couple of functions $\boldsymbol{u}: (0,T) \to V$ and $\boldsymbol{\sigma}: (0,T) \to \mathcal{H}$ which satisfies (7.41) and the constitutive law (7.26) is called a *weak solution* of Problem 7.6. We conclude that, under the hypotheses of Theorem 7.8, the frictional contact problem (7.25)–(7.30) has at least one weak solution. Moreover, using (7.26), (7.35), and arguments similar to those used on page 210 it can be proved that the weak solution has the regularity

$$\boldsymbol{u} \in L^2(0,T;V), \quad \boldsymbol{\sigma} \in L^2(0,T;\mathcal{H}_1).$$

If, in addition, the regularity condition (7.43) holds, then the weak solution of Problem 7.7 is unique.

7.3 An Electro-Elastic Frictional Problem

For the problem studied in this section we assume that the body is piezoelectric and the foundation is conductive. Therefore, we refer to the physical setting described in Sect. 6.4. The classical model for the process is as follows.

Problem 7.9. *Find a displacement field* $\boldsymbol{u}\colon \Omega \to \mathbb{R}^d$, *a stress field* $\boldsymbol{\sigma}\colon \Omega \to \mathbb{S}^d$, *an electric potential* $\varphi\colon \Omega \to \mathbb{R}$ *and an electric displacement field* $\boldsymbol{D}\colon \Omega \to \mathbb{R}^d$ *such that*

$$\operatorname{Div}\boldsymbol{\sigma} + \boldsymbol{f}_0 = \boldsymbol{0} \quad \text{in } \Omega, \tag{7.47}$$

$$\operatorname{div}\boldsymbol{D} - q_0 = 0 \quad \text{in } \Omega, \tag{7.48}$$

$$\boldsymbol{\sigma} = \mathcal{F}\boldsymbol{\varepsilon}(\boldsymbol{u}) - \mathcal{P}^\top \boldsymbol{E}(\varphi) \quad \text{in } \Omega, \tag{7.49}$$

$$\boldsymbol{D} = \mathcal{P}\boldsymbol{\varepsilon}(\boldsymbol{u}) + \boldsymbol{\beta}\boldsymbol{E}(\varphi) \quad \text{in } \Omega, \tag{7.50}$$

$$\boldsymbol{u} = \boldsymbol{0} \quad \text{on } \Gamma_D, \tag{7.51}$$

$$\boldsymbol{\sigma}\boldsymbol{\nu} = \boldsymbol{f}_N \quad \text{on } \Gamma_N, \tag{7.52}$$

$$\varphi = 0 \quad \text{on } \Gamma_a, \tag{7.53}$$

$$\boldsymbol{D}\cdot\boldsymbol{\nu} = q_b \quad \text{on } \Gamma_b, \tag{7.54}$$

$$-\sigma_\nu \in h_\nu(\varphi - \varphi_0)\,\partial j_\nu(u_\nu - g_0) \quad \text{on } \Gamma_C, \tag{7.55}$$

$$-\boldsymbol{\sigma}_\tau \in h_\tau(\varphi - \varphi_0, u_\nu - g_0)\,\partial j_\tau(\boldsymbol{u}_\tau) \quad \text{on } \Gamma_C, \tag{7.56}$$

$$\boldsymbol{D}\cdot\boldsymbol{\nu} \in h_e(u_\nu - g_0)\,\partial j_e(\varphi - \varphi_0) \quad \text{on } \Gamma_C. \tag{7.57}$$

We describe now problem (7.47)–(7.57) and provide explanation of the equations and the boundary conditions.

First, equations (7.47) and (7.48) are the equilibrium equations for the stress and electric displacement fields, see (6.3) and (6.74), respectively. Next, equations (7.49) and (7.50) represent the electro-elastic constitutive laws introduced in Sect. 6.4, see (6.78) and (6.80). Recall that here $\boldsymbol{\varepsilon}(\boldsymbol{u})$ denotes the linearized strain tensor, $\mathcal{F}$ is the elasticity operator, assumed to be nonlinear, $\mathcal{P} = (p_{ijk})$ represents the third-order piezoelectric tensor, $\mathcal{P}^\top$ is its transpose, $\boldsymbol{\beta} = (\beta_{ij})$ denotes the electric permittivity tensor, and $\boldsymbol{E}(\varphi)$ is the electric field.

Conditions (7.51) and (7.52) are the displacement and traction boundary conditions, whereas (7.53) and (7.54) represent the electric boundary conditions. These conditions show that the displacement field and the electric potential vanish on Γ_D and Γ_a, respectively, while the forces and free electric charges are prescribed on Γ_N and Γ_b, respectively.

The boundary conditions (7.55), (7.56), and (7.57) describe the contact, the frictional, and the electrical conductivity conditions on the potential contact surface Γ_C, respectively, and were introduced in Sect. 6.4, see (6.82), (6.85), and (6.89). Recall that here, as usual, $h_\nu, h_\tau, h_e, j_\nu, j_\tau$, and j_e are prescribed functions, $\partial j_\nu, \partial j_\tau$, and ∂j_e denote the Clarke subdifferentials of the functions j_e, j_ν, j_τ, respectively, g_0 is the gap function, and φ_0 represents the electric potential of the foundation.

In the study of Problem 7.9 we assume that the elasticity operator $\mathcal{F}$ satisfies condition (7.7). Moreover, the piezoelectric tensor $\mathcal{P}$ and the electric permittivity tensor $\boldsymbol{\beta}$ satisfy

$$\left.\begin{array}{l}\mathcal{P}\colon \Omega \times \mathbb{S}^d \to \mathbb{R}^d \text{ is such that}\\ \quad\text{(a) } \mathcal{P}(\boldsymbol{x},\boldsymbol{\varepsilon}) = p(\boldsymbol{x})\,\boldsymbol{\varepsilon} \text{ for all } \boldsymbol{\varepsilon} \in \mathbb{S}^d, \text{ a.e. } \boldsymbol{x} \in \Omega.\\ \quad\text{(b) } p(\boldsymbol{x}) = (p_{ijk}(\boldsymbol{x})) \text{ with } p_{ijk} \in L^\infty(\Omega).\end{array}\right\} \qquad (7.58)$$

$$\left.\begin{array}{l}\boldsymbol{\beta}\colon \Omega \times \mathbb{R}^d \to \mathbb{R}^d \text{ is such that}\\ \quad\text{(a) } \boldsymbol{\beta}(\boldsymbol{x},\boldsymbol{\xi}) = \beta(\boldsymbol{x})\,\boldsymbol{\xi} \text{ for all } \boldsymbol{\xi} \in \mathbb{R}^d, \text{ a.e. } \boldsymbol{x} \in \Omega.\\ \quad\text{(b) } \beta(\boldsymbol{x}) = (\beta_{ij}(\boldsymbol{x})) \text{ with } \beta_{ij} = \beta_{ji} \in L^\infty(\Omega).\\ \quad\text{(c) } \beta_{ij}(\boldsymbol{x})\xi_i\xi_j \ge m_\beta \|\boldsymbol{\xi}\|^2_{\mathbb{R}^d} \text{ for all } \boldsymbol{\xi} = (\xi_i) \in \mathbb{R}^d,\\ \qquad\quad \text{a.e. } \boldsymbol{x} \in \Omega \text{ with } m_\beta > 0.\end{array}\right\} \qquad (7.59)$$

The functions j_ν, and j_τ satisfy (7.8) and (7.9), respectively, and, in addition j_e satisfies

$$\left.\begin{array}{l}j_e\colon \Gamma_C \times \mathbb{R} \to \mathbb{R} \text{ is such that}\\ \quad\text{(a) } j_e(\cdot,r) \text{ is measurable on } \Gamma_C \text{ for all } r \in \mathbb{R} \text{ and there}\\ \qquad\quad \text{exists } e_3 \in L^2(\Gamma_C) \text{ such that } j_e(\cdot,e_3(\cdot)) \in L^1(\Gamma_C).\\ \quad\text{(b) } j_e(\boldsymbol{x},\cdot) \text{ is locally Lipschitz on } \mathbb{R} \text{ for a.e. } \boldsymbol{x} \in \Gamma_C.\\ \quad\text{(c) } |\partial j_e(\boldsymbol{x},r)| \le c_{0e} + c_{1e}|r| \text{ for all } r \in \mathbb{R}, \text{ a.e. } \boldsymbol{x} \in \Gamma_C\\ \qquad\quad \text{with } c_{0e}, c_{1e} \ge 0.\\ \quad\text{(d) } j_e^0(\boldsymbol{x},r;-r) \le d_e\,(1+|r|) \text{ for all } r \in \mathbb{R}, \text{ a.e. } \boldsymbol{x} \in \Gamma_C\\ \qquad\quad \text{with } d_e \ge 0.\end{array}\right\} \qquad (7.60)$$

The functions h_ν, h_τ, and h_e satisfy

$$\left.\begin{array}{l}h_\nu\colon \Gamma_C \times \mathbb{R} \to \mathbb{R} \text{ is such that}\\ \quad\text{(a) } h_\nu(\cdot,r) \text{ is measurable on } \Gamma_C \text{ for all } r \in \mathbb{R}.\\ \quad\text{(b) } h_\nu(\boldsymbol{x},\cdot) \text{ is continuous on } \mathbb{R} \text{ for a.e. } \boldsymbol{x} \in \Gamma_C.\\ \quad\text{(c) } 0 \le h_\nu(\boldsymbol{x},r) \le \overline{h}_\nu \text{ for all } r \in \mathbb{R}, \text{ a.e. } \boldsymbol{x} \in \Gamma_C \text{ with } \overline{h}_\nu > 0.\end{array}\right\} \qquad (7.61)$$

$$\left.\begin{array}{l}h_\tau\colon \Gamma_C \times \mathbb{R} \times \mathbb{R} \to \mathbb{R} \text{ is such that}\\ \quad\text{(a) } h_\tau(\cdot,r_1,r_2) \text{ is measurable on } \Gamma_C \text{ for all } r_1,\, r_2 \in \mathbb{R}.\\ \quad\text{(b) } h_\tau(\boldsymbol{x},\cdot,\cdot) \text{ is continuous on } \mathbb{R}\times\mathbb{R} \text{ for a.e. } \boldsymbol{x} \in \Gamma_C.\\ \quad\text{(c) } 0 \le h_\tau(\boldsymbol{x},r_1,r_2) \le \overline{h}_\tau \text{ for all } r_1,\, r_2 \in \mathbb{R}, \text{ a.e. } \boldsymbol{x} \in \Gamma_C\\ \qquad\quad \text{with } \overline{h}_\tau > 0.\end{array}\right\} \qquad (7.62)$$

$$\left.\begin{array}{l} h_e\colon \Gamma_C \times \mathbb{R} \to \mathbb{R} \text{ is such that} \\ \quad \text{(a) } h_e(\cdot, r) \text{ is measurable on } \Gamma_C \text{ for all } r \in \mathbb{R}. \\ \quad \text{(b) } h_e(\boldsymbol{x}, \cdot) \text{ is continuous on } \mathbb{R} \text{ for a.e. } \boldsymbol{x} \in \Gamma_C. \\ \quad \text{(c) } 0 \le h_e(\boldsymbol{x}, r) \le \overline{h}_e \text{ for all } r \in \mathbb{R}, \text{ a.e. } \boldsymbol{x} \in \Gamma_C \text{ with } \overline{h}_e > 0. \end{array}\right\} \tag{7.63}$$

The forces, tractions, volume, and surface free charge densities have the regularity

$$\boldsymbol{f}_0 \in L^2(\Omega;\mathbb{R}^d), \quad \boldsymbol{f}_N \in L^2(\Gamma_N;\mathbb{R}^d), \quad q_0 \in L^2(\Omega), \quad q_b \in L^2(\Gamma_b), \tag{7.64}$$

whereas the gap and the potential of the foundation satisfy

$$g_0 \in L^\infty(\Gamma_C), \quad g_0 \ge 0 \text{ a.e. on } \Gamma_C, \quad \varphi_0 \in L^\infty(\Gamma_C). \tag{7.65}$$

In the analysis of Problem 7.9, for the mechanical unknowns $\boldsymbol{u}$ and $\boldsymbol{\sigma}$ we use the spaces H, $\mathcal{H}$, $\mathcal{H}_1$, V, and V^* introduced in Sect. 7.1. For the electrical unknowns φ and $\boldsymbol{D}$ we need the spaces

$$W = \{\, \boldsymbol{D} \in H \mid \operatorname{div} \boldsymbol{D} \in L^2(\Omega)\,\}, \quad \Phi = \left\{\, \varphi \in H^1(\Omega) \,|\, \varphi = 0 \text{ on } \Gamma_a \,\right\},$$

which are Hilbert spaces equipped with the standard inner products. Moreover, since meas (Γ_a) is positive, it can be shown that Φ is a Hilbert space with the inner product and the corresponding norm given by

$$(\varphi, \psi)_\Phi = (\nabla\varphi, \nabla\psi)_H, \quad \|\psi\|_\Phi = \|\nabla\psi\|_H \text{ for all } \varphi, \psi \in \Phi.$$

In addition, it is well known that the inclusions $\Phi \subset L^2(\Omega) \subset \Phi^*$ are continuous and compact where Φ^* denotes the dual space of Φ.

We turn now to the variational formulation of the contact problem (7.47)–(7.57). To this end, we assume in what follows that $\boldsymbol{u}$, $\boldsymbol{\sigma}$, φ, $\boldsymbol{D}$ are smooth functions which solve (7.47)–(7.57). Let $\boldsymbol{v} \in V$. We use standard arguments based on the equilibrium equation (7.47), the boundary conditions (7.51), (7.52), and the subdifferential inclusions (7.55), (7.56) to see that

$$\begin{aligned} (\boldsymbol{\sigma}, \boldsymbol{\varepsilon}(\boldsymbol{v}))_{\mathcal{H}} + \int_{\Gamma_C} \Big(& h_\nu(\varphi - \varphi_0)\, j_\nu^0(u_\nu - g_0; v_\nu) \\ & + h_\tau(\varphi - \varphi_0, u_\nu - g_0)\, j_\tau^0(\boldsymbol{u}_\tau; \boldsymbol{v}_\tau)\Big)\, d\Gamma \\ \ge \int_\Omega \boldsymbol{f}_0 \cdot \boldsymbol{v}\, dx & + \int_{\Gamma_N} \boldsymbol{f}_N \cdot \boldsymbol{v}\, d\Gamma. \end{aligned}$$

Consider now the element $\boldsymbol{f} \in V^*$ given by (7.16). Then, the previous inequality yields

$$\langle \boldsymbol{\sigma}, \boldsymbol{\varepsilon}(\boldsymbol{v})\rangle_{\mathcal{H}} + \int_{\Gamma_C} \Big(h_\nu(\varphi - \varphi_0)\, j_\nu^0(u_\nu - g_0; v_\nu) \\ + h_\tau(\varphi - \varphi_0, u_\nu - g_0)\, j_\tau^0(\boldsymbol{u}_\tau; \boldsymbol{v}_\tau)\Big)\, d\Gamma \geq \langle \boldsymbol{f}, \boldsymbol{v}\rangle_{V^*\times V}. \quad (7.66)$$

Similarly, for every $\psi \in \Phi$, from (7.48) and the Green formula (2.6) we deduce that

$$\langle \boldsymbol{D}, \nabla\psi\rangle_H + \int_\Omega q_0 \psi\, dx = \int_\Gamma \boldsymbol{D}\cdot\boldsymbol{\nu}\,\psi\, d\Gamma$$

and by (7.54), (7.57), we get

$$-\langle \boldsymbol{D}, \nabla\psi\rangle_H + \int_{\Gamma_C} h_e(u_\nu - g_0)\, j_e^0(\varphi - \varphi_0; \psi)\, d\Gamma \geq \langle q, \psi\rangle_{\Phi^*\times\Phi} \quad (7.67)$$

for all $\psi \in \Phi$, where q is the element of Φ^* given by

$$\langle q, \psi\rangle_{\Phi^*\times\Phi} = \langle q_0, \psi\rangle_{L^2(\Omega)} - \langle q_b, \psi\rangle_{L^2(\Gamma_b)} \quad \text{for all } \psi \in \Phi. \quad (7.68)$$

Also, remark that (7.53) implies that $\varphi \in \Phi$. We substitute (7.49) in (7.66), (7.50) in (7.67) and use the equality $\boldsymbol{E}(\varphi) = -\nabla\varphi$ to derive the following variational formulation of Problem 7.9, in terms of displacement and electric potential fields.

Problem 7.10. *Find a displacement field $\boldsymbol{u} \in V$ and an electric potential $\varphi \in \Phi$ such that*

$$\langle \mathcal{F}\boldsymbol{\varepsilon}(\boldsymbol{u}), \boldsymbol{\varepsilon}(\boldsymbol{v})\rangle_{\mathcal{H}} + \langle \mathcal{P}^\top\nabla\varphi, \boldsymbol{\varepsilon}(\boldsymbol{v})\rangle_{\mathcal{H}} \\ + \int_{\Gamma_C} \Big(h_\nu(\varphi - \varphi_0)\, j_\nu^0(u_\nu - g_0; v_\nu) + h_\tau(\varphi - \varphi_0, u_\nu - g_0)\, j_\tau^0(\boldsymbol{u}_\tau; \boldsymbol{v}_\tau)\Big) d\Gamma \\ \geq \langle \boldsymbol{f}, \boldsymbol{v}\rangle_{V^*\times V} \quad \text{for all } \boldsymbol{v} \in V, \quad (7.69)$$

$$\langle \boldsymbol{\beta}\nabla\varphi, \nabla\psi\rangle_H - \langle \mathcal{P}\boldsymbol{\varepsilon}(\boldsymbol{u}), \nabla\psi\rangle_H \\ + \int_{\Gamma_C} h_e(u_\nu - g_0)\, j_e^0(\varphi - \varphi_0; \psi)\, d\Gamma \geq \langle q, \psi\rangle_{\Phi^*\times\Phi} \quad \text{for all } \psi \in \Phi. \quad (7.70)$$

Note that, in contrast with the variational formulations of the frictional contact problems studied in Sects. 7.1 and 7.2, Problem 7.10 represents a *system of hemivariational inequalities*. One of the main features of this system arises in the strong coupling between the unknowns $\boldsymbol{u}$ and φ, which appears both in the terms containing the piezoelectricity tensor $\mathcal{P}$ as well as in the terms related to the contact conditions.

This last coupling represents a consequence of the assumption that the foundation is conductive; it makes the model more interesting and its mathematical analysis more difficult.

Our main results in the study of Problem 7.10 are presented in Theorems 7.11 and 7.13 . In order to state these results, as in Chap. 4, we consider the space $Z = H^{\delta}(\Omega;\mathbb{R}^{d+1})$ with $\delta \in (1/2, 1)$, the trace operator $\gamma: Z \to L^2(\Gamma;\mathbb{R}^{d+1})$ and we denote by $\|\gamma\|$ its norm in $\mathcal{L}(Z, L^2(\Gamma;\mathbb{R}^{d+1}))$, and by $c_e > 0$ the embedding constant of $V \times \Phi \subset H^1(\Omega;\mathbb{R}^{d+1})$ into Z.

Theorem 7.11. *Assume that* (7.7), (7.58), (7.59), *and* (7.61)–(7.65) *hold. Moreover, assume that one of the following hypotheses:*

(i) (7.8)(a)–(c), (7.9)(a)–(c), (7.60)(a)–(c) *and*

$$\min\{\widehat{\alpha}, m_{\beta}\} > \sqrt{3}\left(c_{1\nu}\,\overline{h}_{\nu} + c_{1\tau}\,\overline{h}_{\tau} + c_{1e}\,\overline{h}_{e}\right) c_e^2\,\|\gamma\|^2$$

(ii) (7.8), (7.9) *and* (7.60)

is satisfied and

$$\left.\begin{array}{l} \text{either } j_\nu(\boldsymbol{x},\cdot),\ j_\tau(\boldsymbol{x},\cdot) \text{ and } j_e(\boldsymbol{x},\cdot) \text{ are regular} \\ \text{or } -j_\nu(\boldsymbol{x},\cdot), -j_\tau(\boldsymbol{x},\cdot) \text{ and } -j_e(\boldsymbol{x},\cdot) \text{ are regular} \end{array}\right\} \tag{7.71}$$

for a.e. $\boldsymbol{x} \in \Gamma_C$. *Then Problem* 7.10 *has at least one solution.*

Proof. We apply Corollary 4.18 with a suitable choice of the functional framework. We work in the product space $Y = V \times \Phi \subset H^1(\Omega;\mathbb{R}^{d+1})$, which is a Hilbert space endowed with the inner product

$$\langle \boldsymbol{y}, \boldsymbol{z}\rangle_Y = \langle \boldsymbol{u}, \boldsymbol{v}\rangle_V + \langle \varphi, \psi\rangle_\Phi \ \text{ for all } \boldsymbol{y} = (\boldsymbol{u}, \varphi) \in Y \text{ and } \boldsymbol{z} = (\boldsymbol{v}, \psi) \in Y$$

and the associated norm $\|\cdot\|_Y$. We introduce the operator $A: Y \to Y^*$ defined by

$$\begin{aligned} \langle A\boldsymbol{y}, \boldsymbol{z}\rangle_{Y^*\times Y} &= \langle \mathcal{F}\boldsymbol{\varepsilon}(\boldsymbol{u}), \boldsymbol{\varepsilon}(\boldsymbol{v})\rangle_{\mathcal{H}} + \langle \mathcal{P}^{\top}\nabla\varphi, \boldsymbol{\varepsilon}(\boldsymbol{v})\rangle_{\mathcal{H}} \\ &\quad + \langle \boldsymbol{\beta}\nabla\varphi, \nabla\psi\rangle_H - \langle \mathcal{P}\boldsymbol{\varepsilon}(\boldsymbol{u}), \nabla\psi\rangle_H \end{aligned} \tag{7.72}$$

for all $\boldsymbol{y}, \boldsymbol{z} \in Y$, $\boldsymbol{y} = (\boldsymbol{u}, \varphi)$, and $\boldsymbol{z} = (\boldsymbol{v}, \psi)$. Consider the functions $h_i, j_i: \Gamma_C \times \mathbb{R}^{d+1} \to \mathbb{R}$ for $i = 1, 2, 3$ given by

$$\begin{aligned} h_1(\boldsymbol{x}, \boldsymbol{\xi}, r) &= h_\nu(\boldsymbol{x}, r - \varphi_0(\boldsymbol{x})), \\ h_2(\boldsymbol{x}, \boldsymbol{\xi}, r) &= h_\tau(\boldsymbol{x}, r - \varphi_0(\boldsymbol{x}), \xi_\nu - g_0(\boldsymbol{x})), \\ h_3(\boldsymbol{x}, \boldsymbol{\xi}, r) &= h_e(\boldsymbol{x}, \xi_\nu - g_0(\boldsymbol{x})), \\ j_1(\boldsymbol{x}, \boldsymbol{\xi}, r) &= j_\nu(\boldsymbol{x}, \xi_\nu - g_0(\boldsymbol{x})), \\ j_2(\boldsymbol{x}, \boldsymbol{\xi}, r) &= j_\tau(\boldsymbol{x}, \boldsymbol{\xi}_\tau), \\ j_3(\boldsymbol{x}, \boldsymbol{\xi}, r) &= j_e(\boldsymbol{x}, r - \varphi_0(\boldsymbol{x})) \end{aligned}$$

for all $(\boldsymbol{\xi}, r) \in \mathbb{R}^d \times \mathbb{R}$ and a.e. $\boldsymbol{x} \in \Gamma_C$. Under the notation above, we associate with Problem 7.10 the following hemivariational inequality:

$$\left.\begin{aligned}
&\text{find } \boldsymbol{y} = (\boldsymbol{u}, \varphi) \in Y \text{ such that} \\
&\langle A\boldsymbol{y}, \boldsymbol{z}\rangle_{Y^*\times Y} + \int_{\Gamma_C} \sum_{i=1}^{3} h_i(\gamma\boldsymbol{y})\, j_i^0(\gamma\boldsymbol{y}; \gamma\boldsymbol{z})\, d\Gamma \ge \langle(\boldsymbol{f}, q), \boldsymbol{z}\rangle_{Y^*\times Y} \\
&\qquad\qquad\qquad\qquad\qquad\qquad\qquad\qquad \text{for all } \boldsymbol{z} \in Y,
\end{aligned}\right\} \tag{7.73}$$

where $(\boldsymbol{f}, q) \in V^* \times \Phi^* = Y^*$ is given by (7.16) and (7.68).

We claim that under hypotheses (7.7), (7.58), and (7.59), the operator A is pseudomonotone and coercive. Indeed, by (7.7)(c), (7.58), (7.59), and the Hölder inequality, we have

$$\begin{aligned}
|\langle A\boldsymbol{y}, \boldsymbol{z}\rangle_{Y^*\times Y}| &\le \int_\Omega \Big(\|\mathcal{F}\boldsymbol{\varepsilon}(\boldsymbol{u})\|_{\mathbb{S}^d} + \|\mathcal{P}^\top\nabla\varphi\|_{\mathbb{S}^d}\Big)\|\boldsymbol{\varepsilon}(\boldsymbol{v})\|_{\mathbb{S}^d}\,dx \\
&\quad + \int_\Omega \Big(\|\boldsymbol{\beta}\nabla\varphi\|_{\mathbb{R}^d} + \|\mathcal{P}\boldsymbol{\varepsilon}(\boldsymbol{u})\|_{\mathbb{R}^d}\Big)\|\nabla\psi\|_{\mathbb{R}^d}\,dx \\
&\le \Big(\int_\Omega \big(a_0(\boldsymbol{x}) + a_1\|\boldsymbol{\varepsilon}(\boldsymbol{u})\|_{\mathbb{S}^d} + c_p\|\nabla\varphi\|_{\mathbb{R}^d}\big)^2\,dx\Big)^{1/2}\|\boldsymbol{v}\|_V \\
&\quad + \Big(\int_\Omega \big(c_b\|\nabla\varphi\|_{\mathbb{R}^d} + c_p\|\boldsymbol{\varepsilon}(\boldsymbol{u})\|_{\mathbb{S}^d}\big)^2\,dx\Big)^{1/2}\|\psi\|_\Phi \\
&\le 2\left(\|a_0\|_{L^2(\Omega)} + a_1\|\boldsymbol{u}\|_V + c_p\|\varphi\|_\Phi\right)\|\boldsymbol{v}\|_V \\
&\quad + \sqrt{2}\left(c_b\|\varphi\|_\Phi + c_p\|\boldsymbol{u}\|_V\right)\|\psi\|_\Phi \le \widehat{c}\,(1 + \|\boldsymbol{y}\|_Y)\|\boldsymbol{z}\|_Y
\end{aligned}$$

for all $\boldsymbol{y} = (\boldsymbol{u}, \varphi)$, $\boldsymbol{z} = (\boldsymbol{v}, \psi)$, $\boldsymbol{y}, \boldsymbol{z} \in Y$ with $c_p, c_b, \widehat{c} > 0$. This gives $\|A\boldsymbol{y}\|_{Y^*} \le \widehat{c}\,(1 + \|\boldsymbol{y}\|_Y)$ for all $\boldsymbol{y} \in Y$ and implies the boundedness of A. To show the monotonicity of A we use (6.79) to see that

$$\langle\mathcal{P}^\top\nabla\varphi, \boldsymbol{\varepsilon}(\boldsymbol{u})\rangle_{\mathcal{H}} = \langle\mathcal{P}\boldsymbol{\varepsilon}(\boldsymbol{u}), \nabla\varphi\rangle_H \tag{7.74}$$

and, therefore, the hypotheses (7.7)(e) and (7.59)(c) yield

$$\begin{aligned}
\langle A\boldsymbol{y}_1 - A\boldsymbol{y}_2, \boldsymbol{y}_1 - \boldsymbol{y}_2\rangle_{Y^*\times Y} &= \langle\mathcal{F}\boldsymbol{\varepsilon}(\boldsymbol{u}_1) - \mathcal{F}\boldsymbol{\varepsilon}(\boldsymbol{u}_2), \boldsymbol{\varepsilon}(\boldsymbol{u}_1) - \boldsymbol{\varepsilon}(\boldsymbol{u}_2)\rangle_{\mathcal{H}} \\
&\quad + \langle\boldsymbol{\beta}\nabla\varphi_1 - \boldsymbol{\beta}\nabla\varphi_2, \nabla\varphi_1 - \nabla\varphi_2\rangle_H \\
&\ge m_\beta\|\nabla\varphi_1 - \nabla\varphi_2\|_H^2 \ge 0
\end{aligned}$$

for all $\boldsymbol{y}_i = (\boldsymbol{u}_i, \varphi_i) \in Y$, $i = 1, 2$. From the hypotheses on $\mathcal{F}$, $\mathcal{P}$, and $\boldsymbol{\beta}$, proceeding anologously as in the proof of Theorem 7.3, we infer that the operator A is continuous. Since the operator A is bounded, monotone, and hemicontinuous, from Theorem 3.69(i), we deduce that A is pseudomonotone.

Next, exploiting the relation (7.74) and the hypotheses (7.7)(d) and (7.59)(c), we get

$$\begin{aligned}\langle A\boldsymbol{y}, \boldsymbol{y}\rangle_{Y^*\times Y} &= \int_\Omega \mathcal{F}\boldsymbol{\varepsilon}(\boldsymbol{u}) : \boldsymbol{\varepsilon}(\boldsymbol{u})\, dx + \int_\Omega \boldsymbol{\beta}\nabla\varphi\cdot\nabla\varphi\, dx \\ &\geq \widehat{\alpha}\, \|\boldsymbol{u}\|_V^2 + m_\beta\, \|\varphi\|_\Phi^2 \geq \min\{\widehat{\alpha}, m_\beta\}\, \|\boldsymbol{y}\|_Y^2\end{aligned}$$

for all $\boldsymbol{y} = (\boldsymbol{u}, \varphi) \in Y$. It follows from here that the operator A is coercive with constant $\alpha = \min\{\widehat{\alpha}, m_\beta\} > 0$ and, therefore, we conclude that it satisfies the hypothesis (4.1).

Now we verify the hypotheses on h_i and j_i. It is obvious to see that h_i, $i = 1$, 2, 3 satisfy (4.40). Moreover, we note that

$$j_1(\boldsymbol{x}, \boldsymbol{\xi}, r) = j_\nu(\boldsymbol{x}, N_\nu\boldsymbol{\xi} - g_0(\boldsymbol{x})), \qquad j_2(\boldsymbol{x}, \boldsymbol{\xi}, r) = j_\tau(\boldsymbol{x}, N_\tau\boldsymbol{\xi})$$

for all $(\boldsymbol{\xi}, r) \in \mathbb{R}^d \times \mathbb{R}$, a.e. $\boldsymbol{x} \in \Gamma_C$, with the operators $N_\nu \in L^\infty(\Gamma; \mathcal{L}(\mathbb{R}^d, \mathbb{R}))$, $N_\tau \in L^\infty(\Gamma; \mathcal{L}(\mathbb{R}^d, \mathbb{R}^d))$ given by $N_\nu\boldsymbol{\xi} = \xi_\nu = \boldsymbol{\xi}\cdot\boldsymbol{\nu}(\boldsymbol{x})$ and $N_\tau\boldsymbol{\xi} = \boldsymbol{\xi}_\tau = \boldsymbol{\xi} - \xi_\nu\boldsymbol{\nu}(\boldsymbol{x})$, respectively, for all $\boldsymbol{\xi} \in \mathbb{R}^d$. Recall that the operators N_ν, N_τ, and $\boldsymbol{\nu}$ depend on the spatial variable $\boldsymbol{x} \in \Gamma$ but, for simplicity of notation, we do not indicate explicitly this dependence.

It is easy to observe that $j_i(\cdot, \boldsymbol{\xi}, r)$ are measurable on Γ_C for all $(\boldsymbol{\xi}, r) \in \mathbb{R}^d \times \mathbb{R}$ and $j_i(\boldsymbol{x}, \cdot, \cdot)$ are locally Lipschitz on $\mathbb{R}^{d+1}$ for a.e. $\boldsymbol{x} \in \Gamma_C$, $i = 1, 2, 3$. Furthermore, by Corollary 3.48 applied to the functions j_ν, j_τ, and j_e we get that $j_i(\cdot, \mathbf{e}(\cdot)) \in L^1(\Gamma_C)$ for all $\mathbf{e} \in L^2(\Gamma_C; \mathbb{R}^{d+1})$, $i = 1, 2, 3$. Hence we deduce that (4.41)(a) holds. Using the definition of the generalized directional derivative of $j_i(\boldsymbol{x}, \cdot, \cdot)$ and Proposition 3.37, we have

$$\left.\begin{aligned} j_1^0(\boldsymbol{x}, \boldsymbol{\xi}, r; \boldsymbol{\varrho}, s) &\leq j_\nu^0(\boldsymbol{x}, \xi_\nu - g_0(\boldsymbol{x}); \varrho_\nu), \\ j_2^0(\boldsymbol{x}, \boldsymbol{\xi}, r; \boldsymbol{\varrho}, s) &\leq j_\tau^0(\boldsymbol{x}, \boldsymbol{\xi}_\tau; \boldsymbol{\varrho}_\tau), \\ j_3^0(\boldsymbol{x}, \boldsymbol{\xi}, r; \boldsymbol{\varrho}, s) &\leq j_e^0(\boldsymbol{x}, r - \varphi_0(\boldsymbol{x}); s) \end{aligned}\right\} \tag{7.75}$$

for all $(\boldsymbol{\xi}, r)$, $(\boldsymbol{\varrho}, s) \in \mathbb{R}^d \times \mathbb{R}$, and a.e. $\boldsymbol{x} \in \Gamma_C$. Also, using the properties of the generalized directional derivative, Proposition 3.23(i) and (iii), analogously as in the proof of Theorem 7.3, we find that

$$\begin{aligned} j_1^0(\boldsymbol{x}, \boldsymbol{\xi}, r; -\boldsymbol{\xi}, -r) &\leq j_\nu^0(\boldsymbol{x}, \xi_\nu - g_0(\boldsymbol{x}); -\xi_\nu) \leq d_{01}\left(1 + \|(\boldsymbol{\xi}, r)\|_{\mathbb{R}^{d+1}}\right), \\ j_2^0(\boldsymbol{x}, \boldsymbol{\xi}, r; -\boldsymbol{\xi}, -r) &\leq j_\tau^0(\boldsymbol{x}, \boldsymbol{\xi}_\tau; -\boldsymbol{\xi}_\tau) \leq d_\tau\left(1 + \|(\boldsymbol{\xi}, r)\|_{\mathbb{R}^{d+1}}\right), \\ j_3^0(\boldsymbol{x}, \boldsymbol{\xi}, r; -\boldsymbol{\xi}, -r) &\leq j_e^0(\boldsymbol{x}, r - \varphi_0(\boldsymbol{x}); -r) \leq d_{03}\left(1 + \|(\boldsymbol{\xi}, r)\|_{\mathbb{R}^{d+1}}\right) \end{aligned}$$

for all $(\boldsymbol{\xi}, r) \in \mathbb{R}^d \times \mathbb{R}$ and a.e. $\boldsymbol{x} \in \Gamma_C$ with $d_{01}, d_{03} \geq 0$. From (7.71) we infer that either $j_i(\boldsymbol{x}, \cdot, \cdot)$ or $-j_i(\boldsymbol{x}, \cdot, \cdot)$ for $i = 1, 2, 3$, are regular for a.e. $\boldsymbol{x} \in \Gamma_C$, so by Proposition 3.38, we have

$$\partial j_i(\boldsymbol{x}, \boldsymbol{\xi}, r) \subseteq \partial_\xi j_i(\boldsymbol{x}, \boldsymbol{\xi}, r) \times \partial_r j_i(\boldsymbol{x}, \boldsymbol{\xi}, r) \tag{7.76}$$

for all $(\boldsymbol{\xi}, r) \in \mathbb{R}^d \times \mathbb{R}$ and a.e. $\boldsymbol{x} \in \Gamma_C$, where ∂j_i denotes the generalized gradient of j_i with respect to the pair $(\boldsymbol{\xi}, r)$ and $\partial_\xi j_i$, $\partial_r j_i$ are the partial generalized gradients of $j_i(\boldsymbol{x}, \cdot, r)$ and $j_i(\boldsymbol{x}, \boldsymbol{\xi}, \cdot)$, respectively. From Proposition 3.37, Lemma 3.39, and (7.76), we obtain

$$\partial j_1(\boldsymbol{x}, \boldsymbol{\xi}, r) \subseteq \partial j_\nu(\boldsymbol{x}, \xi_\nu - g_0(\boldsymbol{x}))\, \boldsymbol{\nu} \times \{0\},$$

$$\partial j_2(\boldsymbol{x}, \boldsymbol{\xi}, r) \subseteq [\partial j_\tau(\boldsymbol{x}, \boldsymbol{\xi}_\tau)]_\tau \times \{0\},$$

$$\partial j_3(\boldsymbol{x}, \boldsymbol{\xi}, r) \subseteq \{\boldsymbol{0}\} \times \partial j_e(\boldsymbol{x}, r - \varphi_0(\boldsymbol{x}))$$

for all $(\boldsymbol{\xi}, r) \in \mathbb{R}^d \times \mathbb{R}$, a.e. $\boldsymbol{x} \in \Gamma_C$. Moreover, we have the estimates

$$\begin{aligned} \|\partial j_1(\boldsymbol{x}, \boldsymbol{\xi}, r)\|_{\mathbb{R}^{d+1}} &\leq |\partial j_\nu(\boldsymbol{x}, \xi_\nu - g_0(\boldsymbol{x}))| \leq c_{0\nu} + c_{1\nu}|\xi_\nu - g_0(\boldsymbol{x})| \\ &\leq c_{0\nu} + c_{1\nu}|g_0(\boldsymbol{x})| + c_{1\nu}\|(\boldsymbol{\xi}, r)\|_{\mathbb{R}^{d+1}}, \end{aligned}$$

$$\begin{aligned} \|\partial j_2(\boldsymbol{x}, \boldsymbol{\xi}, r)\|_{\mathbb{R}^{d+1}} &\leq \|\partial j_\tau(\boldsymbol{x}, \boldsymbol{\xi}_\tau)\|_{\mathbb{R}^d} \leq c_{0\tau} + c_{1\tau}\|\boldsymbol{\xi}_\tau\|_{\mathbb{R}^d} \\ &\leq c_{0\tau} + c_{1\tau}\|(\boldsymbol{\xi}, r)\|_{\mathbb{R}^{d+1}}, \end{aligned}$$

$$\begin{aligned} \|\partial j_3(\boldsymbol{x}, \boldsymbol{\xi}, r)\|_{\mathbb{R}^{d+1}} &\leq |\partial j_e(\boldsymbol{x}, r - \varphi_0(\boldsymbol{x}))| \leq c_{0e} + c_{1e}|r - \varphi_0(\boldsymbol{x})| \\ &\leq c_{0e} + c_{1e}|\varphi_0(\boldsymbol{x})| + c_{1e}\|(\boldsymbol{\xi}, r)\|_{\mathbb{R}^{d+1}} \end{aligned}$$

for all $(\boldsymbol{\xi}, r) \in \mathbb{R}^d \times \mathbb{R}$, a.e. $\boldsymbol{x} \in \Gamma_C$. We conclude from above that the functions j_i, $i = 1, 2, 3$, satisfy the hypothesis (4.41).

Next, we apply Corollary 4.18 to the hemivariational inequality (7.73) in which, clearly, the role of the abstract space V is played by the space Y and $s = d + 1$. We deduce in this way that (7.73) has at least one solution $\boldsymbol{y} = (\boldsymbol{u}, \varphi) \in Y$. It remains to show that the pair of functions $(\boldsymbol{u}, \varphi)$ represents a solution to the system (7.69)–(7.70). To this end, we choose $\boldsymbol{z} = (\boldsymbol{v}, 0) \in Y$ in (7.73) and take into account (7.75) and the definition (7.72) to obtain the inequality (7.69). Then, we choose $\boldsymbol{z} = (\boldsymbol{0}, \psi) \in Y$ in (7.73) and use again (7.75) to obtain (7.70). Therefore, it follows from above that the couple of functions $(\boldsymbol{u}, \varphi) \in V \times \Phi$ represents a solution to the system (7.69)–(7.70). We deduce that Problem 7.10 has at least one solution, which completes the proof. □

The uniqueness of a solution to Problem 7.10 under the general assumptions (7.61), (7.62), and (7.63) represents an open problem which, clearly, deserves more investigation. However, it can be obtained in a particular case, when h_ν, h_τ, and h_e are nonnegative constant functions. To present this uniqueness result we consider the statement of Problem 7.10 in the particular case above.

Problem 7.12. *Find a displacement field* $\boldsymbol{u} \in V$ *and an electric potential* $\varphi \in \Phi$ *such that*

$$\langle \mathcal{F}\boldsymbol{\varepsilon}(\boldsymbol{u}), \boldsymbol{\varepsilon}(\boldsymbol{v})\rangle_{\mathcal{H}} + \langle \mathcal{P}^\top \nabla\varphi, \boldsymbol{\varepsilon}(\boldsymbol{v})\rangle_{\mathcal{H}} + \int_{\Gamma_C} \Big(k_\nu\, j_\nu^0(u_\nu - g_0; v_\nu) + k_\tau\, j_\tau^0(\boldsymbol{u}_\tau; \boldsymbol{v}_\tau)\Big)\, d\Gamma$$

$$\geq \langle \boldsymbol{f}, \boldsymbol{v}\rangle_{V^*\times V} \quad \text{for all } \boldsymbol{v} \in V,$$

$$\langle \boldsymbol{\beta}\nabla\varphi, \nabla\psi\rangle_H - \langle \mathcal{P}\boldsymbol{\varepsilon}(\boldsymbol{u}), \nabla\psi\rangle_H + \int_{\Gamma_C} k_e\, j_e^0(\varphi - \varphi_0; \psi)\, d\Gamma$$

$$\geq \langle q, \psi\rangle_{\Phi^*\times\Phi} \quad \text{for all } \psi \in \Phi.$$

We have the following existence and uniqueness result.

Theorem 7.13. *Assume that* (7.7)(a)–(d), (7.58), (7.59), (7.64), *and* (7.65) *hold,* k_ν, k_τ, $k_e \geq 0$ *and, moreover, assume that*

$$\left.\begin{array}{l}
\text{(a) } (\mathcal{F}(\boldsymbol{x},\boldsymbol{\varepsilon}_1) - \mathcal{F}(\boldsymbol{x},\boldsymbol{\varepsilon}_2)) : (\boldsymbol{\varepsilon}_1 - \boldsymbol{\varepsilon}_2) \geq m_{\mathcal{F}} \|\boldsymbol{\varepsilon}_1 - \boldsymbol{\varepsilon}_2\|^2_{\mathbb{S}^d} \\
\quad \textit{for all } \boldsymbol{\varepsilon}_1, \boldsymbol{\varepsilon}_2 \in \mathbb{S}^d, \textit{ a.e. } \boldsymbol{x} \in \Omega \textit{ with } m_{\mathcal{F}} > 0. \\
\text{(b) } (\zeta_1 - \zeta_2)(r_1 - r_2) \geq -m_\nu |r_1 - r_2|^2 \textit{ for all } \zeta_i \in \partial j_\nu(\boldsymbol{x}, r_i), \\
\quad r_i \in \mathbb{R},\ i = 1,2, \textit{ a.e. } \boldsymbol{x} \in \Gamma_C \textit{ with } m_\nu \geq 0. \\
\text{(c) } (\boldsymbol{\zeta}_1 - \boldsymbol{\zeta}_2)\cdot(\boldsymbol{\xi}_1 - \boldsymbol{\xi}_2) \geq -m_\tau \|\boldsymbol{\xi}_1 - \boldsymbol{\xi}_2\|^2_{\mathbb{R}^d} \textit{ for all} \\
\quad \boldsymbol{\zeta}_i \in \partial j_\tau(\boldsymbol{x},\boldsymbol{\xi}_i), \boldsymbol{\xi}_i \in \mathbb{R}^d, i = 1,2, \textit{ a.e. } \boldsymbol{x} \in \Gamma_C \\
\quad \textit{with } m_\tau \geq 0. \\
\text{(d) } (\zeta_1 - \zeta_2)(r_1 - r_2) \geq -m_e |r_1 - r_2|^2 \textit{ for all } \zeta_i \in \partial j_e(\boldsymbol{x}, r_i), \\
\quad r_i \in \mathbb{R}, i = 1,2, \textit{ a.e. } \boldsymbol{x} \in \Gamma_C \textit{ with } m_e \geq 0. \\
\text{(e) } \min\{m_{\mathcal{F}}, m_\beta\} > \max\{m_\nu k_\nu, m_\tau k_\tau, m_e k_e\}\, c_e^2\, \|\gamma\|^2.
\end{array}\right\} \quad (7.77)$$

Assume, in addition, that one of the following hypotheses

(i) (7.8)(a)–(c), (7.9)(a)–(c), (7.60)(a) –(c) *and*

$$\min\{\widehat{\alpha}, m_\beta\} > \sqrt{3}\,(c_{1\nu}k_\nu + c_{1\tau}k_\tau + c_{1e}k_e)\, c_e^2\, \|\gamma\|^2$$

(ii) (7.8), (7.9) *and* (7.60)

is satisfied and (7.71) *holds. Then Problem* 7.12 *has a unique solution.*

Proof. In order to obtain the existence and uniqueness of solution to Problem 7.12, we apply Theorem 4.20 in a suitable framework. To this end, we consider the product space $Y = V \times \Phi \subset H^1(\Omega;\mathbb{R}^{d+1})$ with the inner product defined on page 220 and the operator $A\colon Y \to Y^*$ defined by (7.72). We define the potential $j\colon \Gamma_C \times \mathbb{R}^{d+1} \to \mathbb{R}$ by

$$j(\boldsymbol{x},\boldsymbol{\xi},r) = k_\nu j_\nu(\boldsymbol{x},\xi_\nu - g_0(\boldsymbol{x})) + k_\tau j_\tau(\boldsymbol{x},\boldsymbol{\xi}_\tau) + k_e j_e(\boldsymbol{x},r - \varphi_0(\boldsymbol{x})) \tag{7.78}$$

for all $(\boldsymbol{\xi},r) \in \mathbb{R}^d \times \mathbb{R}$, a.e. $\boldsymbol{x} \in \Gamma_C$, and we consider the element $(\boldsymbol{f},q) \in V^* \times \Phi^* = Y^*$ defined by (7.16) and (7.68). Under the notation above, using arguments similar to those presented in the proof of Theorem 7.11, it is easy to see that Problem 7.12 can be formulated, equivalently, as follows:

$$\left.\begin{array}{l} \text{find } \boldsymbol{y} = (\boldsymbol{u},\varphi) \in Y \text{ such that} \\[2mm] \langle A\boldsymbol{y},\boldsymbol{z}\rangle_{Y^*\times Y} + \displaystyle\int_{\Gamma_C} j^0(\gamma\boldsymbol{y};\gamma\boldsymbol{z})\,d\Gamma \ge \langle(\boldsymbol{f},q),\boldsymbol{z}\rangle_{Y^*\times Y} \\[2mm] \hfill \text{for all } \boldsymbol{z} \in Y. \end{array}\right\} \tag{7.79}$$

From the proof of Theorem 7.11, we know that the operator A is pseudomonotone and coercive with constant $\alpha = \min\{\widehat{\alpha}, m_\beta\} > 0$. It is also strongly monotone since by (7.74) and the hypotheses (7.77)(a) and (7.59)(c), we have

$$\begin{aligned} &\langle A\boldsymbol{y}_1 - A\boldsymbol{y}_2, \boldsymbol{y}_1 - \boldsymbol{y}_2\rangle_{Y^*\times Y} \\ &\quad = \langle \mathcal{F}\boldsymbol{\varepsilon}(\boldsymbol{u}_1) - \mathcal{F}\boldsymbol{\varepsilon}(\boldsymbol{u}_2), \boldsymbol{\varepsilon}(\boldsymbol{u}_1) - \boldsymbol{\varepsilon}(\boldsymbol{u}_2)\rangle_{\mathcal{H}} \\ &\quad\quad + \langle \boldsymbol{\beta}\nabla\varphi_1 - \boldsymbol{\beta}\nabla\varphi_2, \nabla\varphi_1 - \nabla\varphi_2\rangle_H \ge m_{\mathcal{F}}\,\|\boldsymbol{u}_1 - \boldsymbol{u}_2\|_V^2 \\ &\quad\quad + m_\beta\|\varphi_1 - \varphi_2\|_\Phi^2 \ge \min\{\,m_{\mathcal{F}}, m_\beta\,\}\,\|\boldsymbol{y}_1 - \boldsymbol{y}_2\|_Y^2 \end{aligned}$$

for all $\boldsymbol{y}_i = (\boldsymbol{u}_i,\varphi_i) \in Y$, $i = 1, 2$. Hence, the hypothesis (4.12) holds with $m_1 = \min\{\,m_{\mathcal{F}}, m_\beta\,\} > 0$.

Next, we study the properties of the function j defined by (7.78). We use arguments similar to those used in the proof of Theorem 7.11 and, for this reason, we skip the details.

First, note that from Propositions 3.37 and 3.38, we have

$$\begin{aligned} j^0(\boldsymbol{x},\boldsymbol{\xi},r;\boldsymbol{\varrho},s) &\le k_\nu j_\nu^0(\boldsymbol{x},\xi_\nu - g_0(\boldsymbol{x});\varrho_\nu) \\ &\quad + k_\tau j_\tau^0(\boldsymbol{x},\boldsymbol{\xi}_\tau;\boldsymbol{\varrho}_\tau) + k_e j_e^0(\boldsymbol{x},r - \varphi_0(\boldsymbol{x});s), \end{aligned} \tag{7.80}$$

$$\begin{aligned} \partial j(\boldsymbol{x},\boldsymbol{\xi},r) \subseteq &\Big(k_\nu \partial j_\nu(\boldsymbol{x},\xi_\nu - g_0(\boldsymbol{x}))\,\boldsymbol{\nu} + k_\tau\,[\partial j_\tau(\boldsymbol{x},\boldsymbol{\xi}_\tau)]_\tau\Big) \\ &\times k_e \partial j_e(\boldsymbol{x},r - \varphi_0(\boldsymbol{x})) \end{aligned} \tag{7.81}$$

for all $(\boldsymbol{\xi}, r)$, $(\boldsymbol{\varrho}, s) \in \mathbb{R}^d \times \mathbb{R}$, a.e. $\boldsymbol{x} \in \Gamma_C$.

In order to verify (4.42)(c), let $(\boldsymbol{x}, \boldsymbol{\xi}, r) \in \Gamma_C \times \mathbb{R}^{d+1}$ and $(\boldsymbol{\zeta}, s) \in \mathbb{R}^d \times \mathbb{R}$ be such that $(\boldsymbol{\zeta}, s) \in \partial j(\boldsymbol{x}, \boldsymbol{\xi}, r)$. From (7.81), we have $\boldsymbol{\zeta} = \overline{\zeta}\, \boldsymbol{\nu} + [\overline{\overline{\boldsymbol{\zeta}}}]_\tau$ with $\overline{\zeta} \in k_\nu \partial j_\nu(\boldsymbol{x}, \xi_\nu - g_0(\boldsymbol{x}))$, $\overline{\overline{\boldsymbol{\zeta}}} \in k_\tau \partial j_\tau(\boldsymbol{x}, \boldsymbol{\xi}_\tau)$ and $s \in k_e \partial j_e(\boldsymbol{x}, r - \varphi_0(\boldsymbol{x}))$. Using assumptions (7.8)(c), (7.9)(c), and (7.60)(c), we obtain

$$\|(\boldsymbol{\zeta}, s)\|_{\mathbb{R}^{d+1}} \le |\overline{\zeta}| + \|[\overline{\overline{\boldsymbol{\zeta}}}]_\tau\|_{\mathbb{R}^d} + |s| \le k_\nu\, (c_{0\nu} + c_{1\nu}|g_0(\boldsymbol{x})| + c_{1\nu}|\xi_\nu|)$$

$$+k_\tau\, (c_{0\tau} + c_{1\tau}\|\boldsymbol{\xi}_\tau\|_{\mathbb{R}^d}) + k_e\, (c_{0e} + c_{1e}|\varphi_0(\boldsymbol{x})| + c_{1e}|r|)\,.$$

Thus, by (7.65) we deduce $\|\partial j(\boldsymbol{x}, \boldsymbol{\xi}, r)\|_{\mathbb{R}^{d+1}} \le \overline{c}_0 + \overline{c}_1\|(\boldsymbol{\xi}, r)\|_{\mathbb{R}^{d+1}}$ for all $(\boldsymbol{\xi}, r) \in \mathbb{R}^d \times \mathbb{R}$, a.e. $\boldsymbol{x} \in \Gamma_C$ with $\overline{c}_0 > 0$ and $\overline{c}_1 = c_{1\nu}k_\nu + c_{1\tau}k_\tau + c_{1e}k_e \ge 0$.

Next, using (7.80) and hypotheses (7.8)(e), (7.9)(e), and (7.60)(e), we have

$$j^0(\boldsymbol{x}, \boldsymbol{\xi}, r; -\boldsymbol{\xi}, -r) \le d_\nu k_\nu(1 + |\xi_\nu| + |g_0(\boldsymbol{x})|) + d_\tau k_\tau(1 + \|\boldsymbol{\xi}_\tau\|_{\mathbb{R}^d})$$

$$+d_e k_e(1 + |r| + |\varphi_0(\boldsymbol{x})|) \le \overline{d}_0\, (1 + \|(\boldsymbol{\xi}, r)\|_{\mathbb{R}^{d+1}})$$

for all $(\boldsymbol{\xi}, r) \in \mathbb{R}^d \times \mathbb{R}$, a.e. $\boldsymbol{x} \in \Gamma_C$ with $\overline{d}_0 \ge 0$. It remains to check the relaxed monotonicity condition for j. We consider $(\boldsymbol{\zeta}_i, s_i) \in \partial j(\boldsymbol{x}, \boldsymbol{\xi}_i, r_i)$ where $\boldsymbol{\zeta}_i, \boldsymbol{\xi}_i \in \mathbb{R}^d$, $s_i, r_i \in \mathbb{R}$, $i = 1, 2$. By the formula (7.81), we get $\boldsymbol{\zeta}_i = \overline{\zeta}_i\, \boldsymbol{\nu} + [\overline{\overline{\boldsymbol{\zeta}}}_i]_\tau$ with $\overline{\zeta}_i \in \mathbb{R}$, $\overline{\overline{\boldsymbol{\zeta}}}_i \in \mathbb{R}^d$, $\overline{\zeta}_i \in k_\nu \partial j_\nu(\boldsymbol{x}, \xi_\nu - g_0(\boldsymbol{x}))$, $\overline{\overline{\boldsymbol{\zeta}}}_i \in k_\tau \partial j_\tau(\boldsymbol{x}, \boldsymbol{\xi}_\tau)$ and $s_i \in k_e \partial j_e(\boldsymbol{x}, r_i - \varphi_0(\boldsymbol{x}))$, $i = 1, 2$. Using (7.77)(b)–(d) and (7.23), we have

$$\begin{aligned}
&((\boldsymbol{\zeta}_1, s_1) - (\boldsymbol{\zeta}_2, s_2)) \cdot ((\boldsymbol{\xi}_1, r_1) - (\boldsymbol{\xi}_2, r_2)) \\
&= (\boldsymbol{\zeta}_1 - \boldsymbol{\zeta}_2, s_1 - s_2) \cdot (\boldsymbol{\xi}_1 - \boldsymbol{\xi}_2, r_1 - r_2) \\
&= (\boldsymbol{\zeta}_1 - \boldsymbol{\zeta}_2) \cdot (\boldsymbol{\xi}_1 - \boldsymbol{\xi}_2) + (s_1 - s_2)(r_1 - r_2) \\
&\ge -\max\{\, m_\nu k_\nu, m_\tau k_\tau \,\}\, \|\boldsymbol{\xi}_1 - \boldsymbol{\xi}_2\|^2_{\mathbb{R}^d} - m_e k_e |r_1 - r_2|^2 \\
&\ge -\max\{\, m_\nu k_\nu, m_\tau k_\tau, m_e k_e \,\}\, \|(\boldsymbol{\xi}_1, r_1) - (\boldsymbol{\xi}_2, r_2)\|^2_{\mathbb{R}^{d+1}},
\end{aligned}$$

which proves (4.42)(d) with $m_2 = \max\{\, m_\nu k_\nu, m_\tau k_\tau, m_e k_e \,\} \ge 0$. We conclude from here that the function j given by (7.78) satisfies condition (4.42). Finally, we note that (4.43) is satisfied and, moreover, the inequality in hypothesis (i) of Theorem 4.20 also holds.

We are now in a position to apply Theorem 4.20 by choosing the space Y as playing the role of the abstract space V and $s = d + 1$. It follows from here that problem (7.79) has a unique solution $\boldsymbol{y} = (\boldsymbol{u}, \varphi) \in Y$, which ends the proof. □

A quadruple of functions $(\boldsymbol{u}, \boldsymbol{\sigma}, \varphi, \boldsymbol{D})$ which satisfies (7.49), (7.50), (7.69), and (7.70) is called a *weak solution* of Problem 7.9. It follows from above that, under the assumptions of Theorem 7.11, there exists at least one weak solution of

Problem 7.9. To describe precisely the regularity of the weak solution, we note that the constitutive relations (7.49) and (7.50), assumptions (7.7), (7.58), (7.59), and regularity $\boldsymbol{u} \in V$, $\varphi \in \Phi$ show that $\boldsymbol{\sigma} \in \mathcal{H}$ and $\boldsymbol{D} \in H$. Moreover, using (7.49), (7.50), (7.69), and (7.70), it follows that (7.66) and (7.67) hold for all $\boldsymbol{v} \in V$ and $\psi \in \Phi$. Then using arguments similar to those used on page 210 we deduce that

$$\operatorname{Div} \boldsymbol{\sigma} + \boldsymbol{f}_0 = \boldsymbol{0}, \quad \operatorname{div} \boldsymbol{D} - q_0 = 0 \quad \text{in } \Omega.$$

It follows now from (7.64) that $\operatorname{Div} \boldsymbol{\sigma} \in H$ and $\operatorname{div} \boldsymbol{D} \in L^2(\Omega)$. We conclude that the weak solution of Problem 7.9 satisfies $(\boldsymbol{u}, \boldsymbol{\sigma}, \varphi, \boldsymbol{D}) \in V \times \mathcal{H}_1 \times \Phi \times W$. In addition, under the assumptions of Theorem 7.13, the weak solution is unique.

7.4 Examples

We end this chapter with some examples of constitutive laws of the form (7.2) for which assumption (7.7) is satisfied. Then, we present basic examples of frictional contact conditions of the forms (7.5) and (7.6) for which assumptions (7.8), (7.9), and (7.10) are satisfied, too. These examples lead to mathematical models for which the existence and uniqueness results presented in Sect. 7.1 work. Also, we provide examples of friction law of the form (7.30) for which assumption (7.34) holds. Describing these examples we develop arguments which can be used to obtain various concrete models of contact for which the results presented both in the rest of Chap. 7 and in Chap. 8 are valid. Nevertheless, since the modifications are straightforward, we restrict ourselves to the examples described in this section.

Constitutive laws. We consider an elasticity operator $\mathcal{F}$ which satisfies

$$\left.\begin{array}{l}
\mathcal{F}: \Omega \times \mathbb{S}^d \to \mathbb{S}^d \text{ is such that} \\
\quad \text{(a) } \mathcal{F}(\cdot, \boldsymbol{\varepsilon}) \text{ is measurable on } \Omega \text{ for all } \boldsymbol{\varepsilon} \in \mathbb{S}^d. \\
\quad \text{(b) } \|\mathcal{F}(\boldsymbol{x}, \boldsymbol{\varepsilon}_1) - \mathcal{F}(\boldsymbol{x}, \boldsymbol{\varepsilon}_2)\|_{\mathbb{S}^d} \le L_{\mathcal{F}} \|\boldsymbol{\varepsilon}_1 - \boldsymbol{\varepsilon}_2\|_{\mathbb{S}^d} \\
\qquad \text{for all } \boldsymbol{\varepsilon}_1, \boldsymbol{\varepsilon}_2 \in \mathbb{S}^d, \text{ a.e. } \boldsymbol{x} \in \Omega \text{ with } L_{\mathcal{F}} > 0. \\
\quad \text{(c) } (\mathcal{F}(\boldsymbol{x}, \boldsymbol{\varepsilon}_1) - \mathcal{F}(\boldsymbol{x}, \boldsymbol{\varepsilon}_2)) : (\boldsymbol{\varepsilon}_1 - \boldsymbol{\varepsilon}_2) \ge m_{\mathcal{F}} \|\boldsymbol{\varepsilon}_1 - \boldsymbol{\varepsilon}_2\|^2_{\mathbb{S}^d} \\
\qquad \text{for all } \boldsymbol{\varepsilon}_1, \boldsymbol{\varepsilon}_2 \in \mathbb{S}^d, \text{ a.e. } \boldsymbol{x} \in \Omega \text{ with } m_{\mathcal{F}} > 0. \\
\quad \text{(d) } \mathcal{F}(\boldsymbol{x}, \boldsymbol{0}) = \boldsymbol{0} \text{ for a.e. } \boldsymbol{x} \in \Omega.
\end{array}\right\} \quad (7.82)$$

It is obvious to see that condition (7.82) implies condition (7.7) with $a_0 = 0$, $a_1 = L_{\mathcal{F}}$, and $\widehat{\alpha} = m_{\mathcal{F}}$. This remark leads to several examples of elasticity operators for which condition (7.7) is satisfied, by checking only the validity of condition (7.82).

Example 7.14. Assume that $\mathcal{F}$ is a linear and positive definite operator with respect to the second variable, i.e.,

$$\left.\begin{array}{l}\mathcal{F}: \Omega \times \mathbb{S}^d \to \mathbb{S}^d \text{ is such that}\\[4pt] \text{(a) } \mathcal{F}(\boldsymbol{x}, \boldsymbol{\varepsilon}) = f(\boldsymbol{x})\boldsymbol{\varepsilon} \text{ for all } \boldsymbol{\varepsilon} \in \mathbb{S}^d, \text{a.e. } \boldsymbol{x} \in \Omega.\\[4pt] \text{(b) } f(\boldsymbol{x}) = \big(f_{ijkl}(\boldsymbol{x})\big) \text{ with } f_{ijkl} = f_{jikl} = f_{lkij} \in L^\infty(\Omega).\\[4pt] \text{(c) } f_{ijkl}(\boldsymbol{x})\varepsilon_{ij}\varepsilon_{kl} \ge m_{\mathcal{F}} \|\boldsymbol{\varepsilon}\|^2_{\mathbb{S}^d} \text{ for all } \boldsymbol{\varepsilon} = (\varepsilon_{ij}) \in \mathbb{S}^d,\\ \qquad \text{a.e. } \boldsymbol{x} \in \Omega \text{ with } m_{\mathcal{F}} > 0.\end{array}\right\} \tag{7.83}$$

It is clear that $\mathcal{F}$ satisfies condition (7.82) with $L_{\mathcal{F}} = \|f\|_{L^\infty(\Omega;\mathbb{S}^{2d})}$. In particular, it follows that the operator (6.12) satisfies condition (7.82), if μ, $\lambda > 0$. We conclude from here that the results in Sect. 7.1 are valid for the linear constitutive law (6.10) and, in particular, for the constitutive law of linearly elastic isotropic materials, (6.11).

Example 7.15. Let $\mathcal{E}: \Omega \times \mathbb{S}^d \to \mathbb{S}^d$ be an operator that satisfies (7.83). Consider the operator $\mathcal{F}$ given by (6.15), where $\beta > 0$, K is a closed convex subset of $\mathbb{S}^d$ such that $\boldsymbol{0} \in K$ and $\mathcal{P}_K: \mathbb{S}^d \to K$ is the projection operator. Then, using the properties of the projection map in Proposition 1.24, it is easy to see that the operator (6.15) satisfies condition (7.82). We conclude that the results in Sect. 7.1 are valid for the nonlinear constitutive law (6.14).

Example 7.16. It was shown in [102, p. 125] that, under the assumptions (6.22), the nonlinear operator defined by (6.21) satisfies condition (7.82). This implies that the results in Sect. 7.1 can be used in the study of the Hencky materials (6.20) as well.

We note that Examples 7.14, 7.15, and 7.16 provide operators $\mathcal{B}$ which satisfy assumption (7.31) too and, therefore, they can be used to construct concrete models of viscoelastic contact for which our results in Sect. 7.2 are valid.

Single-valued contact conditions. We turn to examples of single-valued contact conditions which lead to subdifferential conditions of the form (7.5). For simplicity, we consider only the case when there is no gap between the body and the foundation, i.e., $g_0 = 0$. These conditions are obtained in the framework described in Sect. 6.3 and resumed below. Consider the normal compliance contact condition

$$-\sigma_\nu = k_\nu\, p_\nu(u_\nu) \quad \text{on } \Gamma_C, \tag{7.84}$$

where $p_\nu: \mathbb{R} \to \mathbb{R}$ is a prescribed nonnegative continuous function which vanishes when its argument is negative and $k_\nu \in L^\infty(\Gamma_C)$ is a positive function, the stiffness coefficient. Let $g_\nu: \mathbb{R} \to \mathbb{R}$ and $j_\nu: \Gamma_C \times \mathbb{R} \to \mathbb{R}$ be the functions defined by

$$g_\nu(r) = \int_0^r p_\nu(s)\, ds \quad \text{for all } r \in \mathbb{R}, \tag{7.85}$$

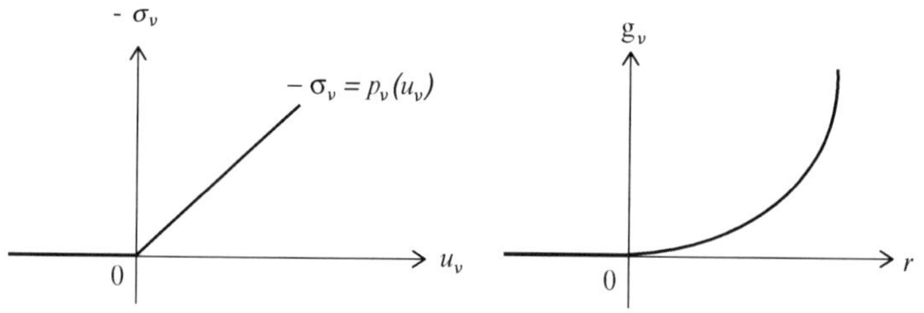

Fig. 7.1 Contact condition in Example 7.17

$$j_\nu(\boldsymbol{x}, r) = k_\nu(\boldsymbol{x})\, g_\nu(r) \quad \text{for all } r \in \mathbb{R}, \text{ a.e. } \boldsymbol{x} \in \Gamma_C. \tag{7.86}$$

Then, as explained on page 186, using Lemma 3.50(iii) we have

$$\partial g_\nu(r) = p_\nu(r), \quad \partial j_\nu(\boldsymbol{x}, r) = k_\nu(\boldsymbol{x})\, \partial g_\nu(r) = k_\nu(\boldsymbol{x})\, p_\nu(r)$$

for all $r \in \mathbb{R}$, a.e. $\boldsymbol{x} \in \Gamma_C$ and, therefore, it is easy to see that the contact condition (7.84) is of the form (7.5). The following concrete examples lead to functions j_ν which satisfy condition (7.8).

Example 7.17. Let $p_\nu : \mathbb{R} \to \mathbb{R}$ be the function given by

$$p_\nu(r) = a\, r_+ = \begin{cases} 0 & \text{if } r < 0, \\ ar & \text{if } r \ge 0, \end{cases} \tag{7.87}$$

with $a > 0$. Then, using (7.85) we have

$$g_\nu(r) = \begin{cases} 0 & \text{if } r < 0, \\ \dfrac{ar^2}{2} & \text{if } r \ge 0. \end{cases} \tag{7.88}$$

Clearly the function p_ν is convex (hence continuous and regular) and increasing. In this case the function j_ν satisfies hypotheses (7.8) with constants $c_{0\nu} = 0$ and $c_{1\nu} = a\, \|k_\nu\|_{L^\infty(\Gamma_C)}$. Moreover, since p_ν is increasing, from Corollary 3.53 on page 80, it follows that the function j_ν satisfies the relaxed monotonicity condition (7.21)(b) with $m_\nu = 0$. The contact condition (7.84) for $k_\nu \equiv 1$ and p_ν given by (7.87), as well as the potential (7.88), is depicted in Fig. 7.1. This contact condition corresponds to a linear dependence of the reactive force with respect to the penetration and, therefore, it models a linearly elastic behavior of the foundation.

Example 7.18. Let $p_\nu : \mathbb{R} \to \mathbb{R}$ be the function given by

$$p_\nu(r) = \begin{cases} 0 & \text{if } r < 0, \\ ar & \text{if } 0 \le r \le l, \\ al & \text{if } r > l, \end{cases} \tag{7.89}$$

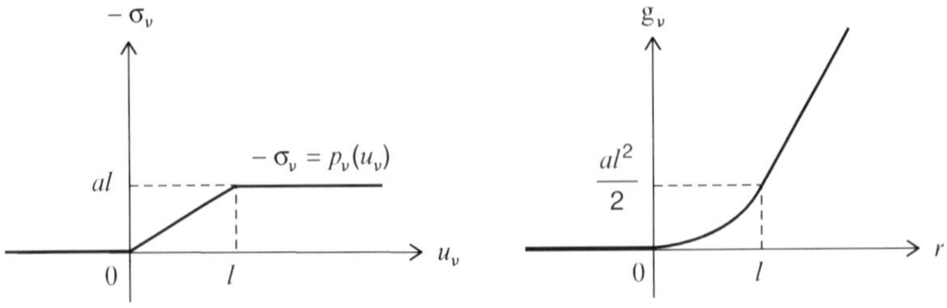

Fig. 7.2 Contact condition in Example 7.18

with $a > 0$ and $l > 0$. Then, using (7.85) we have

$$g_\nu(r) = \begin{cases} 0 & \text{if } r < 0, \\ \dfrac{ar^2}{2} & \text{if } 0 \le r \le l, \\ alr - \dfrac{al^2}{2} & \text{if } r > l. \end{cases} \tag{7.90}$$

Clearly the function p_ν is continuous and satisfies the inequality $|p_\nu(r)| \le al$ for all $r \in \mathbb{R}$. Moreover, the function g_ν defined by (7.85) is continuously differentiable (hence regular), $\partial g_\nu(r) = g'_\nu(r) = p_\nu(r)$ for all $r \in \mathbb{R}$, and by Lemma 3.50(iii), we have

$$\partial j_\nu(\boldsymbol{x}, r) = k_\nu(\boldsymbol{x}) p_\nu(r)$$

for all $r \in \mathbb{R}$, a.e. $\boldsymbol{x} \in \Gamma_C$. In this case the function j_ν satisfies hypotheses (7.8) with constants $c_{0\nu} = al\, \|k_\nu\|_{L^\infty(\Gamma_C)}$ and $c_{1\nu} = 0$. In addition, since p_ν is increasing, Corollary 3.53 implies that the function j_ν satisfies (7.21)(b) with $m_\nu = 0$. The contact condition (7.84) for $k_\nu \equiv 1$ and p_ν given by (7.89), as well as the potential (7.90), is depicted in Fig. 7.2. This contact condition corresponds to an elastic-perfect plastic behavior of the foundation. The plasticity consists in the fact that when the penetration reaches the limit l, then the surface offers no additional resistance.

Example 7.19. Let $p_\nu \colon \mathbb{R} \to \mathbb{R}$ be the function given by

$$p_\nu(r) = \begin{cases} 0 & \text{if } r < 0, \\ \dfrac{a + e^{-b}}{b}\, r & \text{if } 0 \le r \le b, \\ e^{-r} + a & \text{if } r > b, \end{cases} \tag{7.91}$$

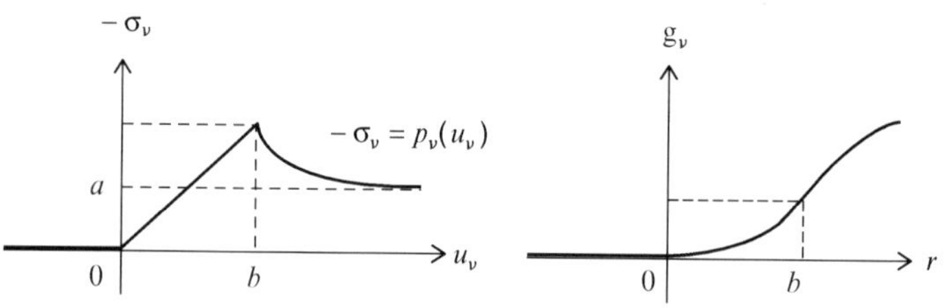

Fig. 7.3 Contact condition in Example 7.19

with $a \geq 0$, $b > 0$. Then, using (7.85) we have

$$g_\nu(r) = \begin{cases} 0 & \text{if } r < 0, \\ \dfrac{a + e^{-b}}{2b}\, r^2 & \text{if } 0 \leq r \leq b, \\ ar - e^{-r} + \dfrac{(b+2)e^{-b} - ab}{2} & \text{if } r > b. \end{cases} \tag{7.92}$$

It is easy to see that p_ν is a continuous function; hence, the function g_ν is continuously differentiable. We infer that $j_\nu(\boldsymbol{x}, \cdot)$ is regular and, from Lemma 3.50(iii), we have $\partial g_\nu(r) = p_\nu(r)$ for all $r \in \mathbb{R}$. Hence

$$|\partial j_\nu(\boldsymbol{x}, r)| = |k_\nu(\boldsymbol{x})|\, |p_\nu(r)| \leq (a + e^{-b})\, \|k_\nu\|_{L^\infty(\Gamma_C)}$$

for all $r \in \mathbb{R}$, a.e. $\boldsymbol{x} \in \Gamma_C$. In this case the function j_ν satisfies hypotheses (7.8) with constants $c_{0\nu} = (a + e^{-b})\, \|k_\nu\|_{L^\infty(\Gamma_C)}$ and $c_{1\nu} = 0$. Again, from Corollary 3.53, performing a direct computation we can show that the relaxed monotonicity condition (7.21)(b) is satisfied with $m_\nu = e^{-b}\, \|k_\nu\|_{L^\infty(\Gamma_C)}$. Note that in contrast to the previous two examples, here the function p_ν is not increasing and, therefore, the potential function j_ν is not a convex function. The contact condition (7.84) for $k_\nu \equiv 1$ and p_ν given by (7.91), as well as the potential (7.92), is depicted in Fig. 7.3. This contact condition corresponds to an elastic-plastic behavior of the foundation, with softening. The softening effect consists in the fact that, when the penetration reach the limit b, then the reactive force decreases.

Multivalued contact conditions. We turn now to examples of multivalued contact conditions which lead to a subdifferential condition of the form (7.5). They are constructed by using the "filling in a gap procedure" described on page 78 that we resume below, for convenience of the reader. Let $p_\nu \in L^\infty_{loc}(\mathbb{R})$ be such that for all $r \in \mathbb{R}$

$$\text{there exist } \lim_{\xi \to r-} p_\nu(\xi) = p_\nu(r-) \in \mathbb{R} \text{ and } \lim_{\xi \to r+} p_\nu(\xi) = p_\nu(r+) \in \mathbb{R}.$$

We define the multivalued function $\widehat{p}_\nu\colon \mathbb{R} \to 2^{\mathbb{R}}$ by

$$\widehat{p}_\nu(r) = \begin{cases} [p_\nu(r-), p_\nu(r+)] \text{ if } & p_\nu(r-) \le p_\nu(r+) \\ [p_\nu(r+), p_\nu(r-)] \text{ if } & p_\nu(r+) \le p_\nu(r-) \end{cases} \quad \text{for all } r \in \mathbb{R}.$$

Also, we consider the functions $g_\nu\colon \mathbb{R} \to \mathbb{R}$ and $j_\nu\colon \Gamma_C \times \mathbb{R} \to \mathbb{R}$ given by (7.85) and (7.86), respectively, where, again, $k_\nu \in L^\infty(\Gamma_C)$ is a positive function. Then, using Lemma 3.50(iii), we have

$$\partial g_\nu(r) = \widehat{p}_\nu(r), \quad \partial j_\nu(\boldsymbol{x}, r) = k_\nu(\boldsymbol{x})\, \partial g_\nu(r) = k_\nu(\boldsymbol{x})\, \widehat{p}_\nu(r)$$

for all $r \in \mathbb{R}$, a.e. $\boldsymbol{x} \in \Gamma_C$.

Assume now that the normal stress σ_ν satisfies the multivalued condition

$$-\sigma_\nu \in k_\nu\, \widehat{p}_\nu(u_\nu) \quad \text{on } \Gamma_C. \tag{7.93}$$

Then, it is easy to see that this contact condition is of the form (7.5), with a zero gap function. The following concrete examples lead to functions j_ν which satisfy condition (7.8).

Example 7.20. Let $p_\nu\colon \mathbb{R} \to \mathbb{R}$ be the function given by

$$p_\nu(r) = \begin{cases} 0 & \text{if } r < 0, \\ M & \text{if } r \ge 0, \end{cases} \tag{7.94}$$

with $M \ge 0$. It is clear that $p_\nu \in L^\infty(\mathbb{R})$ and, moreover, there exists $\lim_{\xi \to r\pm} p_\nu(\xi) = p_\nu(r\pm) \in \mathbb{R}$ for all $r \in \mathbb{R}$. Then, using the filling in a gap procedure, we have

$$\widehat{p}_\nu(r) = \begin{cases} 0 & \text{if } r < 0, \\ [0, M] & \text{if } r = 0, \\ M & \text{if } r > 0, \end{cases} \tag{7.95}$$

and, in addition,

$$g_\nu(r) = \begin{cases} 0 & \text{if } r < 0, \\ Mr & \text{if } r \ge 0. \end{cases} \tag{7.96}$$

It follows that $|p_\nu(r)| \le M$ and $\partial g_\nu(r) = \widehat{p}_\nu(r)$ for all $r \in \mathbb{R}$. Also,

$$|\partial j_\nu(\boldsymbol{x}, r)| = |k_\nu(\boldsymbol{x})|\, |\widehat{p}_\nu(r)| \le M\, \|k_\nu\|_{L^\infty(\Gamma_C)}$$

for all $r \in \mathbb{R}$, a.e. $\boldsymbol{x} \in \Gamma_C$. In this case the function j_ν satisfies hypotheses (7.8) with constants $c_{0\nu} = M\, \|k_\nu\|_{L^\infty(\Gamma_C)}$ and $c_{1\nu} = 0$. Moreover, since $\widehat{p}_\nu$ has a monotone graph, it follows that (7.21)(b) holds with $m_\nu = 0$. The contact condition (7.93) for $k_\nu \equiv 1$ and $\widehat{p}_\nu$ given by (7.95), as well as the potential (7.96), is depicted in

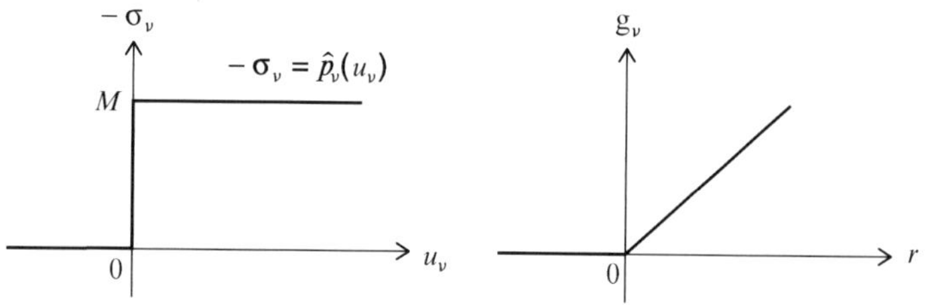

Fig. 7.4 Contact condition in Example 7.20

Fig. 7.4. This contact condition corresponds to a rigid-perfect plastic behavior of the foundation. The rigid behavior corresponds to the fact that, as far as the magnitude of the normal force is less than the critical limit M, there is no penetration. The perfect plastic behavior is given by the fact that, when the normal force reaches the limit M, then the foundation offers no additional resistance.

Example 7.21. Let $p_\nu : \mathbb{R} \to \mathbb{R}$ be the function given by

$$p_\nu(r) = \begin{cases} 0 & \text{if } r < 0, \\ -\dfrac{b}{a} r + b & \text{if } 0 \le r \le a, \\ 0 & \text{if } r > a \end{cases} \tag{7.97}$$

with $a,\ b > 0$. It is clear that $p_\nu \in L^\infty(\mathbb{R})$ and, moreover, there exists $\lim_{\xi \to r^\pm} p_\nu(\xi) = p_\nu(r\pm) \in \mathbb{R}$ for all $r \in \mathbb{R}$. Then, using the filling in a gap procedure, we have

$$\widehat{p}_\nu(r) = \begin{cases} 0 & \text{if } r < 0, \\ [0, b] & \text{if } r = 0, \\ -\dfrac{b}{a} r + b & \text{if } 0 < r \le a, \\ 0 & \text{if } r > a. \end{cases} \tag{7.98}$$

In this case the function g_ν is given by

$$g_\nu(r) = \begin{cases} 0 & \text{if } r < 0, \\ -\dfrac{b}{2a} r^2 + br & \text{if } 0 \le r \le a, \\ \dfrac{ab}{2} & \text{if } r > a. \end{cases} \tag{7.99}$$

We have

$$|\partial j_\nu(\boldsymbol{x}, r)| = |k_\nu(\boldsymbol{x})|\, |\widehat{p}_\nu(r)| \le b\, \|k_\nu\|_{L^\infty(\Gamma_C)}$$

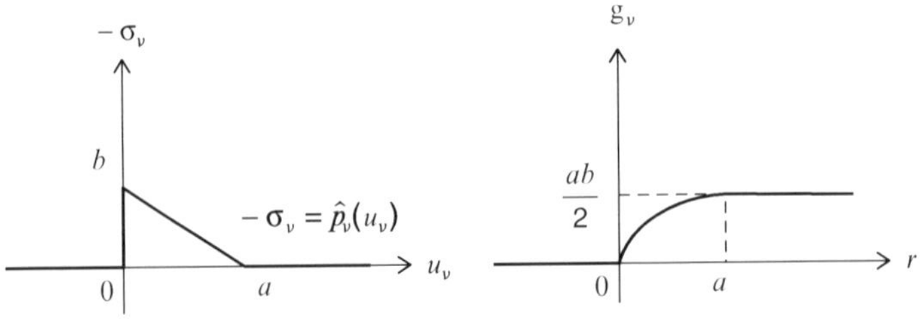

Fig. 7.5 Contact condition in Example 7.21

for all $r \in \mathbb{R}$, a.e. $\boldsymbol{x} \in \Gamma_C$. Therefore, the function j_ν satisfies hypotheses (7.8) with constants $c_{0\nu} = b\,\|k_\nu\|_{L^\infty(\Gamma_C)}$ and $c_{1\nu} = 0$. From Corollary 3.53, by direct computation we know that $j_\nu(\boldsymbol{x},\cdot)$ satisfies the relaxed monotonicity condition (7.21)(b) with $m_\nu = \frac{b}{a}\,\|k_\nu\|_{L^\infty(\Gamma_C)}$. Moreover, we show that $j_\nu(\boldsymbol{x},\cdot)$ is regular for a.e. $\boldsymbol{x} \in \Gamma_C$, i.e. (7.19) is also satisfied. Indeed, $j_\nu(\boldsymbol{x},\cdot)$ can be represented as the difference of convex functions, i.e.,

$$j_\nu(\boldsymbol{x},r) = \varphi_1(\boldsymbol{x},r) - \varphi_2(\boldsymbol{x},r)$$

with $\varphi_1(\boldsymbol{x},r) = k_\nu(\boldsymbol{x})h_1(r)$, where

$$h_1(r) = \begin{cases} \dfrac{b}{2a}\,r^2 - br + ab & \text{if } r < 0, \\[2ex] ab & \text{if } 0 \le r \le a, \\[2ex] \dfrac{b}{2a}\,r^2 - br + \dfrac{3ab}{2} & \text{if } r > a \end{cases}$$

and

$$\varphi_2(\boldsymbol{x},r) = k_\nu(\boldsymbol{x})\left(\frac{b}{2a}\,r^2 - br + ab\right)$$

for all $r \in \mathbb{R}$, a.e. $\boldsymbol{x} \in \Gamma_C$. Since $\varphi_1(\boldsymbol{x},\cdot)$ and $\varphi_2(\boldsymbol{x},\cdot)$ are convex functions and $\partial\varphi_2(\boldsymbol{x},\cdot)$ is single-valued, from Proposition 3.42, we deduce that $j_\nu(\boldsymbol{x},\cdot)$ is regular on $\mathbb{R}$ for a.e. $\boldsymbol{x} \in \Gamma_C$ and, in addition,

$$\partial j_\nu(\boldsymbol{x},r) = \partial\varphi_1(\boldsymbol{x},r) - \partial\varphi_2(\boldsymbol{x},r)$$

for all $r \in \mathbb{R}$, a.e. $\boldsymbol{x} \in \Gamma_C$. The contact condition (7.93) for $k_\nu \equiv 1$ and $\widehat{p}_\nu$ given by (7.98), as well as the potential (7.99), is depicted in Fig. 7.5. This contact condition corresponds to a rigid-plastic behavior of the foundation, with softening. The softening effect is in such a way that, if the penetration a is reached, the surface is completely disintegrate and offers no resistance to penetration.

Example 7.22. Let $p_\nu : \mathbb{R} \to \mathbb{R}$ be the function given by

$$p_\nu(r) = \begin{cases} 0 & \text{if } r < 0, \\ e^{-r} + a & \text{if } r \geq 0 \end{cases} \tag{7.100}$$

with $a \geq 0$. Obviously $p_\nu \in L^\infty(\mathbb{R})$ and, moreover, there exists $\lim_{\xi \to r\pm} p_\nu(\xi) = p_\nu(r\pm) \in \mathbb{R}$ for all $r \in \mathbb{R}$. Then, using the filling in a gap procedure, we have

$$\widehat{p}_\nu(r) = \begin{cases} 0 & \text{if } r < 0, \\ [0, 1 + a] & \text{if } r = 0, \\ e^{-r} + a & \text{if } r > 0. \end{cases} \tag{7.101}$$

In this case the function g_ν is given by

$$g_\nu(r) = \begin{cases} 0 & \text{if } r < 0, \\ -e^{-r} + ar + 1 & \text{if } r \geq 0. \end{cases} \tag{7.102}$$

We have

$$|\partial j_\nu(\boldsymbol{x}, r)| = |k_\nu(\boldsymbol{x})|\, |\widehat{p}_\nu(r)| \leq (1 + a)\, \|k_\nu\|_{L^\infty(\Gamma_C)}$$

for all $r \in \mathbb{R}$, a.e. $\boldsymbol{x} \in \Gamma_C$. Therefore, the function j_ν satisfies hypotheses (7.8) with constants $c_{0\nu} = (1+a)\, \|k_\nu\|_{L^\infty(\Gamma_C)}$ and $c_{1\nu} = 0$. Moreover, by a direct computation we can show that the condition (7.21)(b) holds with $m_\nu = 1$. Finally, we observe that $j_\nu(\boldsymbol{x}, \cdot)$ can be written as the difference of convex functions, i.e.,

$$j_\nu(\boldsymbol{x}, r) = \varphi_1(\boldsymbol{x}, r) - \varphi_2(\boldsymbol{x}, r)$$

with $\varphi_1(\boldsymbol{x}, r) = k_\nu(\boldsymbol{x}) h_1(r)$, where

$$h_1(r) = \begin{cases} e^{-r} - ar - 1 & \text{if } r < 0, \\ 0 & \text{if } r \geq 0 \end{cases}$$

and

$$\varphi_2(\boldsymbol{x}, r) = k_\nu(\boldsymbol{x})(e^{-r} - ar - 1)$$

for all $r \in \mathbb{R}$, a.e. $\boldsymbol{x} \in \Gamma_C$. Since $\varphi_1(\boldsymbol{x}, \cdot)$ and $\varphi_2(\boldsymbol{x}, \cdot)$ are convex functions and $\partial\varphi_2(\boldsymbol{x}, \cdot)$ is single-valued, from Proposition 3.42 we have that $j_\nu(\boldsymbol{x}, \cdot)$ is regular on $\mathbb{R}$ for a.e. $\boldsymbol{x} \in \Gamma_C$ and, in addition,

$$\partial j_\nu(\boldsymbol{x}, r) = \partial\varphi_1(\boldsymbol{x}, r) - \partial\varphi_2(\boldsymbol{x}, r)$$

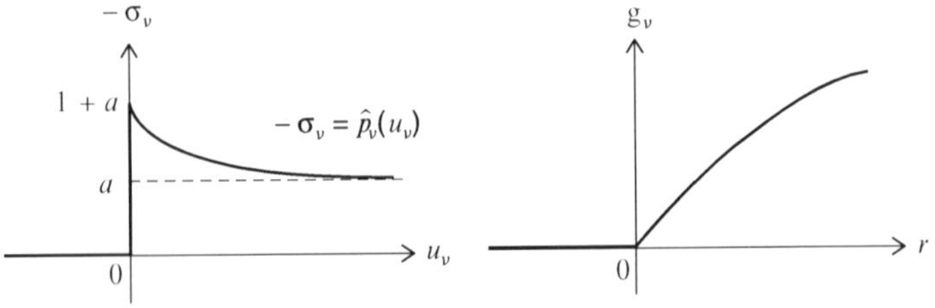

Fig. 7.6 Contact condition in Example 7.22

for all $r \in \mathbb{R}$, a.e. $\boldsymbol{x} \in \Gamma_C$. The contact condition (7.93) for $k_\nu \equiv 1$ and $\widehat{p}_\nu$ given by (7.101), as well as the potential (7.102), is depicted in Fig. 7.6. And, again, this contact condition corresponds to a rigid-plastic behavior of the foundation, with softening. Nevertheless, in contrast with the contact condition in Example 7.21 where the resistance of the foundation vanishes for large penetrations, here it remains larger than the limit a.

Next, we note that the arguments presented above can be used in order to provide examples of functions j_ν which satisfy assumption (7.33). Indeed, assume that the stiffness coefficient depends both on $\boldsymbol{x}$ and t, i.e., it is a positive function which satisfies $k_\nu \in L^\infty(\Sigma_C)$, and let p_ν be one of the functions in Examples 7.17–7.22. Let j_ν be the potential function defined by

$$j_\nu(\boldsymbol{x}, t, r) = k_\nu(\boldsymbol{x}, t) g_\nu(r) \quad \text{for all } r \in \mathbb{R}, \text{ a.e. } (\boldsymbol{x}, t) \in \Sigma_C$$

where, again, g_ν is given by (7.85). Then, it can be easily proved that j_ν satisfies assumption (7.33). Since the modifications are straightforward, in order to avoid repetitions, we omit the details. Nevertheless, we remark that in this way we can construct concrete models of viscoelastic contact for which our results in Sect. 7.2 are valid.

Frictional contact conditions with normal compliance. We proceed in what follows with an example of frictional condition (7.6). To this end, we consider the static version of Coulomb's law of dry friction, (6.71), with the friction bound (6.54). The choice (7.84) in (6.54) leads to the friction bound $F_b = \mu\, k_\nu\, p_\nu(u_\nu)$. Therefore, using the arguments on page 188, it is easy to check that (6.71) leads to the subdifferential condition (6.69) in which

$$h_\tau(r) = \mu\, k_\nu\, p_\nu(r) \quad \text{for all } r \in \mathbb{R}, \tag{7.103}$$

$$j_\tau(\boldsymbol{\xi}) = \|\boldsymbol{\xi}\|_{\mathbb{R}^d} \quad \text{for all } \boldsymbol{\xi} \in \mathbb{R}^d. \tag{7.104}$$

The following concrete example leads to functions h_τ and j_τ which satisfy conditions (7.10) and (7.9), respectively.

Example 7.23. We assume that $\mu \in L^\infty(\Gamma_C)$ and $k_\nu \in L^\infty(\Gamma_C)$ are positive functions. Let p_ν be the normal compliance function considered in Example 7.18. Then, it is easy to see that the function h_τ satisfies condition (7.10) with

$$\overline{h}_\tau = al\, \|\mu\|_{L^\infty(\Gamma_C)} \|k_\nu\|_{L^\infty(\Gamma_C)}.$$

Also, if we choose the normal compliance function p_ν in Example 7.19, then the corresponding function h_τ satisfies condition (7.10) with

$$\overline{h}_\tau = (a + e^{-b})\, \|\mu\|_{L^\infty(\Gamma_C)} \|k_\nu\|_{L^\infty(\Gamma_C)}.$$

And, obviously, if in the two examples above the coefficient of friction vanish, then $h_\tau \equiv 0$ and, therefore, condition (7.10) is satisfied. This situation arises in the case of the frictionless condition, since in this case $\boldsymbol{\sigma}_\tau = \mathbf{0}$ on Γ_C.

Next, we verify that the function j_τ given by (7.104) satisfies condition (7.9). We have

$$\partial j_\tau(\boldsymbol{\xi}) = \begin{cases} \overline{B}(\mathbf{0}, 1) & \text{if } \boldsymbol{\xi} = \mathbf{0}, \\ \dfrac{\boldsymbol{\xi}}{\|\boldsymbol{\xi}\|_{\mathbb{R}^d}} & \text{if } \boldsymbol{\xi} \neq \mathbf{0} \end{cases}$$

for all $\boldsymbol{\xi} \in \mathbb{R}^d$, where $\overline{B}(\mathbf{0}, 1)$ denotes the closed unit ball in $\mathbb{R}^d$, i.e.,

$$\overline{B}(\mathbf{0}, 1) = \{\, \boldsymbol{\xi} \in \mathbb{R}^d \mid \|\boldsymbol{\xi}\|_{\mathbb{R}^d} \leq 1 \,\}.$$

The function j_τ is convex, regular, its subdifferential ∂j_τ is monotone (hence the relaxed monotonicity condition (7.21)(c) holds with $m_\tau = 0$) and $\|\partial j_\tau(\boldsymbol{\xi})\|_{\mathbb{R}^d} \leq 1$ for all $\boldsymbol{\xi} \in \mathbb{R}^d$. It is easy to see that the function j_τ satisfies hypotheses (7.9) with constants $c_{0\tau} = 1$ and $c_{1\tau} = 0$ and (7.77)(c) with $m_\tau = 0$.

Other frictional conditions. We pass now to examples of frictional conditions of the form (7.30) for which assumption (7.34) holds. These examples can be used to construct concrete models of viscoelastic contact for which our results in Sect. 7.2 are valid.

Example 7.24. Consider the function $j_\tau \colon \Sigma_C \times \mathbb{R}^d \to \mathbb{R}$ given by

$$j_\tau(\boldsymbol{x}, t, \boldsymbol{\xi}) = F_b(\boldsymbol{x}, t)\Big(\sqrt{\|\boldsymbol{\xi}\|^2_{\mathbb{R}^d} + \rho^2} - \rho\Big) \quad \text{for all } \boldsymbol{\xi} \in \mathbb{R}^d, \text{ a.e. } (\boldsymbol{x}, t) \in \Sigma_C.$$

Here $\rho > 0$ and F_b is a positive function such that $F_b \in L^\infty(\Sigma_C)$. The function $j_\tau(\boldsymbol{x}, t, \cdot)$ is convex, regular, continuously differentiable for a.e. $(\boldsymbol{x}, t) \in \Sigma_C$ and its subdifferential is monotone. We conclude from here that the relaxed monotonicity condition holds with $m_\tau = 0$. In this case the function j_τ satisfies hypotheses (7.34) with constants $c_{0\tau} = \|F_b\|_{L^\infty(\Sigma_C)}$, $c_{1\tau} = 0$ and $m_\tau = 0$. A simple computation shows that

$$\partial j_\tau(\boldsymbol{x},t,\boldsymbol{\xi}) = F_b(\boldsymbol{x},t)\,\frac{\boldsymbol{\xi}}{\sqrt{\|\boldsymbol{\xi}\|^2_{\mathbb{R}^d}+\rho^2}} \quad \text{for all } \boldsymbol{\xi}\in\mathbb{R}^d,\ \text{a.e. } (\boldsymbol{x},t)\in\Sigma_C.$$

Therefore, the corresponding frictional condition (7.30) corresponds to the static version of the regularized Coulomb friction law (6.65).

Example 7.25. Consider the function $j_\tau\colon \Sigma_C\times\mathbb{R}^d\to\mathbb{R}$ given by

$$j_\tau(\boldsymbol{x},t,\boldsymbol{\xi}) = \frac{F_b(\boldsymbol{x},t)}{\rho+1}\,\|\boldsymbol{\xi}\|^{\rho+1}_{\mathbb{R}^d} \quad \text{for all } \boldsymbol{\xi}\in\mathbb{R}^d,\ \text{a.e. } (\boldsymbol{x},t)\in\Sigma_C.$$

with $\rho\in(0,1]$. Here, again, F_b is a positive function which belongs to $L^\infty(\Sigma_C)$. The function $j_\tau(\boldsymbol{x},t,\cdot)$ is convex, regular and continuously differentiable, for a.e. $(\boldsymbol{x},t)\in\Sigma_C$. Moreover, by the chain rule, we have

$$\partial j_\tau(\boldsymbol{x},t,\boldsymbol{\xi}) = \begin{cases} F_b(\boldsymbol{x},t)\,\|\boldsymbol{\xi}\|^{\rho-1}_{\mathbb{R}^d}\,\boldsymbol{\xi} & \text{if } \boldsymbol{\xi}\neq\boldsymbol{0},\\ \boldsymbol{0} & \text{if } \boldsymbol{\xi}=\boldsymbol{0}\end{cases}$$

for all $\boldsymbol{\xi}\in\mathbb{R}^d$, a.e. $(\boldsymbol{x},t)\in\Sigma_C$. It is clear that

$$\|\partial j_\tau(\boldsymbol{x},t,\boldsymbol{\xi})\|_{\mathbb{R}^d} \le \|F_b\|_{L^\infty(\Sigma_C)}\,\|\boldsymbol{\xi}\|^{\rho}_{\mathbb{R}^d} \le \|F_b\|_{L^\infty(\Sigma_C)}\,(\|\boldsymbol{\xi}\|_{\mathbb{R}^d}+1)$$

for all $\boldsymbol{\xi}\subset\mathbb{R}^d$, a.e. $(\boldsymbol{x},t)\in\Sigma_C$, which implies that this function satisfies hypotheses (7.34)(c) with constants $c_{0\tau}=c_{1\tau}=\|F_b\|_{L^\infty(\Sigma_C)}$. Moreover, since $j_\tau(\boldsymbol{x},t,\cdot)$ is convex for a.e. $(\boldsymbol{x},t)\in\Sigma_C$, it satisfies the relaxed monotonicity condition (7.34)(d) with $m_\tau=0$. The rest of the conditions in (7.34) are obviously satisfied. Note that the corresponding frictional condition (7.30) corresponds to the static version of the regularized Coulomb friction law (6.66), called also the power-law friction.

Example 7.26. Consider the function $j_\tau\colon \Sigma_C\times\mathbb{R}^d\to\mathbb{R}$ given by

$$j_\tau(\boldsymbol{x},t,\boldsymbol{\xi}) = \big(a(\boldsymbol{x},t)-1\big)\,e^{-\|\boldsymbol{\xi}\|_{\mathbb{R}^d}} + a(\boldsymbol{x},t)\,\|\boldsymbol{\xi}\|_{\mathbb{R}^d} \tag{7.105}$$

for all $\boldsymbol{\xi}\in\mathbb{R}^d$, a.e. $(\boldsymbol{x},t)\in\Sigma_C$ with $a\in L^\infty(\Sigma_C)$, $0\le a(\boldsymbol{x},t)<1$, for a.e. $(\boldsymbol{x},t)\in\Sigma_C$. Then, using the chain rule and the generalized gradient formula (see Proposition 3.34), we have

$$\partial j_\tau(\boldsymbol{x},t,\boldsymbol{\xi}) = \begin{cases} \overline{B}(\boldsymbol{0},1) & \text{if } \boldsymbol{\xi}=\boldsymbol{0},\\ \Big((1-a(\boldsymbol{x},t))\,e^{-\|\boldsymbol{\xi}\|_{\mathbb{R}^d}} + a(\boldsymbol{x},t)\Big)\,\dfrac{\boldsymbol{\xi}}{\|\boldsymbol{\xi}\|_{\mathbb{R}^d}} & \text{if } \boldsymbol{\xi}\neq\boldsymbol{0}\end{cases}$$

for all $\boldsymbol{\xi}\in\mathbb{R}^d$, a.e. $(\boldsymbol{x},t)\in\Sigma_C$. The function $j_\tau(\boldsymbol{x},t,\cdot)$ is nonconvex and $\|\partial j_\tau(\boldsymbol{x},t,\boldsymbol{\xi})\|_{\mathbb{R}^d}\le 1$ for all $\boldsymbol{\xi}\in\mathbb{R}^d$, a.e. $(\boldsymbol{x},t)\in\Sigma_C$. In this case the function j_τ satisfies hypotheses (7.34) with constants $c_{0\tau}=1$, $c_{1\tau}=0$ and $m_\tau=1$. Note also that in the particular case $a\equiv 1$ the function (7.105) becomes the function (7.104) associated with the classical version of Coulomb's law of dry friction.

We turn now to some comments on the frictional condition (7.30) with j_τ given by (7.105). To this end, we assume that a is a constant such that $0 \le a < 1$ and $d = 2$, i.e., we deal with the case of a two-dimensional body. Then, at each point of the contact surface there is only one tangential direction. Denoting by $\boldsymbol{\tau}$ its unit vector, we have $\boldsymbol{\sigma}_\tau = \sigma_\tau \boldsymbol{\tau}$ and $\boldsymbol{u}_\tau = u_\tau \boldsymbol{\tau}$ where σ_τ and u_τ are real-valued functions. It is easy to see that in this special case, the friction law $-\boldsymbol{\sigma}_\tau \in \partial j_\tau(\boldsymbol{u}_\tau)$ leads to the multivalued relation

$$-\sigma_\tau = \begin{cases} (a-1)\, e^{u_\tau} - a & \text{if } u_\tau < 0, \\ [-1, 1] & \text{if } u_\tau = 0, \\ (1-a)\, e^{-u_\tau} + a & \text{if } u_\tau > 0. \end{cases} \tag{7.106}$$

Consider now the function (7.105) in the case $d = 1$, $a(\boldsymbol{x}, t) \equiv a \in [0, 1)$, i.e., $j_\tau \colon \mathbb{R} \to \mathbb{R}$,

$$j_\tau(r) = (a-1)\, e^{-|r|} + a\, |r| \tag{7.107}$$

for all $r \in \mathbb{R}$. The subdifferential of this function is given by

$$\partial j_\tau(r) = \begin{cases} (a-1)\, e^{r} - a & \text{if } r < 0, \\ [-1, 1] & \text{if } r = 0, \\ (1-a)\, e^{-r} + a & \text{if } r > 0. \end{cases}$$

Therefore, we note that equality (7.106) can be written in the subdifferential form $-\sigma_\tau \in \partial j_\tau(u_\tau)$. Moreover, it can be written in the equivalent form

$$-\sigma_\tau \in F_b(|u_\tau|)\partial |u_\tau|$$

where

$$F_b(r) = (1-a)\, e^{-r} + a \quad \text{for all } r \in \mathbb{R}_+ \tag{7.108}$$

represents the friction bound and $\partial |r|$ is the convex subdifferential of the function $r \mapsto |r|$, i.e.,

$$\partial |r| = \begin{cases} -1 & \text{if } r < 0, \\ [-1, 1] & \text{if } r = 0, \\ 1 & \text{if } r > 0. \end{cases}$$

The friction law (7.106) and the potential function (7.107) are depicted in Fig. 7.7. One of the main features of this friction law arises in the fact that the friction bound (7.108) decreases with the slip, from the value one to the limit value a. We conclude that (7.106) describes the slip weakening phenomenon which appears in the study of geophysical problems, see [36, 37, 228] for details.

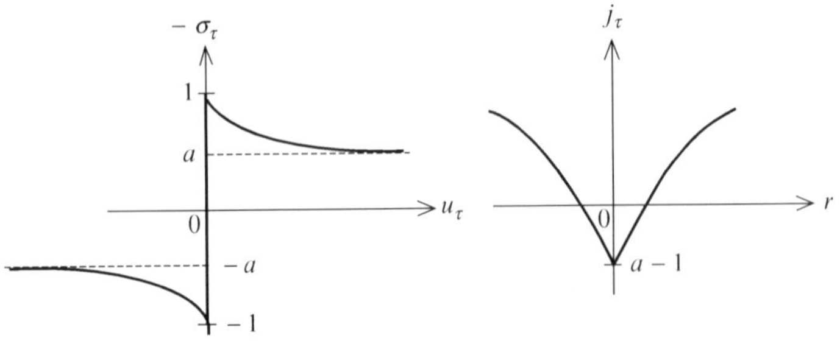

Fig. 7.7 Contact condition in Example 7.26

We end this section with the remark that examples of contact conditions of the forms (7.55), (7.56), and (7.57) in which the functions j_ν, j_τ, j_e, h_ν, h_τ and h_e satisfy assumptions (7.8), (7.9), (7.60), (7.61), (7.62), and (7.63), respectively, can be constructed by using arguments similar to those presented on pages 228–236. The only difference arises in the fact that here we have to take into account the dependence of some functions on the difference $\varphi - \varphi_0$.

For instance, assuming the normal compliance contact condition (6.83) in which $k_\nu : \mathbb{R} \to \mathbb{R}$ is a continuous bounded function and p_ν is one of the functions in Examples 7.17–7.22, we obtain a contact condition of the form (7.55) in which (7.8) and (7.61) are satisfied. Moreover, considering $\mu > 0$ and k_ν as above, the arguments of Example 7.23 lead to a contact condition of the form (7.56) in which (7.9) and (7.62) hold. Note also that the functions h_e in (6.94) and (6.95) satisfy assumptions (7.63) with $\overline{h}_e = k_e$. And, concrete examples of functions j_e such that (7.60) holds are provided in Examples 7.17–7.22.

We conclude that in this way we can construct concrete models of electro-elastic contact for which our results in Sect. 7.3 are valid.

Chapter 8
Analysis of Dynamic Contact Problems

In this chapter we apply the abstract results of Chap. 5 in the study of three dynamic frictional contact problems. In the first two problems we model the material's behavior with a nonlinear viscoelastic constitutive laws with short and long memory, respectively. In the third problem the body is supposed to be piezoelectric and, therefore, the process is mechanically dynamic and electrically static. In all problems under investigation we describe frictional contact with subdifferential boundary conditions. For each problem we deliver a variational formulation. For the first two problems it is in a form of a hemivariational inequality for the displacement field and, for the third one, it is in a form of a system of hemivariational inequalities in which the unknowns are the displacement and electric potential fields. Next, we use the abstract existence and uniqueness results presented in Chap. 5 to prove the weak solvability of the corresponding contact problems and, under additional assumptions, their unique weak solvability. Everywhere in this chapter we use the notation introduced in Chap. 6.

8.1 A First Viscoelastic Frictional Problem

For the physical setting of the first problem under consideration we refer to Sect. 6.1. We suppose that the process is dynamic and the material is viscoelastic with short memory. We recall that, in the reference configuration, the viscoelastic body occupies an open bounded connected set $\Omega \subset \mathbb{R}^d$ with boundary $\Gamma = \partial\Omega$, assumed to be Lipschitz continuous. Also, it is assumed that Γ consists of three sets $\overline{\Gamma}_D$, $\overline{\Gamma}_N$, and $\overline{\Gamma}_C$, with mutually disjoint relatively open sets Γ_D, Γ_N, and Γ_C, such that meas $(\Gamma_D) > 0$. Let $[0, T]$ denotes the time interval, $T > 0$. Everywhere in this chapter we use the notation $Q = \Omega \times (0, T)$, $\Sigma_D = \Gamma_D \times (0, T)$, $\Sigma_N = \Gamma_N \times (0, T)$, and $\Sigma_C = \Gamma_C \times (0, T)$. Then, the classical model of the frictional problem can be stated as follows.

S. Migórski et al., *Nonlinear Inclusions and Hemivariational Inequalities*, Advances in Mechanics and Mathematics 26, DOI 10.1007/978-1-4614-4232-5_8,

Problem 8.1. *Find a displacement field* $\boldsymbol{u} : Q \to \mathbb{R}^d$ *and a stress field* $\boldsymbol{\sigma} : Q \to \mathbb{S}^d$ *such that*

$$\boldsymbol{u}''(t) - \operatorname{Div} \boldsymbol{\sigma}(t) = \boldsymbol{f}_0(t) \quad \text{in } Q, \tag{8.1}$$

$$\boldsymbol{\sigma}(t) = \mathcal{A}(t, \boldsymbol{\varepsilon}(\boldsymbol{u}'(t))) + \mathcal{B}\boldsymbol{\varepsilon}(\boldsymbol{u}(t)) \quad \text{in } Q, \tag{8.2}$$

$$\boldsymbol{u}(t) = \boldsymbol{0} \quad \text{on } \Sigma_D, \tag{8.3}$$

$$\boldsymbol{\sigma}(t)\boldsymbol{\nu} = \boldsymbol{f}_N(t) \quad \text{on } \Sigma_N, \tag{8.4}$$

$$-\sigma_\nu(t) \in \partial j_\nu(t, u_\nu(t) - g_0) \quad \text{on } \Sigma_C, \tag{8.5}$$

$$-\boldsymbol{\sigma}_\tau(t) \in h_\tau(u_\nu(t) - g_0)\, \partial j_\tau(t, \boldsymbol{u}'_\tau(t)) \quad \text{on } \Sigma_C, \tag{8.6}$$

$$\boldsymbol{u}(0) = \boldsymbol{u}_0, \;\; \boldsymbol{u}'(0) = \boldsymbol{v}_0 \quad \text{in } \Omega. \tag{8.7}$$

We now provide some comments on equations and conditions in (8.1)–(8.7). Equation (8.1) is the equation of motion, (6.1), in which "Div" denotes the divergence operator for tensor-valued functions and, for simplicity, we assume that the density of mass $\rho \equiv 1$. Equation (8.2) represents the viscoelastic constitutive law with short memory, (6.23), while conditions (8.3) and (8.4) are the displacement and the traction boundary conditions, respectively. Conditions (8.5) and (8.6) represent the contact and friction conditions, respectively, in which j_ν, h_τ, and j_τ are given functions and, recall, g_0 denotes the gap function. Note that the explicit dependence of the functions j_ν and j_τ on the time variable allows to model situation when the frictional contact conditions depend on the temperature, which plays the role of a parameter, i.e., its evolution in time is prescribed. Finally, conditions (8.7) represent the initial conditions where $\boldsymbol{u}_0$ and $\boldsymbol{v}_0$ denote the initial displacement and the initial velocity, respectively.

In the study of Problem 8.1 we assume that the viscosity operator $\mathcal{A}$ and the elasticity operator $\mathcal{B}$ satisfy the following hypotheses.

$$\left.\begin{array}{l}
\mathcal{A}\colon Q \times \mathbb{S}^d \to \mathbb{S}^d \text{ is such that}\\
\text{(a) } \mathcal{A}(\cdot,\cdot,\boldsymbol{\varepsilon}) \text{ is measurable on } Q \text{ for all } \boldsymbol{\varepsilon} \in \mathbb{S}^d.\\
\text{(b) } \mathcal{A}(\boldsymbol{x},t,\cdot) \text{ is continuous on } \mathbb{S}^d \text{ for a.e. } (\boldsymbol{x},t) \in Q.\\
\text{(c) } \|\mathcal{A}(\boldsymbol{x},t,\boldsymbol{\varepsilon})\|_{\mathbb{S}^d} \le a_0(\boldsymbol{x},t) + a_1\|\boldsymbol{\varepsilon}\|_{\mathbb{S}^d} \text{ for all } \boldsymbol{\varepsilon} \in \mathbb{S}^d,\\
\qquad \text{a.e. } (\boldsymbol{x},t) \in Q \text{ with } a_0 \in L^2(Q), a_0 \ge 0 \text{ and } a_1 > 0.\\
\text{(d) } \mathcal{A}(\boldsymbol{x},t,\boldsymbol{\varepsilon}) : \boldsymbol{\varepsilon} \ge \alpha\,\|\boldsymbol{\varepsilon}\|^2_{\mathbb{S}^d} \text{ for all } \boldsymbol{\varepsilon} \in \mathbb{S}^d, \text{a.e. } (\boldsymbol{x},t) \in Q\\
\qquad \text{with } \alpha > 0.\\
\text{(e) } (\mathcal{A}(\boldsymbol{x},t,\boldsymbol{\varepsilon}_1) - \mathcal{A}(\boldsymbol{x},t,\boldsymbol{\varepsilon}_2)) : (\boldsymbol{\varepsilon}_1 - \boldsymbol{\varepsilon}_2) \ge 0 \text{ for all}\\
\qquad \boldsymbol{\varepsilon}_1, \boldsymbol{\varepsilon}_2 \in \mathbb{S}^d, \text{ a.e. } (\boldsymbol{x},t) \in Q.
\end{array}\right\} \tag{8.8}$$

$$
\left.\begin{array}{l}
\mathcal{B}\colon \Omega \times \mathbb{S}^d \to \mathbb{S}^d \text{ is such that} \\
\quad \text{(a) } \mathcal{B}(\boldsymbol{x}, \boldsymbol{\varepsilon}) = b(\boldsymbol{x})\boldsymbol{\varepsilon} \text{ for all } \boldsymbol{\varepsilon} \in \mathbb{S}^d, \text{ a.e. } \boldsymbol{x} \in \Omega. \\
\quad \text{(b) } b(\boldsymbol{x}) = \big(b_{ijkl}(\boldsymbol{x})\big) \text{ with} \\
\qquad\qquad b_{ijkl} = b_{jikl} = b_{lkij} \in L^\infty(\Omega). \\
\quad \text{(c) } b_{ijkl}(\boldsymbol{x})\varepsilon_{ij}\varepsilon_{kl} \geq 0 \text{ for all } \boldsymbol{\varepsilon} = (\varepsilon_{ij}) \in \mathbb{S}^d, \text{ a.e. } \boldsymbol{x} \in \Omega.
\end{array}\right\} \tag{8.9}
$$

The contact and frictional potentials j_ν and j_τ and the function h_τ satisfy

$$
\left.\begin{array}{l}
j_\nu\colon \Sigma_C \times \mathbb{R} \to \mathbb{R} \text{ is such that} \\
\quad \text{(a) } j_\nu(\cdot,\cdot,r) \text{ is measurable on } \Sigma_C \text{ for all } r \in \mathbb{R} \text{ and there} \\
\qquad\quad \text{exists } e_1 \in L^2(\Gamma_C) \text{ such that } j_\nu(\cdot,\cdot,e_1(\cdot)) \in L^1(\Sigma_C). \\
\quad \text{(b) } j_\nu(\boldsymbol{x},t,\cdot) \text{ is locally Lipschitz on } \mathbb{R} \text{ for a.e. } (\boldsymbol{x},t) \in \Sigma_C. \\
\quad \text{(c) } |\partial j_\nu(\boldsymbol{x},t,r)| \leq c_{0\nu}(t) + c_{1\nu}|r| \text{ for all } r \in \mathbb{R}, \text{ a.e. } (\boldsymbol{x},t) \in \Sigma_C \\
\qquad\quad \text{with } c_{0\nu} \in L^\infty(0,T),\, c_{0\nu}, c_{1\nu} \geq 0. \\
\quad \text{(d) } (\zeta_1 - \zeta_2)(r_1 - r_2) \geq -m_\nu|r_1 - r_2|^2 \text{ for all } \zeta_i \in \partial j_\nu(\boldsymbol{x},t,r_i), \\
\qquad\quad r_i \in \mathbb{R}, i = 1,2, \text{ a.e. } (\boldsymbol{x},t) \in \Sigma_C \text{ with } m_\nu \geq 0.
\end{array}\right\} \tag{8.10}
$$

$$
\left.\begin{array}{l}
j_\tau\colon \Sigma_C \times \mathbb{R}^d \to \mathbb{R} \text{ is such that} \\
\quad \text{(a) } j_\tau(\cdot,\cdot,\boldsymbol{\xi}) \text{ is measurable on } \Sigma_C \text{ for all } \boldsymbol{\xi} \in \mathbb{R}^d \text{ and there} \\
\qquad\quad \text{exists } \mathbf{e}_2 \in L^2(\Gamma_C;\mathbb{R}^d) \text{ such that } j_\tau(\cdot,\cdot,\mathbf{e}_2(\cdot)) \in L^1(\Sigma_C). \\
\quad \text{(b) } j_\tau(\boldsymbol{x},t,\cdot) \text{ is locally Lipschitz on } \mathbb{R}^d \text{ for a.e. } (\boldsymbol{x},t) \in \Sigma_C. \\
\quad \text{(c) } \|\partial j_\tau(\boldsymbol{x},t,\boldsymbol{\xi})\|_{\mathbb{R}^d} \leq c_{0\tau}(t) + c_{1\tau}\|\boldsymbol{\xi}\|_{\mathbb{R}^d} \text{ for all } \boldsymbol{\xi} \in \mathbb{R}^d, \\
\qquad\quad \text{a.e. } (\boldsymbol{x},t) \in \Sigma_C \text{ with } c_{0\tau} \in L^\infty(0,T),\, c_{0\tau}, c_{1\tau} \geq 0. \\
\quad \text{(d) } (\boldsymbol{\zeta}_1 - \boldsymbol{\zeta}_2)\cdot(\boldsymbol{\xi}_1 - \boldsymbol{\xi}_2) \geq -m_\tau\|\boldsymbol{\xi}_1 - \boldsymbol{\xi}_2\|^2_{\mathbb{R}^d} \text{ for all} \\
\qquad\quad \boldsymbol{\zeta}_i \in \partial j_\tau(\boldsymbol{x},t,\boldsymbol{\xi}_i), \boldsymbol{\xi}_i \in \mathbb{R}^d, i = 1,2, \text{ a.e. } (\boldsymbol{x},t) \in \Sigma_C \\
\qquad\quad \text{with } m_\tau \geq 0.
\end{array}\right\} \tag{8.11}
$$

$$
\left.\begin{array}{l}
h_\tau\colon \Gamma_C \times \mathbb{R} \to \mathbb{R} \text{ is such that} \\
\quad \text{(a) } h_\tau(\cdot,r) \text{ is measurable on } \Gamma_C \text{ for all } r \in \mathbb{R}. \\
\quad \text{(b) } h_\tau(\boldsymbol{x},\cdot) \text{ is continuous on } \mathbb{R} \text{ for a.e. } \boldsymbol{x} \in \Gamma_C. \\
\quad \text{(c) } 0 \leq h_\tau(\boldsymbol{x},r) \leq \overline{h}_\tau \text{ for all } r \in \mathbb{R}, \text{ a.e. } \boldsymbol{x} \in \Gamma_C \text{ with } \overline{h}_\tau > 0.
\end{array}\right\} \tag{8.12}
$$

The volume force and traction densities satisfy

$$
\boldsymbol{f}_0 \in L^2(0,T;L^2(\Omega;\mathbb{R}^d)), \qquad \boldsymbol{f}_N \in L^2(0,T;L^2(\Gamma_N;\mathbb{R}^d)), \tag{8.13}
$$

the gap function satisfies

$$
g_0 \in L^\infty(\Gamma_C), \;\; g_0 \geq 0 \quad \text{a.e. on } \Gamma_C. \tag{8.14}
$$

and, finally, the initial data have the regularity

$$\boldsymbol{u}_0 \in V, \quad \boldsymbol{v}_0 \in H. \tag{8.15}$$

Here and everywhere in this chapter we use the Hilbert spaces

$$H = L^2(\Omega;\mathbb{R}^d), \qquad \mathcal{H} = \big\{ \boldsymbol{\tau} = (\tau_{ij}) \mid \tau_{ij} = \tau_{ji} \in L^2(\Omega) \big\} = L^2(\Omega;\mathbb{S}^d),$$

$$\mathcal{H}_1 = \{ \boldsymbol{\tau} \in \mathcal{H} \mid \mathrm{Div}\, \boldsymbol{\tau} \in H \}, \qquad V = \big\{ \boldsymbol{v} \in H^1(\Omega;\mathbb{R}^d) \mid \boldsymbol{v} = \boldsymbol{0} \text{ on } \Gamma_D \big\},$$

introduced in Sect. 2.3. Also, V^* denotes the dual space of V and $\langle \cdot,\cdot \rangle_{V^*\times V}$ represents the duality paring. Finally, as usual in the study of evolutionary problems, we need the spaces

$$\mathcal{V} = L^2(0,T;V), \quad \mathcal{V}^* = L^2(0,T;V^*),$$
$$\mathcal{W} = \{ \boldsymbol{v} \in \mathcal{V} \mid \boldsymbol{v}' \in \mathcal{V}^* \}.$$

Next, we present the variational formulation of Problem 8.1. We consider the function $\boldsymbol{f}\colon (0,T) \to V^*$ given by

$$\langle \boldsymbol{f}(t), \boldsymbol{v} \rangle_{V^*\times V} = \langle \boldsymbol{f}_0(t), \boldsymbol{v} \rangle_H + \langle \boldsymbol{f}_N(t), \boldsymbol{v} \rangle_{L^2(\Gamma_N;\mathbb{R}^d)} \tag{8.16}$$

for $\boldsymbol{v} \in V$, a.e. $t \in (0,T)$. Assume that $(\boldsymbol{u}, \boldsymbol{\sigma})$ is a couple of sufficiently smooth functions which solve (8.1)–(8.7), $\boldsymbol{v} \in V$ and $t \in (0,T)$. We use the equation of motion (8.1) and the Green formula (2.7) to find that

$$\langle \boldsymbol{u}''(t), \boldsymbol{v} \rangle_{V^*\times V} + \langle \boldsymbol{\sigma}(t), \boldsymbol{\varepsilon}(\boldsymbol{v}) \rangle_{\mathcal{H}} = \langle \boldsymbol{f}_0(t), \boldsymbol{v} \rangle_H + \int_\Gamma \boldsymbol{\sigma}(t)\boldsymbol{\nu} \cdot \boldsymbol{v}\, d\Gamma. \tag{8.17}$$

We now take into account the boundary condition (8.4) to see that

$$\int_\Gamma \boldsymbol{\sigma}(t)\boldsymbol{\nu} \cdot \boldsymbol{v}\, d\Gamma = \int_{\Gamma_N} \boldsymbol{f}_N(t) \cdot \boldsymbol{v}\, d\Gamma + \int_{\Gamma_C} (\sigma_\nu(t) v_\nu + \boldsymbol{\sigma}_\tau(t) \cdot \boldsymbol{v}_\tau)\, d\Gamma. \tag{8.18}$$

On the other hand, from the definition of the Clarke subdifferential combined with (8.5) and (8.6), we have

$$-\sigma_\nu(t) v_\nu \le j_\nu^0(t, u_\nu(t) - g_0; v_\nu),$$
$$-\boldsymbol{\sigma}_\tau(t) \cdot \boldsymbol{v}_\tau \le h_\tau(u_\nu(t) - g_0)\, j_\tau^0(t, \boldsymbol{u}_\tau'(t); \boldsymbol{v}_\tau)$$

on Σ_C, which implies that

$$\int_{\Gamma_C} (\sigma_\nu(t)v_\nu + \boldsymbol{\sigma}_\tau(t)\cdot \boldsymbol{v}_\tau)\, d\Gamma$$

$$\geq -\int_{\Gamma_C} \Big(j_\nu^0(t, u_\nu(t) - g_0; v_\nu) + h_\tau(u_\nu(t) - g_0)\, j_\tau^0(t, \boldsymbol{u}_\tau'(t); \boldsymbol{v}_\tau)\Big)\, d\Gamma. \quad (8.19)$$

We now combine (8.16)–(8.19) to see that

$$\langle \boldsymbol{u}''(t), \boldsymbol{v}\rangle_{V^*\times V} + \langle \boldsymbol{\sigma}(t), \boldsymbol{\varepsilon}(\boldsymbol{v})\rangle_{\mathcal{H}}$$

$$+ \int_{\Gamma_C} \Big(j_\nu^0(t, u_\nu(t) - g_0; v_\nu) + h_\tau(u_\nu(t) - g_0)\, j_\tau^0(t, \boldsymbol{u}_\tau'(t); \boldsymbol{v}_\tau)\Big)\, d\Gamma$$

$$\geq \langle \boldsymbol{f}(t), \boldsymbol{v}\rangle_{V^*\times V} \quad \text{for all } \boldsymbol{v} \in V \text{ and a.e. } t \in (0, T). \quad (8.20)$$

Finally, we use (8.20), the constitutive law (8.2) and the initial conditions (8.7) to obtain the following variational formulation of the mechanical problem (8.1)–(8.7).

Problem 8.2. *Find a displacement field* $\boldsymbol{u}\colon (0, T) \to V$ *such that* $\boldsymbol{u} \in \mathcal{V}$, $\boldsymbol{u}' \in \mathcal{W}$ *and*

$$\langle \boldsymbol{u}''(t), \boldsymbol{v}\rangle_{V^*\times V} + \langle \mathcal{A}(t, \boldsymbol{\varepsilon}(\boldsymbol{u}'(t))), \boldsymbol{\varepsilon}(\boldsymbol{v})\rangle_{\mathcal{H}} + \langle \mathcal{B}\boldsymbol{\varepsilon}(\boldsymbol{u}(t)), \boldsymbol{\varepsilon}(\boldsymbol{v})\rangle_{\mathcal{H}}$$

$$+ \int_{\Gamma_C} \Big(j_\nu^0(t, u_\nu(t) - g_0; v_\nu) + h_\tau(u_\nu(t) - g_0)\, j_\tau^0(t, \boldsymbol{u}_\tau'(t); \boldsymbol{v}_\tau)\Big)\, d\Gamma$$

$$\geq \langle \boldsymbol{f}(t), \boldsymbol{v}\rangle_{V^*\times V} \quad \text{for all } \boldsymbol{v} \in V \text{ and a.e. } t \in (0, T), \quad (8.21)$$

$$\boldsymbol{u}(0) = \boldsymbol{u}_0, \quad \boldsymbol{u}'(0) = \boldsymbol{v}_0.$$

In what follows we need the space $Z = H^\delta(\Omega; \mathbb{R}^d)$ with a fixed $\delta \in (\frac{1}{2}, 1)$ and the trace operator $\gamma\colon Z \to L^2(\Gamma_C; \mathbb{R}^d)$. We denote by $\|\gamma\|$ the norm of the trace operator in $\mathcal{L}(Z, L^2(\Gamma_C; \mathbb{R}^d))$ and by $c_e > 0$ the embedding constant of V into Z.

The solvability of Problem 8.2 is given by the following result.

Theorem 8.3. *Assume that hypotheses* (8.8), (8.9), (8.10)(a)–(c), (8.11)(a)–(c), (8.12)–(8.15) *are satisfied. Moreover, assume that*

$$\left.\begin{array}{l} \text{either } j_\nu(\boldsymbol{x}, t, \cdot) \text{ and } j_\tau(\boldsymbol{x}, t, \cdot) \text{ are regular} \\ \text{or } -j_\nu(\boldsymbol{x}, t, \cdot) \text{ and } -j_\tau(\boldsymbol{x}, t, \cdot) \text{ are regular} \end{array}\right\} \quad (8.22)$$

for a.e. $(\boldsymbol{x}, t) \in \Sigma_C$ *and, in addition,*

$$\alpha > 6 \max\{ c_{1\nu}, c_{1\tau}\overline{h}_\tau \}\, c_e^2\, \|\gamma\|^2\, T. \quad (8.23)$$

Then Problem 8.2 has at least one solution.

Proof. The proof is based on Theorem 5.19. We introduce the operators $A\colon (0,T)\times V\to V^*$ and $B\colon V\to V^*$ defined by

$$\langle A(t,\boldsymbol{u}),\boldsymbol{v}\rangle_{V^*\times V} = \langle \mathcal{A}(t,\boldsymbol{\varepsilon}(\boldsymbol{u})),\boldsymbol{\varepsilon}(\boldsymbol{v})\rangle_{\mathcal{H}} \tag{8.24}$$

$$\langle B\boldsymbol{u},\boldsymbol{v}\rangle_{V^*\times V} = \langle \mathcal{B}\boldsymbol{\varepsilon}(\boldsymbol{u}),\boldsymbol{\varepsilon}(\boldsymbol{v})\rangle_{\mathcal{H}} \tag{8.25}$$

for $\boldsymbol{u},\boldsymbol{v}\in V$, a.e. $t\in(0,T)$. We need to check that the operators A, B, and the function $\boldsymbol{f}$ given by (8.24), (8.25), and (8.16) satisfy the conditions (5.1), (5.2), and (5.4), respectively.

First, we observe that by (8.8) and Hölder's inequality, we have

$$\begin{aligned}
|\langle A(t,\boldsymbol{v}),\boldsymbol{w}\rangle_{V^*\times V}| &\le \int_\Omega \|\mathcal{A}(\boldsymbol{x},t,\boldsymbol{\varepsilon}(\boldsymbol{v}))\|_{\mathbb{S}^d}\,\|\boldsymbol{\varepsilon}(\boldsymbol{w})\|_{\mathbb{S}^d}\,dx\\
&\le \int_\Omega (a_0(\boldsymbol{x},t)+a_1\,\|\boldsymbol{\varepsilon}(\boldsymbol{v})\|_{\mathbb{S}^d})\|\boldsymbol{\varepsilon}(\boldsymbol{w})\|_{\mathbb{S}^d}\,dx\\
&\le \sqrt{2}\left(\|a_0(t)\|_{L^2(\Omega)}+a_1\,\|\boldsymbol{v}\|_V\right)\|\boldsymbol{w}\|_V
\end{aligned} \tag{8.26}$$

for all $\boldsymbol{v},\boldsymbol{w}\in V$, a.e. $t\in(0,T)$. Therefore, the function

$$(\boldsymbol{x},t)\mapsto \mathcal{A}(\boldsymbol{x},t,\boldsymbol{\varepsilon}(\boldsymbol{v}))\colon\varepsilon(\boldsymbol{w})$$

is integrable on Q, for all $\boldsymbol{v},\boldsymbol{w}\in V$. Next, by Fubini's theorem (Theorem 1.69 on p. 22), we have that

$$t\mapsto \int_\Omega \mathcal{A}(\boldsymbol{x},t,\boldsymbol{\varepsilon}(\boldsymbol{v}))\colon\varepsilon(\boldsymbol{w})\,dx = \langle A(t,\boldsymbol{v}),\boldsymbol{w}\rangle_{V^*\times V}$$

is measurable for all $\boldsymbol{v},\boldsymbol{w}\in V$. Therefore, the function $t\mapsto A(t,\boldsymbol{v})$ is weakly measurable from $(0,T)$ into V^*, for all $\boldsymbol{v}\in V$. Since the latter is separable, from the Pettis measurability theorem, (Theorem 2.28 on p. 38), it follows that $t\mapsto A(t,\boldsymbol{v})$ is measurable for all $\boldsymbol{v}\in V$, i.e., (5.1)(a) holds. The properties of A in (5.1)(b)–(d) follow from the hypothesis (8.8) by a reasoning analogous to the one of the proofs of Theorem 7.3. Therefore, we conclude that the operator A given by (8.24) satisfies condition (5.1).

It is also clear that, under assumption (8.9), the operator B given by (8.25) satisfies condition (5.2). Moreover, from (8.13), (8.15), and the definition of $\boldsymbol{f}$, it follows that (5.4) holds, too.

To proceed, we define the functions $j_i\colon \Sigma_C\times\mathbb{R}^d\to\mathbb{R}$ for $i=1,2$ and $h\colon\Gamma_C\times\mathbb{R}^d\times\mathbb{R}^d\to\mathbb{R}$ by

$$j_1(\boldsymbol{x},t,\boldsymbol{\xi}) = j_\nu(\boldsymbol{x},t,N_\nu\boldsymbol{\xi}-g_0(\boldsymbol{x}))$$

$$j_2(\boldsymbol{x},t,\boldsymbol{\xi}) = j_\tau(\boldsymbol{x},t,N_\tau\boldsymbol{\xi})$$

$$h(\boldsymbol{x}, \boldsymbol{\eta}_1, \boldsymbol{\eta}_2) = h_\tau(\boldsymbol{x}, N_\nu \boldsymbol{\eta}_1 - g_0(\boldsymbol{x}))$$

for all $\boldsymbol{\xi}$, $\boldsymbol{\eta}_1$, $\boldsymbol{\eta}_2 \in \mathbb{R}^d$, a.e. $(\boldsymbol{x}, t) \in \Sigma_C$, where $N_\nu \in L^\infty(\Gamma; \mathcal{L}(\mathbb{R}^d, \mathbb{R}))$, $N_\tau \in L^\infty(\Gamma; \mathcal{L}(\mathbb{R}^d, \mathbb{R}^d))$ are given by $N_\nu \boldsymbol{\xi} = \xi_\nu = \boldsymbol{\xi} \cdot \boldsymbol{\nu}$ and $N_\tau \boldsymbol{\xi} = \boldsymbol{\xi}_\tau = \boldsymbol{\xi} - \xi_\nu \boldsymbol{\nu}$, respectively. By the arguments presented in the proof of Theorem 7.3, we deduce that j_i for $i = 1, 2$ and h satisfy (5.90) and (5.89), respectively.

Finally, the condition (8.23) implies (5.91). We use now Theorem 5.19 to complete the proof. □

We underline that the result of Theorem 8.3 represents a local existence result, since the length of the time interval has to satisfy the smallness condition (8.23). Nevertheless, given a time interval $(0, T)$, if the constant α which appears in the coercivity condition (8.8)(d) is large enough, then (8.23) holds and, therefore, Theorem 8.3 provides a global existence result.

A couple of functions $\boldsymbol{u}: (0, T) \to V$ and $\boldsymbol{\sigma}: (0, T) \to \mathcal{H}$ which satisfies (8.21), (8.2) and (8.7) is called a *weak solution* of Problem 8.1. We remark that under the assumptions of Theorem 8.3 there exists at least one weak solution of Problem 8.1.

To describe precisely the regularity of the weak solution, we note that the regularity $\boldsymbol{u} \in \mathcal{V}$ and $\boldsymbol{u}' \in \mathcal{W}$ combined with the continuity of the embedding $\mathcal{W} \subset C(0, T; H)$, implies that

$$\boldsymbol{u} \in W^{1,2}(0, T; V), \quad \boldsymbol{u}' \in C(0, T; H), \quad \boldsymbol{u}'' \in L^2(0, T; V^*). \tag{8.27}$$

Moreover, the constitutive law (8.2), assumptions (8.8), (8.9), and (8.27) show that $\boldsymbol{\sigma} \in L^2(0, T; \mathcal{H})$. In addition, using (8.2) and (8.21), it follows that (8.20) holds which implies that

$$\boldsymbol{u}''(t) - \operatorname{Div} \boldsymbol{\sigma}(t) = \boldsymbol{f}_0(t) \quad \text{in } Q.$$

This equality combined with (8.13) and (8.27) implies that Div $\boldsymbol{\sigma} \in L^2(0, T; V^*)$. Therefore, the regularity of the stress field is given by

$$\boldsymbol{\sigma} \in L^2(0, T; \mathcal{H}), \quad \operatorname{Div} \boldsymbol{\sigma} \in L^2(0, T; V^*). \tag{8.28}$$

We conclude from above that the weak solution of Problem 8.1 satisfies (8.27), (8.28). The question of the uniqueness of the weak solution is left open and, clearly, this question deserves more investigation in the future.

We end this section with the remark that examples of viscoelastic constitutive laws of the form (8.2) in which the operators $\mathcal{A}$ and $\mathcal{B}$ satisfy conditions (8.8) and (8.9) as well as examples of contact conditions of the forms (8.5), (8.6) in which the functions j_ν, j_τ, and h_τ satisfy assumptions (8.10), (8.11), and (8.12), respectively, can be constructed by using arguments similar to those presented in Sect. 7.4. We conclude that our results presented in this section are valid for the corresponding frictional contact models.

8.2 A Second Viscoelastic Frictional Problem

In this section we describe a dynamic viscoelastic frictional contact problem which leads to a hemivariational inequality involving the Volterra-type integral term. We refer to Sect. 6.1 for the physical setting of the problem under consideration. In contrast to Sect. 8.1, the material behavior is now described with a viscoelastic constitutive law with both short and long memory, (6.29). The classical formulation of frictional contact problem is stated as follows.

Problem 8.4. *Find a displacement field* $\boldsymbol{u}\colon Q \to \mathbb{R}^d$ *and a stress field* $\boldsymbol{\sigma}\colon Q \to \mathbb{S}^d$ *such that*

$$\boldsymbol{u}''(t) - \operatorname{Div}\boldsymbol{\sigma}(t) = \boldsymbol{f}_0(t) \quad \text{in } Q, \tag{8.29}$$

$$\boldsymbol{\sigma}(t) = \mathcal{A}(t, \boldsymbol{\varepsilon}(\boldsymbol{u}'(t))) + \mathcal{B}\boldsymbol{\varepsilon}(\boldsymbol{u}(t)) + \int_0^t \mathcal{C}(t-s, \boldsymbol{\varepsilon}(\boldsymbol{u}(s)))\,ds \quad \text{in } Q, \tag{8.30}$$

$$\boldsymbol{u}(t) = \boldsymbol{0} \quad \text{on } \Sigma_D, \tag{8.31}$$

$$\boldsymbol{\sigma}(t)\boldsymbol{\nu} = \boldsymbol{f}_N(t) \quad \text{on } \Sigma_N, \tag{8.32}$$

$$-\sigma_\nu(t) \in \partial j_\nu(t, u_\nu'(t)) \quad \text{on } \Sigma_C, \tag{8.33}$$

$$-\boldsymbol{\sigma}_\tau(t) \in \partial j_\tau(t, \boldsymbol{u}_\tau'(t)) \quad \text{on } \Sigma_C, \tag{8.34}$$

$$\boldsymbol{u}(0) = \boldsymbol{u}_0, \ \ \boldsymbol{u}'(0) = \boldsymbol{v}_0 \quad \text{in } \Omega. \tag{8.35}$$

The dynamic contact problem (8.29)–(8.35) is similar to the dynamic problem (8.1)–(8.7) studied in Sect. 8.1. The difference arises in the fact that now we replace the constitutive law (8.2) with the viscoelastic constitutive law (8.30) and the frictional contact conditions (8.5), (8.6) with the frictional contact conditions (8.33), (8.34). Here j_ν and j_τ are given functions and, recall, the symbols ∂j_ν and ∂j_τ denote the Clarke subdifferential of j_ν and j_τ with respect to their last variables.

In the study of Problem 8.4 we use the function spaces H, $\mathcal{H}$, $\mathcal{H}_1$, V, V^*, Z, $\mathcal{V}$, $\mathcal{V}^*$, and $\mathcal{W}$ considered in Sect. 8.1. Also, as usual, we denote by $\|\gamma\|$ the norm of the trace operator in $\mathcal{L}(Z, L^2(\Gamma_C;\mathbb{R}^d))$ and by $c_e > 0$ the embedding constant of V into Z.

We assume that the viscosity operator $\mathcal{A}$ satisfies

$$\left.\begin{array}{l}
\mathcal{A}\colon Q \times \mathbb{S}^d \to \mathbb{S}^d \text{ is such that}\\
\quad \text{(a) } \mathcal{A}(\cdot,\cdot,\boldsymbol{\varepsilon}) \text{ is measurable on } Q \text{ for all } \boldsymbol{\varepsilon} \in \mathbb{S}^d.\\
\quad \text{(b) } \mathcal{A}(\boldsymbol{x},t,\cdot) \text{ is continuous on } \mathbb{S}^d \text{ for a.e. } (\boldsymbol{x},t) \in Q.\\
\quad \text{(c) } \|\mathcal{A}(\boldsymbol{x},t,\boldsymbol{\varepsilon})\|_{\mathbb{S}^d} \le a_0(\boldsymbol{x},t) + a_1\|\boldsymbol{\varepsilon}\|_{\mathbb{S}^d} \text{ for all } \boldsymbol{\varepsilon} \in \mathbb{S}^d,\\
\qquad\quad \text{a.e. } (\boldsymbol{x},t) \in Q \text{ with } a_0 \in L^2(Q), a_0 \ge 0 \text{ and } a_1 > 0.\\
\quad \text{(d) } (\mathcal{A}(\boldsymbol{x},t,\boldsymbol{\varepsilon}_1) - \mathcal{A}(\boldsymbol{x},t,\boldsymbol{\varepsilon}_2)) : (\boldsymbol{\varepsilon}_1 - \boldsymbol{\varepsilon}_2) \ge m_{\mathcal{A}}\|\boldsymbol{\varepsilon}_1 - \boldsymbol{\varepsilon}_2\|^2_{\mathbb{S}^d}\\
\qquad\quad \text{for all } \boldsymbol{\varepsilon}_1, \boldsymbol{\varepsilon}_2 \in \mathbb{S}^d, \text{ a.e. } (\boldsymbol{x},t) \in Q \text{ with } m_{\mathcal{A}} > 0.\\
\quad \text{(e) } \mathcal{A}(\boldsymbol{x},t,\boldsymbol{0}) = \boldsymbol{0} \text{ for a.e. } (\boldsymbol{x},t) \in Q.
\end{array}\right\} \tag{8.36}$$

Also, we suppose that the elasticity operator $\mathcal{B}$ satisfies (8.9) and the relaxation operator $\mathcal{C}$ is following

$$\left.\begin{array}{l} \mathcal{C}\colon Q\times\mathbb{S}^d\to\mathbb{S}^d \text{ is such that} \\ \quad \text{(a) } \mathcal{C}(\boldsymbol{x},t,\boldsymbol{\varepsilon})=c(\boldsymbol{x},t)\boldsymbol{\varepsilon} \text{ for all } \boldsymbol{\varepsilon}\in\mathbb{S}^d, \text{ a.e. } (\boldsymbol{x},t)\in Q. \\ \quad \text{(b) } c(\boldsymbol{x},t)=\big(c_{ijkl}(\boldsymbol{x},t)\big) \\ \qquad\quad \text{with } c_{ijkl}=c_{jikl}=c_{lkij}\in L^2(0,T;L^\infty(\Omega)). \end{array}\right\} \tag{8.37}$$

Furthermore, the contact and frictional potentials j_ν and j_τ satisfy (8.10) and (8.11), respectively. The volume force and traction densities satisfy (8.13) and, finally, the initial data have the regularity (8.15).

We present now the variational formulation of Problem 8.4 and, to this end, we consider the function $\boldsymbol{f}\colon(0,T)\to V^*$ given by (8.16). Assume that $(\boldsymbol{u},\boldsymbol{\sigma})$ is a couple of sufficiently smooth functions which solve (8.29)–(8.35), $\boldsymbol{v}\in V$ and $t\in(0,T)$. We use the equation of motion (8.29) and the Green formula (2.7) on page 37 to find that

$$\langle\boldsymbol{u}''(t),\boldsymbol{v}\rangle_{V^*\times V}+\langle\boldsymbol{\sigma}(t),\boldsymbol{\varepsilon}(\boldsymbol{v})\rangle_{\mathcal{H}}=\langle\boldsymbol{f}_0(t),\boldsymbol{v}\rangle_H+\int_\Gamma\boldsymbol{\sigma}(t)\boldsymbol{\nu}\cdot\boldsymbol{v}\,d\Gamma. \tag{8.38}$$

Now, we take into account the boundary condition (8.32) as well as the definition of the Clarke subdifferential in (8.33), (8.34) and, analogously as in Sect. 8.1, we have

$$\begin{aligned}&\langle\boldsymbol{u}''(t),\boldsymbol{v}\rangle_{V^*\times V}+\langle\boldsymbol{\sigma}(t),\boldsymbol{\varepsilon}(\boldsymbol{v})\rangle_{\mathcal{H}}\\&\quad+\int_{\Gamma_C}\Big(j_\nu^0(t,u'_\nu(t);v_\nu)+j_\tau^0(t,\boldsymbol{u}'_\tau(t);\boldsymbol{v}_\tau)\Big)\,d\Gamma\\&\quad\ge\langle\boldsymbol{f}(t),\boldsymbol{v}\rangle_{V^*\times V}\quad\text{for all }\boldsymbol{v}\in V\text{ and a.e. }t\in(0,T).\end{aligned} \tag{8.39}$$

We use now (8.39), the constitutive law (8.30), and the initial conditions (8.35) to obtain the following variational formulation of the mechanical problem (8.29)–(8.35).

Problem 8.5. *Find a displacement field $\boldsymbol{u}\colon(0,T)\to V$ such that $\boldsymbol{u}\in\mathcal{V}$, $\boldsymbol{u}'\in\mathcal{W}$ and*

$$\begin{aligned}&\langle\boldsymbol{u}''(t),\boldsymbol{v}\rangle_{V^*\times V}+\langle\mathcal{A}(t,\boldsymbol{\varepsilon}(\boldsymbol{u}'(t))),\boldsymbol{\varepsilon}(\boldsymbol{v})\rangle_{\mathcal{H}}\\&\quad+\langle\mathcal{B}\boldsymbol{\varepsilon}(\boldsymbol{u}(t)),\boldsymbol{\varepsilon}(\boldsymbol{v})\rangle_{\mathcal{H}}+\left\langle\int_0^t\mathcal{C}(t-s,\boldsymbol{\varepsilon}(\boldsymbol{u}(s)))\,ds,\boldsymbol{\varepsilon}(\boldsymbol{v})\right\rangle_{\mathcal{H}}\\&\quad+\int_{\Gamma_C}\Big(j_\nu^0(t,u'_\nu(t);v_\nu)+j_\tau^0(t,\boldsymbol{u}'_\tau(t);\boldsymbol{v}_\tau)\Big)\,d\Gamma\ge\langle\boldsymbol{f}(t),\boldsymbol{v}\rangle_{V^*\times V}\\&\qquad\text{for all }\boldsymbol{v}\in V\text{ and a.e. }t\in(0,T),\end{aligned} \tag{8.40}$$

$$\boldsymbol{u}(0)=\boldsymbol{u}_0,\quad\boldsymbol{u}'(0)=\boldsymbol{v}_0.$$

The unique solvability of Problem 8.5 is given by the following result.

Theorem 8.6. *Assume that* (8.9)–(8.13), (8.15), (8.22), (8.36), (8.37) *hold and the following conditions are satisfied:*

$$m_{\mathcal{A}} \geq \max\{m_\nu, m_\tau\}\, c_e^2\, \|\gamma\|^2, \tag{8.41}$$

$$m_{\mathcal{A}} > 6\max\{c_{1\nu}, c_{1\tau}\}\, c_e^2\, \|\gamma\|^2. \tag{8.42}$$

Then Problem 8.5 *has a unique solution.*

Proof. The proof follows from Theorem 5.25. We introduce the operators $A\colon (0,T)\times V \to V^*$ and $B\colon V \to V^*$ by (8.24) and (8.25), respectively, and the operator $C\colon (0,T)\times V \to V^*$ defined by

$$\langle C(t)\boldsymbol{u}, \boldsymbol{v}\rangle_{V^*\times V} = \langle \mathcal{C}(t, \boldsymbol{\varepsilon}(\boldsymbol{u})), \boldsymbol{\varepsilon}(\boldsymbol{v})\rangle_{\mathcal{H}} \tag{8.43}$$

for all $\boldsymbol{u}$, $\boldsymbol{v} \in V$, a.e. $t \in (0,T)$. The operator A given by (8.24) satisfies the hypothesis (5.67)(a)–(d), as it results from the proof of Theorem 8.3. Moreover, (8.36)(d) implies the strong monotonicity of $A(t,\cdot)$ for a.e. $t \in (0,T)$, thus (5.67) is satisfied with $\alpha = m_1 = m_{\mathcal{A}} > 0$. It is obvious to see that, under condition (8.9), the operator B given by (8.25) satisfies (5.2). Also, from (8.13), (8.15), and the definition (8.16) of $\boldsymbol{f}$, we know that (5.4) holds. In addition, condition (8.37) implies that the operator C defined by (8.43) satisfies (5.74).

Next, we define the function $j\colon \Sigma_C \times \mathbb{R}^d \to \mathbb{R}$ by

$$j(\boldsymbol{x}, t, \boldsymbol{\xi}) = j_\nu(\boldsymbol{x}, t, \xi_\nu) + j_\tau(\boldsymbol{x}, t, \boldsymbol{\xi}_\tau) \ \text{ for all } \boldsymbol{\xi} \in \mathbb{R}^d, \text{ a.e. } (\boldsymbol{x}, t) \in \Sigma_C.$$

Applying Corollary 3.48 to the functions j_ν and j_τ, it is easy to see that $j(\cdot,\cdot,\mathbf{e}(\cdot)) \in L^1(\Sigma_C)$ for all $\mathbf{e} \in L^2(\Gamma_C;\mathbb{R}^d)$. Under the assumptions (8.10), (8.11), and (8.22), the function j satisfies (5.96) with $\overline{c}_0 \in L^\infty(0,T)$, $\overline{c}_0 \geq 0$, $\overline{c}_1 = \max\{c_{1\nu}, c_{1\tau}\} \geq 0$, and $m_2 = \max\{m_\nu, m_\tau\}$. Finally, the conditions (8.41) and (8.42) guarantee (5.97) and (5.98), respectively. We apply Theorem 5.25 which gives that Problem 8.5 has a unique solution. This completes the proof. □

A couple of functions $\boldsymbol{u}\colon (0,T) \to V$ and $\boldsymbol{\sigma}\colon (0,T) \to \mathcal{H}$ which satisfies (8.40), (8.30) and (8.35) is called a *weak solution* of Problem 8.4. We conclude that, under the assumptions of Theorem 8.6, there exists a unique weak solution of Problem 8.4. Moreover, following arguments similar to those presented in Sect. 8.2 it follows that the weak solution has the regularity (8.27) and (8.28).

We provide now a brief comparison between the results presented in Theorems 8.3 and 8.6. First, we note that the regularity of the solutions provided by the two theorems is the same. Also, note that in the two theorems the assumptions on the elasticity operator $\mathcal{B}$, on the potential functions j_ν, j_τ, on the forces and tractions as well as on the initial data are the same. Nevertheless, we remark that condition (8.36) implies (8.8) and, therefore, it follows that Theorem 8.6 holds under stronger assumption on the viscosity operator $\mathcal{A}$. Second, we note that Theorem 8.3 states the existence of at least one solution to Problem 8.2 while Theorem 8.6 states the existence of a unique solution to Problem 8.5. And, finally, note that Theorem 8.3

holds if the time interval T satisfies the smallness assumption (8.23) and, for this reason, the solutions provided by this theorem can be considered as a local solution to Problem 8.2. In contrast, the result provided by Theorem 8.6 holds without any restriction on T, i.e., the solution to Problem 8.5 provided by this theorem is global in time.

We end this section with the remark that examples of contact conditions of the forms (8.33) and (8.34) in which the functions j_ν and j_τ satisfy assumptions (8.10) and (8.11), respectively, can be constructed by using arguments similar to those presented in Sect. 7.4. We conclude that our results in this section are valid for the corresponding frictional contact models.

8.3 An Electro-Viscoelastic Frictional Problem

For the problem studied in this section we assume that the body is piezoelectric and the foundation is conductive. Therefore, we refer to the physical setting described in Sect. 6.4. As a consequence, besides the partition of Γ into three sets $\overline{\Gamma}_D$, $\overline{\Gamma}_N$, and $\overline{\Gamma}_C$, which corresponds to the mechanical boundary conditions, we consider a partition of $\overline{\Gamma}_D \cup \overline{\Gamma}_N$ into two sets $\overline{\Gamma}_a$ and $\overline{\Gamma}_b$ with mutually disjoint relatively open sets Γ_a and Γ_b, which corresponds to the electrical boundary conditions. We assume that meas $(\Gamma_D) > 0$ and meas $(\Gamma_a) > 0$. We denote by $[0,T]$ the time interval of interest, with $T > 0$ and, besides the notation Q, Σ_D, Σ_N, Σ_C introduced in Sect. 8.1, we also use the notation $\Sigma_a = \Gamma_a \times (0,T)$ and $\Sigma_b = \Gamma_b \times (0,T)$. The classical model for the process is as follows.

Problem 8.7. *Find a displacement field* $\boldsymbol{u}\colon Q \to \mathbb{R}^d$, *a stress field* $\boldsymbol{\sigma}\colon Q \to \mathbb{S}^d$, *an electric potential* $\varphi\colon Q \to \mathbb{R}$, *and an electric displacement field* $\boldsymbol{D}\colon Q \to \mathbb{R}^d$ *such that*

$$\boldsymbol{u}''(t) - \text{Div}\,\boldsymbol{\sigma}(t) = \boldsymbol{f}_0(t) \quad \text{in } Q, \tag{8.44}$$

$$\text{div}\,\boldsymbol{D}(t) = q_0(t) \quad \text{in } Q, \tag{8.45}$$

$$\boldsymbol{\sigma}(t) = \mathcal{A}(t, \varepsilon(\boldsymbol{u}'(t))) + \mathcal{B}\varepsilon(\boldsymbol{u}(t)) - \mathcal{P}^\top \boldsymbol{E}(\varphi(t)) \quad \text{in } Q, \tag{8.46}$$

$$\boldsymbol{D}(t) = \mathcal{P}\boldsymbol{\varepsilon}(\boldsymbol{u}(t)) + \boldsymbol{\beta}\boldsymbol{E}(\varphi(t)) \quad \text{in } Q, \tag{8.47}$$

$$\boldsymbol{u}(t) = \boldsymbol{0} \quad \text{on } \Sigma_D, \tag{8.48}$$

$$\boldsymbol{\sigma}(t)\,\boldsymbol{\nu} = \boldsymbol{f}_N(t) \quad \text{on } \Sigma_N, \tag{8.49}$$

$$\varphi(t) = 0 \quad \text{on } \Sigma_a, \tag{8.50}$$

$$\boldsymbol{D}(t)\cdot\boldsymbol{\nu} = q_b(t) \quad \text{on } \Sigma_b, \tag{8.51}$$

$$-\sigma_\nu(t) \in \partial j_\nu(t, u'_\nu(t)) \quad \text{on } \Sigma_C, \tag{8.52}$$

$$-\boldsymbol{\sigma}_\tau(t) \in \partial j_\tau(t, \boldsymbol{u}'_\tau(t)) \quad \text{on } \Sigma_C, \tag{8.53}$$

$$\boldsymbol{D}(t) \cdot \boldsymbol{\nu} \in \partial j_e(\varphi(t) - \varphi_0) \quad \text{on } \Sigma_C, \tag{8.54}$$

$$\boldsymbol{u}(0) = \boldsymbol{u}_0, \quad \boldsymbol{u}'(0) = \boldsymbol{v}_0 \quad \text{in } \Omega. \tag{8.55}$$

We now provide explanation of the equations and the conditions (8.44)–(8.55) and we send the reader to Sect. 6.4 for more details and comments.

First, (8.44) is the equation of motion for the stress field in which the density of mass has been taken to equal one and (8.45) represents the balance equation for the electric displacement field. We use these equations since the process is assumed to be mechanically dynamic and electrically static. Equations (8.46) and (8.47) represent the electro-viscoelastic constitutive laws, see (6.81) and (6.80), respectively. We recall that $\mathcal{A}$ is the viscosity operator, $\mathcal{B}$ is the elasticity operator, $\mathcal{P}$ represents the third-order piezoelectric tensor with $\mathcal{P}^\top$ its transpose, $\boldsymbol{\beta}$ denotes the electric permittivity tensor and $\boldsymbol{E}(\varphi)$ is the electric field, i.e., $\boldsymbol{E}(\varphi) = -\nabla\varphi$. Also, the tensors $\mathcal{P}$ and $\mathcal{P}^\top$ satisfy the equality (6.79) and, if $\mathcal{P} = (p_{ijk})$, then the components of the tensor $\mathcal{P}^\top$ are given by $p^\top_{ijk} = p_{kij}$.

Next, conditions (8.48) and (8.49) are the displacement and traction boundary conditions, whereas (8.50) and (8.51) represent the electric boundary conditions. Moreover, (8.52), (8.53), and (8.54) are the contact, the frictional and the electrical conductivity conditions on the contact surface Γ_C, respectively, in which the function φ_0 represents the electric potential of the foundation.

Finally, conditions (8.55) represent the initial conditions where $\boldsymbol{u}_0$ and $\boldsymbol{v}_0$ denote the initial displacement and the initial velocity, respectively.

Note that the operator $\mathcal{A}$ and the functions j_ν and j_τ may depend explicitly on the time variable and this is the case when the viscosity properties of the material and the frictional contact conditions depend on the temperature, which plays the role of a parameter, i.e., its evolution in time is prescribed.

In the study of Problem 8.7, we assume that the viscosity operator $\mathcal{A}$ satisfies (8.36) and the elasticity tensor $\mathcal{B}$ satisfies (8.9). Moreover, the piezoelectric tensor $\mathcal{P}$ and the electric permittivity tensor $\boldsymbol{\beta}$ satisfy

$$\left.\begin{array}{l} \mathcal{P}\colon \Omega \times \mathbb{S}^d \to \mathbb{R}^d \text{ is such that} \\ \quad \text{(a) } \mathcal{P}(\boldsymbol{x}, \boldsymbol{\varepsilon}) = p(\boldsymbol{x})\,\boldsymbol{\varepsilon} \text{ for all } \boldsymbol{\varepsilon} \in \mathbb{S}^d, \text{ a.e. } \boldsymbol{x} \in \Omega. \\ \quad \text{(b) } p(\boldsymbol{x}) = (p_{ijk}(\boldsymbol{x})) \text{ with } p_{ijk} \in L^\infty(\Omega). \end{array}\right\} \tag{8.56}$$

$$\left.\begin{array}{l} \boldsymbol{\beta}\colon \Omega \times \mathbb{R}^d \to \mathbb{R}^d \text{ is such that} \\ \quad \text{(a) } \boldsymbol{\beta}(\boldsymbol{x}, \boldsymbol{\xi}) = \boldsymbol{\beta}(\boldsymbol{x})\,\boldsymbol{\xi} \text{ for all } \boldsymbol{\xi} \in \mathbb{R}^d, \text{ a.e. } \boldsymbol{x} \in \Omega. \\ \quad \text{(b) } \boldsymbol{\beta}(\boldsymbol{x}) = (\beta_{ij}(\boldsymbol{x})) \text{ with } \beta_{ij} = \beta_{ji} \in L^\infty(\Omega). \\ \quad \text{(c) } \beta_{ij}(\boldsymbol{x})\xi_i\xi_j \ge m_\beta \|\boldsymbol{\xi}\|^2_{\mathbb{R}^d} \text{ for all } \boldsymbol{\xi} = (\xi_i) \in \mathbb{R}^d, \\ \qquad \text{a.e. } \boldsymbol{x} \in \Omega \text{ with } m_\beta > 0. \end{array}\right\} \tag{8.57}$$

The contact and frictional potentials j_ν and j_τ satisfy conditions (8.10) and (8.11), respectively, and the electric potential function j_e satisfies

$$\left.\begin{array}{l} j_e\colon \Gamma_C \times \mathbb{R} \to \mathbb{R} \text{ is such that} \\ \quad \text{(a) } j_e(\cdot, r) \text{ is measurable on } \Gamma_C \text{ for all } r \in \mathbb{R} \text{ and there} \\ \qquad \text{exists } e_3 \in L^2(\Gamma_C) \text{ such that } j_e(\cdot, e_3(\cdot)) \in L^1(\Gamma_C). \\ \quad \text{(b) } j_e(\boldsymbol{x}, \cdot) \text{ is locally Lipschitz on } \mathbb{R} \text{ for a.e. } \boldsymbol{x} \in \Gamma_C. \\ \quad \text{(c) } |\partial j_e(\boldsymbol{x}, r)| \le c_{0e} + c_{1e}|r| \text{ for all } r \in \mathbb{R}, \text{ a.e. } \boldsymbol{x} \in \Gamma_C \\ \qquad \text{with } c_{0e}, c_{1e} \ge 0. \\ \quad \text{(d) } (\zeta_1 - \zeta_2)(r_1 - r_2) \ge -m_e|r_1 - r_2|^2 \text{ for all } \zeta_i \in \partial j_e(\boldsymbol{x}, t, r_i), \\ \qquad r_i \in \mathbb{R}, i = 1, 2, \text{ a.e.} (\boldsymbol{x}, t) \in \Sigma_C \text{ with } m_e \ge 0. \end{array}\right\} \qquad (8.58)$$

The volume force and traction densities satisfy (8.13), the densities of electric charge and surface free electrical charge have the regularity

$$q_0 \in L^2(0, T; L^2(\Omega)), \quad q_b \in L^2(0, T; L^2(\Gamma_b)), \qquad (8.59)$$

the initial data satisfy (8.15) and, finally, we assume the potential of the foundation is such that

$$\varphi_0 \in L^\infty(\Gamma_C). \qquad (8.60)$$

In order to give the variational formulation of Problem 8.7, for the electric displacement field and for the stress field we use the spaces

$$H = L^2(\Omega; \mathbb{R}^d), \qquad \mathcal{H} = \{\boldsymbol{\tau} = (\tau_{ij}) \mid \tau_{ij} = \tau_{ji} \in L^2(\Omega)\} = L^2(\Omega; \mathbb{S}^d).$$

Recall that these are real Hilbert spaces with the inner products defined on page 35. Also, for the displacement field and the electric potential, we introduce the spaces

$$V = \{\boldsymbol{v} \in H^1(\Omega; \mathbb{R}^d) \mid \boldsymbol{v} = \boldsymbol{0} \text{ on } \Gamma_D\}, \quad \Phi = \{\psi \in H^1(\Omega) \mid \psi = 0 \text{ on } \Gamma_a\}$$

which are Hilbert spaces with the inner products and the corresponding norms given by

$$\langle \boldsymbol{u}, \boldsymbol{v}\rangle_V = \langle \boldsymbol{\varepsilon}(\boldsymbol{u}), \boldsymbol{\varepsilon}(\boldsymbol{v})\rangle_{\mathcal{H}}, \quad \|\boldsymbol{v}\|_V = \|\boldsymbol{\varepsilon}(\boldsymbol{v})\|_{\mathcal{H}} \text{ for all } \boldsymbol{u}, \boldsymbol{v} \in V,$$

$$\langle \varphi, \psi\rangle_\Phi = \langle \nabla\varphi, \nabla\psi\rangle_H, \quad \|\psi\|_\Phi = \|\nabla\psi\|_H \text{ for all } \varphi, \psi \in \Phi.$$

Also, V^* and Φ^* will denote the dual space of V and Φ, respectively, and $\langle\cdot,\cdot\rangle_{V^*\times V}$, $\langle\cdot,\cdot\rangle_{\Phi^*\times\Phi}$ will represent the corresponding duality parings.

We turn now to the variational formulation of the contact problems (8.44)–(8.55). To this end, we assume in what follows that $\boldsymbol{u}$, $\boldsymbol{\sigma}$, φ, $\mathbf{D}$ are sufficiently smooth functions which solve (8.44)–(8.55). We use the equation of motion (8.44), take into account the boundary conditions (8.48), (8.49), (8.52), (8.53) and, analogously as in Sect. 8.1, we obtain

$$\langle \boldsymbol{u}''(t), \boldsymbol{v}\rangle_{V^*\times V} + \langle \boldsymbol{\sigma}(t), \boldsymbol{\varepsilon}(\boldsymbol{v})\rangle_{\mathcal{H}} + \int_{\Gamma_C} \Big(j_\nu^0(t, u_\nu'(t); v_\nu) + j_\tau^0(t, \boldsymbol{u}_\tau'(t); \boldsymbol{v}_\tau)\Big)\, d\Gamma$$

$$\geq \langle \boldsymbol{f}(t), \boldsymbol{v}\rangle_{V^*\times V} \quad \text{for all } \boldsymbol{v} \in V, \text{ a.e. } t \in (0,T), \tag{8.61}$$

where the function $\boldsymbol{f}\colon (0,T) \to V^*$ is given by (8.16).

Similarly, for $\psi \in \Phi$ and $t \in (0,T)$, from (8.45) and the Green formula (2.6) we deduce that

$$\langle \boldsymbol{D}(t), \nabla\psi\rangle_H + \int_\Omega q_0(t)\psi\, dx = \int_\Gamma \boldsymbol{D}(t)\cdot\boldsymbol{\nu}\,\psi\, d\Gamma$$

and then, by (8.51) we get

$$-\langle \boldsymbol{D}(t), \nabla\psi\rangle_H + \int_{\Gamma_C} \boldsymbol{D}(t)\cdot\boldsymbol{\nu}\,\psi\, d\Gamma = \langle q(t), \psi\rangle_{\Phi^*\times\Phi}, \tag{8.62}$$

where $q\colon (0,T) \to \Phi^*$ is the function given by

$$\langle q(t), \psi\rangle_{\Phi^*\times\Phi} = \int_\Omega q_0(t)\psi\, dx - \int_{\Gamma_b} q_b(t)\psi\, d\Gamma$$

for $\psi \in \Phi$ and a.e. $t \in (0,T)$. From the definition of the Clarke subdifferential and (8.54) we have

$$\boldsymbol{D}(t)\cdot\boldsymbol{\nu}\,\psi \leq j_e^0(\varphi(t) - \varphi_0; \psi) \quad \text{on } \Sigma_C,$$

which implies that

$$\int_{\Gamma_C} \boldsymbol{D}(t)\cdot\boldsymbol{\nu}\,\psi\, d\Gamma \leq \int_{\Gamma_C} j_e^0(\varphi(t) - \varphi_0; \psi)\, d\Gamma. \tag{8.63}$$

We combine now (8.62) and (8.63) to obtain

$$-\langle \boldsymbol{D}(t), \nabla\psi\rangle_H + \int_{\Gamma_C} j_e^0(\varphi(t) - \varphi_0; \psi)\, d\Gamma \geq \langle q(t), \psi\rangle_{\Phi^*\times\Phi} \tag{8.64}$$

for all $\psi \in \Phi$, a.e. $t \in (0,T)$. We substitute now (8.46) in (8.61), (8.47) in (8.64), use the equality $\boldsymbol{E}(\varphi) = -\nabla\varphi$ and the initial conditions (8.55) to derive the following variational formulation of Problem 8.7, in terms of displacement field and electric potential.

Problem 8.8. *Find a displacement field* $\boldsymbol{u}\colon (0,T) \to V$ *and an electric potential* $\varphi\colon (0,T) \to \Phi$ *such that* $\boldsymbol{u} \in \mathcal{V}$, $\boldsymbol{u}' \in \mathcal{W}$, $\varphi \in L^2(0,T;\Phi)$, *and*

$$\begin{aligned}
&\langle \boldsymbol{u}''(t), \boldsymbol{v}\rangle_{V^*\times V} + \langle \mathcal{A}(t, \boldsymbol{\varepsilon}(\boldsymbol{u}'(t))), \boldsymbol{\varepsilon}(\boldsymbol{v})\rangle_{\mathcal{H}} + \langle \mathcal{B}\boldsymbol{\varepsilon}(\boldsymbol{u}(t)), \boldsymbol{\varepsilon}(\boldsymbol{v})\rangle_{\mathcal{H}} \\
&\quad + \langle \mathcal{P}^{\top}\nabla\varphi(t), \boldsymbol{\varepsilon}(\boldsymbol{v})\rangle_{\mathcal{H}} + \int_{\Gamma_C} \Big(j_\nu^0(t, u_\nu'(t); v_\nu) + j_\tau^0(t, \boldsymbol{u}_\tau'(t); \boldsymbol{v}_\tau)\Big)\, d\Gamma \\
&\qquad \ge \langle \boldsymbol{f}(t), \boldsymbol{v}\rangle_{V^*\times V} \ \text{ for all } \boldsymbol{v} \in V, \ \text{a.e. } t \in (0,T), \qquad (8.65)
\end{aligned}$$

$$\begin{aligned}
&\langle \boldsymbol{\beta}\nabla\varphi(t), \nabla\psi\rangle_H - \langle \mathcal{P}\boldsymbol{\varepsilon}(\boldsymbol{u}(t)), \nabla\psi\rangle_H + \int_{\Gamma_C} j_e^0(\varphi(t) - \varphi_0; \psi)\, d\Gamma \\
&\qquad \ge \langle q(t), \psi\rangle_{\Phi^*\times\Phi} \ \text{ for all } \psi \in \Phi, \ \text{a.e. } t \in (0,T), \qquad (8.66)
\end{aligned}$$

$$\boldsymbol{u}(0) = \boldsymbol{u}_0, \ \ \boldsymbol{u}'(0) = \boldsymbol{v}_0.$$

Note that, in contrast with the variational formulations of the frictional contact problems studied in Sects. 8.1 and 8.2, Problem 8.8 represents a *system of hemivariational inequalities*. One of the main features of this system arises in the coupling between the unknowns $\boldsymbol{u}$ and φ, which appears in the terms containing the piezoelectricity tensor $\mathcal{P}$.

In order to state the main existence and uniqueness result of this section, we need the spaces $Z = H^\delta(\Omega;\mathbb{R}^d)$ and $Z_1 = H^\delta(\Omega)$ where $\delta \in (1/2, 1)$ is fixed. We denote by $c_e > 0$ the embedding constant of V into Z and also of Φ into Z_1. Moreover, we introduce the trace operators $\gamma\colon Z \to L^2(\Gamma_C;\mathbb{R}^d)$ and $\gamma_1\colon Z_1 \to L^2(\Gamma_C)$ and, finally, we denote by $\|\gamma\|$ and $\|\gamma_1\|$ their norms in $\mathcal{L}(Z, L^2(\Gamma_C;\mathbb{R}^d))$ and $\mathcal{L}(Z_1, L^2(\Gamma_C))$, respectively.

Our main result in the study of Problem 8.8 is the following.

Theorem 8.9. *Assume that* (8.9)–(8.11), (8.13), (8.15), (8.36), (8.41), (8.42), (8.56)–(8.60) *hold. Moreover, assume that*

$$\left.\begin{aligned}
&\text{either } \ j_\nu(\boldsymbol{x},t,\cdot),\, j_\tau(\boldsymbol{x},t,\cdot) \ \text{ and } j_e(\boldsymbol{x},t,\cdot) \text{ are regular} \\
&\text{or } \ -j_\nu(\boldsymbol{x},t,\cdot), -j_\tau(\boldsymbol{x},t,\cdot) \ \text{ and } \ -j_e(\boldsymbol{x},t,\cdot) \text{ are regular}
\end{aligned}\right\} \qquad (8.67)$$

for a.e. $(\boldsymbol{x},t) \in \Sigma_C$ *and, in addition,*

$$m_\beta > \sqrt{3}\, c_{1e}\, c_e^2\, \|\gamma_1\|^2 \ \text{ and } \ m_\beta > m_e\, c_e^2\, \|\gamma_1\|^2. \qquad (8.68)$$

Then Problem 8.8 *has a unique solution.*

Proof. The proof is carried out into several steps. It is based on the study of three intermediate problems combined with a fixed-point argument. We start by considering the operators $A\colon (0,T)\times V \to V^*$ and $B\colon V \to V^*$ defined by (8.24) and (8.25), respectively, and recall that the function $\boldsymbol{f}\colon (0,T) \to V^*$ is defined by (8.16). It is clear that the operators A and B satisfy conditions (5.67) and (5.2), respectively, and $\boldsymbol{f} \in \mathcal{V}^*$.

Step 1. Let $\boldsymbol{\eta} \in \mathcal{V}^*$ be fixed. In this first step we prove the well-posedness of the following intermediate evolutionary problem.

Problem P_η. *Find $\boldsymbol{u}_\eta \in \mathcal{V}$ such that $\boldsymbol{u}'_\eta \in \mathcal{W}$ and*

$$\langle \boldsymbol{u}''_\eta(t) + A(t,\boldsymbol{u}'_\eta(t)) + B\boldsymbol{u}_\eta(t), \boldsymbol{v}\rangle_{V^*\times V} + \langle \boldsymbol{\eta}(t), \boldsymbol{v}\rangle_{V^*\times V}$$

$$+ \int_{\Gamma_C} \Big(j_\nu^0(u'_{\eta\nu}(t); v_\nu) + j_\tau^0(\boldsymbol{u}'_{\eta\tau}(t); \boldsymbol{v}_\tau)\Big)\, d\Gamma \ge \langle \boldsymbol{f}(t), v\rangle_{V^*\times V}$$

$$\text{for all } \boldsymbol{v} \in V, \text{ a.e. } t \in (0,T),$$

$$\boldsymbol{u}_\eta(0) = \boldsymbol{u}_0, \quad \boldsymbol{u}'_\eta(0) = \boldsymbol{v}_0.$$

To solve Problem P_η, we define the function $j\colon \Sigma_C \times \mathbb{R}^d \to \mathbb{R}$ by

$$j(\boldsymbol{x},t,\boldsymbol{\xi}) = j_\nu(\boldsymbol{x},t,\xi_\nu) + j_\tau(\boldsymbol{x},t,\boldsymbol{\xi}_\tau)$$

for all $\boldsymbol{\xi} \in \mathbb{R}^d$, a.e. $(\boldsymbol{x},t) \in \Sigma_C$. Under assumptions (8.10), (8.11), and (8.67), from the proof of Theorem 7.5 it follows that j satisfies (5.96) with $\overline{c}_1 = \max\{c_{1\nu}, c_{1\tau}\}$ and $m_2 = \max\{m_\nu, m_\tau\}$. By Propositions 3.35, 3.37, and (8.67), we have

$$j^0(\boldsymbol{x},t,\boldsymbol{\xi};\boldsymbol{\rho}) = j_\nu^0(\boldsymbol{x},t,\xi_\nu;\rho_\nu) + j_\tau^0(\boldsymbol{x},t,\boldsymbol{\xi}_\tau;\boldsymbol{\rho}_\tau)$$

for all $\boldsymbol{\xi}$, $\boldsymbol{\rho} \in \mathbb{R}^d$, a.e. $(\boldsymbol{x},t) \in \Sigma_C$. Therefore, from Theorem 5.23, it follows that Problem P_η has a unique solution. Moreover, using inequality (5.99) we have

$$\|\boldsymbol{u}_1(t) - \boldsymbol{u}_2(t)\|_V^2 \le c \int_0^t \|\boldsymbol{\eta}_1(s) - \boldsymbol{\eta}_2(s)\|_{V^*}^2\, ds \quad \text{for all } t \in [0,T], \tag{8.69}$$

where $\boldsymbol{u}_i = \boldsymbol{u}_{\eta_i}$ is the unique solution to Problem P_η corresponding to $\boldsymbol{\eta}_i$ for $i = 1$, 2. Note that in (8.69) and below, c represents a positive constant which does not depend on time and whose value can change from line to line.

Step 2. Let $\boldsymbol{w} \in V$ be fixed and consider the bilinear forms $\widetilde{\beta}\colon \Phi \times \Phi \to \mathbb{R}$ and $\widetilde{p}\colon V \times \Phi \to \mathbb{R}$ defined by

$$\widetilde{\beta}(\varphi,\psi) = \langle \boldsymbol{\beta}\nabla\varphi, \nabla\psi\rangle_H, \qquad \widetilde{p}(\boldsymbol{v},\psi) = \langle \mathcal{P}\boldsymbol{\varepsilon}(\boldsymbol{v}), \nabla\psi\rangle_H$$

for all φ, $\psi \in \Phi$ and $\boldsymbol{v} \in V$. In the second step we prove the well-posedness of the following hemivariational inequality.

Problem P_w. *Find $\varphi_w \in \Phi$ such that*

$$\widetilde{\beta}(\varphi_w, \psi) + \int_{\Gamma_C} j_e^0(\varphi_w - \varphi_0; \psi)\, d\Gamma \geq \widetilde{p}(\boldsymbol{w}, \psi) + \langle q, \psi \rangle_{\Phi^* \times \Phi} \quad \text{for all } \psi \in \Phi.$$

To solve Problem P_w we define the operator $\widetilde{B}\colon \Phi \to \Phi^*$ and the functional $\widetilde{q}_w \in \Phi^*$ by equalities

$$\langle \widetilde{B}\varphi, \psi \rangle_{\Phi^* \times \Phi} = \widetilde{\beta}(\varphi, \psi), \quad \langle \widetilde{q}_w, \psi \rangle_{\Phi^* \times \Phi} = \widetilde{p}(\boldsymbol{w}, \psi) + \langle q, \psi \rangle_{\Phi^* \times \Phi}$$

for $\varphi, \psi \in \Phi$. It follows from Theorem 4.20 that there exists a unique element $\varphi_w \in \Phi$ such that

$$\langle \widetilde{B}\varphi_w, \psi \rangle_{\Phi^* \times \Phi} + \int_{\Gamma_C} j_e^0(\varphi_w - \varphi_0; \psi)\, d\Gamma \geq \langle \widetilde{q}_w, \psi \rangle_{\Phi^* \times \Phi} \quad \text{for all } \psi \in \Phi$$

and, clearly, φ_w is the unique solution of Problem P_w. Moreover, using the inequality

$$\begin{aligned} \langle \widetilde{q}_{w_1} - \widetilde{q}_{w_2}, \psi_1 - \psi_2 \rangle_{\Phi^* \times \Phi} &\leq \widetilde{p}(\boldsymbol{w}_1 - \boldsymbol{w}_2, \psi_1 - \psi_2) \\ &\leq c\, \|\boldsymbol{w}_1 - \boldsymbol{w}_2\|_V\, \|\psi_1 - \psi_2\|_\Phi, \end{aligned}$$

valid for all $\boldsymbol{w}_1, \boldsymbol{w}_2 \in V$ and $\psi_1, \psi_2 \in \Phi$, we have

$$\|\widetilde{q}_{w_1} - \widetilde{q}_{w_2}\|_{\Phi^*} \leq c\, \|\boldsymbol{w}_1 - \boldsymbol{w}_2\|_V.$$

Therefore, from (4.44), we obtain

$$\|\varphi_1 - \varphi_2\|_\Phi \leq c\, \|\boldsymbol{w}_1 - \boldsymbol{w}_2\|_V, \tag{8.70}$$

where $c > 0$ and $\varphi_i = \varphi_{w_i}$ is the unique solution to Problem P_w corresponding to $\boldsymbol{w}_i$ for $i = 1, 2$.

Step 3. For $\boldsymbol{\eta} \in \mathcal{V}^*$ we use the solution $\boldsymbol{u}_\eta$ of Problem P_η obtained in Step 1 and consider the following intermediate problem.

Problem Q_η. *Find $\varphi_\eta \in L^2(0, T; \Phi)$ such that*

$$\begin{aligned} &\langle \boldsymbol{\beta} \nabla \varphi_\eta(t), \nabla \psi \rangle_H - \langle \mathcal{P} \boldsymbol{\varepsilon}(\boldsymbol{u}_\eta(t)), \nabla \psi \rangle_H + \int_{\Gamma_C} j^0(\varphi_\eta(t) - \varphi_0; \psi)\, d\Gamma \\ &\quad \geq \langle q(t), \psi \rangle_{\Phi^* \times \Phi} \quad \text{for all } \psi \in \Phi, \text{ a.e. } t \in (0, T). \end{aligned}$$

The aim of the third step is to prove the well-posedness of Problem Q_η. First, from Step 2 it follows that Problem Q_η has a unique solution. Moreover, if $\boldsymbol{u}_1$, $\boldsymbol{u}_2$ represent the solutions to Problem P_η and φ_1, φ_2 represent the solutions to Problem

Q_η, respectively, corresponding to $\boldsymbol{\eta}_1, \boldsymbol{\eta}_2 \in \mathcal{V}^*$, then (8.70) shows that there exists $c > 0$ such that

$$\|\varphi_1(t) - \varphi_2(t)\|_\Phi \leq c\, \|\boldsymbol{u}_1(t) - \boldsymbol{u}_2(t)\|_V \quad \text{for a.e. } t \in (0,T). \tag{8.71}$$

Step 4. For $\boldsymbol{\eta} \in \mathcal{V}^*$, we denote by $\boldsymbol{u}_\eta$ and φ_η the functions obtained in Step 1 and Step 3, respectively. We introduce the operator $\Lambda\colon \mathcal{V}^* \to \mathcal{V}^*$ defined by

$$\langle (\Lambda \boldsymbol{\eta})(t), \boldsymbol{v} \rangle_{V^* \times V} = \langle \mathcal{P}^\top \nabla \varphi_\eta(t), \boldsymbol{\varepsilon}(\boldsymbol{v}) \rangle_{\mathcal{H}}$$

for all $\boldsymbol{v} \in V$, a.e. $t \in (0,T)$. In this step we prove that Λ has a unique fixed point $\boldsymbol{\eta}^* \in \mathcal{V}^*$. Let $\boldsymbol{\eta} \in \mathcal{V}^*$. Since

$$\begin{aligned}\langle \mathcal{P}^\top \nabla \varphi_\eta(t), \boldsymbol{\varepsilon}(\boldsymbol{v}) \rangle_{\mathcal{H}} &= \int_\Omega \mathcal{P}^\top \nabla \varphi_\eta(t) \colon \boldsymbol{\varepsilon}(\boldsymbol{v})\, dx = \int_\Omega \mathcal{P}\boldsymbol{\varepsilon}(\boldsymbol{v}) \cdot \nabla \varphi_\eta(t)\, dx \\ &= \langle \mathcal{P}\boldsymbol{\varepsilon}(\boldsymbol{v}), \nabla \varphi_\eta(t) \rangle_H = \widetilde{p}(\boldsymbol{v}, \varphi_\eta(t)) \leq c\, \|\boldsymbol{v}\|_V \|\varphi_\eta(t)\|_\Phi\end{aligned}$$

for all $\boldsymbol{v} \in V$, a.e. $t \in (0,T)$, we have

$$\|(\Lambda \boldsymbol{\eta})(t)\|_{V^*} = \sup_{\|\boldsymbol{v}\|_V \leq 1} |\langle (\Lambda \boldsymbol{\eta})(t), \boldsymbol{v} \rangle_{V^* \times V}| \leq c\, \|\varphi_\eta(t)\|_\Phi \tag{8.72}$$

for a.e. $t \in (0,T)$. This implies that $\|\Lambda \boldsymbol{\eta}\|_{\mathcal{V}^*} \leq c\, \|\varphi_\eta\|_{L^2(0,T;\Phi)}$ which shows that the operator Λ is well defined.

We prove in what follows that the operator Λ has a unique fixed point. To this end, consider $\boldsymbol{\eta}_1, \boldsymbol{\eta}_2 \in \mathcal{V}^*$. Using arguments similar to those used to obtain (8.72), from (8.69) and (8.71), we have

$$\begin{aligned}\|(\Lambda \boldsymbol{\eta}_1)(t) - (\Lambda \boldsymbol{\eta}_2)(t)\|_{V^*}^2 &\leq c\, \|\varphi_{\eta_1}(t) - \varphi_{\eta_2}(t)\|_\Phi^2 \\ &\leq c\, \|\boldsymbol{u}_{\eta_1}(t) - \boldsymbol{u}_{\eta_2}(t)\|_V^2 \leq c \int_0^t \|\boldsymbol{\eta}_1(s) - \boldsymbol{\eta}_2(s)\|_{V^*}^2\, ds\end{aligned}$$

for a.e. $t \in (0,T)$. Subsequently, we deduce that

$$\begin{aligned}\|(\Lambda^2 \boldsymbol{\eta}_1)(t) - (\Lambda^2 \boldsymbol{\eta}_2)(t)\|_{V^*}^2 &= \|(\Lambda(\Lambda \boldsymbol{\eta}_1))(t) - (\Lambda(\Lambda \boldsymbol{\eta}_2))(t)\|_{V^*}^2 \\ &\leq c \int_0^t \|(\Lambda \boldsymbol{\eta}_1)(s) - (\Lambda \boldsymbol{\eta}_2)(s)\|_{V^*}^2\, ds \\ &\leq c \int_0^t \left(c \int_0^s \|\boldsymbol{\eta}_1(r) - \boldsymbol{\eta}_2(r)\|_{V^*}^2\, dr \right) ds \\ &\leq c^2 \left(\int_0^t \|\boldsymbol{\eta}_1(r) - \boldsymbol{\eta}_2(r)\|_{V^*}^2\, dr \right) \left(\int_0^t ds \right) \\ &= c^2 t \int_0^t \|\boldsymbol{\eta}_1(r) - \boldsymbol{\eta}_2(r)\|_{V^*}^2\, dr\end{aligned}$$

for a.e. $t \in (0, T)$. Here and below, as usual, Λ^k represents the kth power of the operator Λ, for $k \in \mathbb{N}$. Reiterating the previous inequality k times, we find that

$$\|(\Lambda^k \boldsymbol{\eta}_1)(t) - (\Lambda^k \boldsymbol{\eta}_2)(t)\|_{V^*}^2 \leq \frac{c^k t^{k-1}}{(k-1)!} \int_0^t \|\boldsymbol{\eta}_1(r) - \boldsymbol{\eta}_2(r)\|_{V^*}^2 \, dr$$

for a.e. $t \in (0, T)$, which leads to

$$\begin{aligned}
\|\Lambda^k \boldsymbol{\eta}_1 - \Lambda^k \boldsymbol{\eta}_2\|_{\mathcal{V}^*} &= \left(\int_0^T \|(\Lambda^k \boldsymbol{\eta}_1)(t) - (\Lambda^k \boldsymbol{\eta}_2)(t)\|_{V^*}^2 \, dt \right)^{\frac{1}{2}} \\
&\leq \left(\int_0^T \frac{c^k T^{k-1}}{(k-1)!} \left(\int_0^t \|\boldsymbol{\eta}_1(r) - \boldsymbol{\eta}_2(r)\|_{V^*}^2 \, dr \right) dt \right)^{\frac{1}{2}} \\
&= \left(\frac{c^k T^k}{(k-1)!} \right)^{\frac{1}{2}} \|\boldsymbol{\eta}_1 - \boldsymbol{\eta}_2\|_{\mathcal{V}^*}.
\end{aligned}$$

Since

$$\lim_{k \to \infty} \frac{a^k}{k!} = 0 \ \text{ for all } \ a > 0,$$

by the Banach contraction principle (Lemma 1.12 on p. 7), we deduce from above that for k sufficiently large, Λ^k is a contraction on $\mathcal{V}^*$. Therefore, there exists a unique $\boldsymbol{\eta}^* \in \mathcal{V}^*$ such that $\boldsymbol{\eta}^* = \Lambda^k \boldsymbol{\eta}^*$ and, moreover, $\boldsymbol{\eta}^* \in \mathcal{V}^*$ is the unique fixed point of the operator Λ.

Step 5 - Existence. Let $\boldsymbol{\eta}^* \in \mathcal{V}^*$ be the fixed point of the operator Λ. We denote by $\boldsymbol{u}$ the solution of Problem P_η for $\boldsymbol{\eta} = \boldsymbol{\eta}^*$, i.e. $\boldsymbol{u} = \boldsymbol{u}_{\eta^*}$, and by φ the solution of Problem Q_η for $\boldsymbol{u}_{\eta^*}$, i.e., $\varphi = \varphi_{\eta^*}$. The regularity of $\boldsymbol{u}$ and φ follows from the regularity of solutions to Problems P_{η^*} and Q_{η^*}. Furthermore, since $\boldsymbol{\eta}^* = \Lambda \boldsymbol{\eta}^*$, we have

$$\langle \boldsymbol{\eta}^*(t), \boldsymbol{v} \rangle_{V^* \times V} = (\mathcal{P}^\top \nabla \varphi_{\eta^*}(t), \boldsymbol{\varepsilon}(\boldsymbol{v}))_{\mathcal{H}} \quad \text{for all } \boldsymbol{v} \in V, \text{ a.e. } t \in (0, T).$$

We conclude that $(\boldsymbol{u}, \varphi)$ is a solution of Problem 8.8 with the desired regularity.

Step 6 - Uniqueness. The uniqueness of the solution of Problem 8.8 is a consequence of Steps 1 and 3, and the uniqueness of the fixed point of Λ. □

A quadruple of functions $(\boldsymbol{u}, \boldsymbol{\sigma}, \varphi, \boldsymbol{D})$ which satisfies (8.46), (8.47), (8.65), (8.66), and (8.55) is called a *weak solution* of Problem 8.7. It follows from above that, under the assumptions of Theorem 8.9, there exists a unique weak solution of Problem 8.7. And, following arguments similar to those presented in Sects. 7.1 and 8.1 it follows that the weak solution satisfies

$$
\begin{aligned}
&\boldsymbol{u} \in W^{1,2}(0,T;V), \quad \boldsymbol{u}' \in C(0,T;H), \quad \boldsymbol{u}'' \in L^2(0,T;V^*),\\
&\boldsymbol{\sigma} \in L^2(0,T;\mathcal{H}), \quad \mathrm{Div}\,\boldsymbol{\sigma} \in L^2(0,T;V^*),\\
&\varphi \in L^2(0,T;\Phi),\\
&\boldsymbol{D} \in L^2(0,T;H), \quad \mathrm{div}\,\boldsymbol{D} \in L^2(0,T;L^2(\Omega)).
\end{aligned}
$$

We end this section with the remark that examples of contact conditions of the forms (8.52), (8.53), and (8.54) in which the functions j_ν, j_τ, and j_e satisfy assumptions (8.10), (8.11), and (8.58), respectively, can be constructed by using arguments similar to those presented in Sect. 7.4. We conclude that our results presented in this section are valid for the corresponding frictional contact models.

Bibliographical Notes

Contact problems with deformable bodies have been studied by many authors, both for theoretical and numerical point of view. The famous Signorini problem was formulated in [235] to model the unilateral contact between an elastic material and a rigid foundation. Mathematical analysis of the Signorini problem was provided in [81] and its numerical analysis was performed in [129]. A reference concerning the Signorini contact problem for elastic and inelastic materials is [246]. References treating modeling and analysis of various contact problems include [72, 77, 102, 117, 129, 202, 233]. Computational methods for problems in Contact Mechanics can be found in the monographs [144, 253, 255] and in the extensive lists of references therein. The state of the art in the field can also be found in the proceedings [95, 155, 218, 254], the surveys [96, 255] and in the special issues [157, 230, 241] as well.

For a comprehensive treatment of basic aspects of Solid Mechanics used in Sects. 6.1 and 6.2 the reader is referred to [11, 70, 85, 97, 127, 145, 154, 197, 198, 248] and to [44] for an in-depth mathematical treatment of three-dimensional elasticity. More information concerning the elastic and viscoelastic constitutive laws presented in Sect. 6.2 can be found in [71, 216, 250].

The literature concerning the frictional contact conditions, including those presented in Sect. 6.3, is abundant. Experimental background and elements of surface physics which justify such conditions can be found in [96, 125, 129, 158, 200, 201, 217, 251]. The regularization (6.65) of the Coulomb friction law was used in [112, 113] in the study of a dynamic contact problem with slip-dependent coefficient of friction involving viscoelastic and elastic materials, respectively. Its static version was used in [106]; there, the uniqueness of a finite element discretized problem in linear elasticity with unilateral contact was investigated. The versions (6.63) and (6.64) of the Coulomb law were used in many geophysical publications in order to model the motion of tectonic plates; details can be found in [37, 118, 217, 228] and the references therein.

Currently, there is a considerable interest in the study of contact problems involving piezoelectric materials and, therefore, there exist a large number of references which can be used to complete the contents of Sect. 6.4. For instance, general models for electro-elastic materials were studied in [27,110,215]. Problems involving piezoelectric contact arising in smart structures and various device applications can be found in [257] and the references therein. The relative motion of two bodies may be detected by a piezoelectric sensor in frictional contact with them, as stated in [29], and vibration of elastic plates may be obtained by the contact with a piezoelectric actuator under electric voltage, see [258] for details. Also, the contact of a read/write piezoelectric head on a hard disk is based on the mechanical deformation generated by the inverse piezoelectric effect, see for instance [18]. There, error estimates and numerical simulations in the study of the corresponding piezoelectric contact problem have been provided.

The existence and uniqueness results presented in Sect. 7.1 are new and were not published before. Nevertheless, we recall that a similar elastic contact problem was considered in [152]. There, a version of the existence and uniqueness result presented in Theorem 7.5 was obtained and a convergence result was provided, based on arguments of H-convergence.

A large number of contact problems with elastic materials were analyzed by many authors, within the framework of variational inequalities, in the static, the quasistatic and the dynamic case. For instance, the existence of the solution to the static contact problem with nonlocal Coulomb law of friction for linearly elastic materials was obtained in [50, 60, 74]. Static problems with Coulomb's friction law were considered in [75] (the coercive case) and in [76] (the semi-coercive case). The analysis of frictional contact problems with normal compliance was performed in [193, 194]. There, the existence of the solutions was proved by using abstract results on elliptic and evolutionary variational inequalities. Contact problems with slip-dependent coefficient of friction for elastic materials have been considered in [45, 116] in the static case and in [56] in the quasistatic case. Existence and uniqueness results in the study of quasistatic problems with normal compliance and Coulomb friction law were obtained in [7, 8] for the case when the magnitude of the normal compliance term is restricted. This restriction was removed in [9]. The quasistatic Signorini frictional problem has been studied in [51, 52] in the case of nonlocal Coulomb law of friction and in [10, 219, 220] in the case of local Coulomb law. Dual variational formulations for the quasistatic Signorini problem with friction have been considered in [247]. The numerical analysis of an elastic quasistatic contact problem with Tresca friction law was performed in [19]. Dynamic frictional problems for linearly elastic materials in which the coefficient of friction is assumed to depend on the slip velocity have been considered in [114, 115]. There it was shown that the solution to the model is not uniquely determined and presents shocks.

Section 7.2, dealing with a study of a time-dependent viscoelastic contact problem, was written following our paper [188]. Other models of frictionless or frictional contact with viscoelastic materials having the same constitutive law, (7.26), have been studied in [223–225, 240]. There, various existence, uniqueness and convergence results were proved. The numerical approximation of the corresponding contact problems was also considered, based on spatially semi-discrete and fully discrete schemes, and error estimates were derived.

Section 7.3 was written following our paper [184]. Our interest in this section was to describe a physical process in which contact, friction, and piezoelectric effects are involved, and to show that the resulting model leads to a well-posed mathematical problem. Other static frictional contact problems for piezoelectric materials with a constitutive law of the forms (7.49)–(7.50) were studied in [29, 153, 167, 169, 237], under the assumption that the foundation is insulated. The results in [29, 153] concern mainly the numerical simulation of the problems while the results in [167, 237] deal with the variational formulations of the problems and their unique weak solvability. A general method in the study of electro-elastic contact problems involving subdifferential boundary conditions was presented in [167] within the framework of hemivariational inequalities. A contact problem for electro-elastic materials with slip-dependent friction was considered in [169]. Unlike the models considered [29, 153, 167, 169, 237], in the model presented in Sect. 7.3 we assume that the foundation is electrically conductive and, moreover, the frictional contact conditions depend on the difference between the potential on the foundation and the body surface. This assumption leads to a full coupling between the mechanical and the electrical unknowns on the boundary conditions and, as a consequence, the resulting mathematical model is new and nonstandard; its variational analysis requires arguments on subdifferential calculus which are different to those used in [167, 237] and it represents the main novelty of the results presented in Sect. 7.3.

Section 8.1 was written following our paper [174]. There, various existence and uniqueness results in the study of second-order variational inequalities were presented, with emphasis to the study of dynamic contact problems with viscoelastic materials with a constitutive law of the form (8.2). The quasistatic version of the results presented in this section was obtained in [189]. We also recall that a new existence result on hemivariational inequalities with applications to quasistatic viscoelastic frictional contact problems can be found in [179]. There, the solution of the quasistatic problem was obtained in the limit of the solution of the corresponding dynamic problem, as the acceleration term converges to zero.

Dynamic frictional contact problems for viscoelastic materials with constitutive law of the form (8.2) were considered by many authors, within the framework of variational inequalities. For instance, a dynamic problem with unilateral conditions formulated in velocities has been considered in [119, 121]. There, the coefficient of friction was assumed to depend on the solution and the solvability of the problem was proved by arguments of penalization and regularization. Dynamic contact problems with normal compliance and Coulomb friction for linearly viscoelastic materials of the form (8.2) have been considered in [112, 138, 140]. In [112] the coefficient of friction was assumed to depend on the slip and the existence of a weak solution was proved by using the Galerkin method. In [140] the coefficient of friction was assumed to depend on the slip velocity and the weak solution of the model was obtained by using arguments of evolutionary inclusions with multivalued pseudomonotone operators. A similar method was used in [141] in the study of a bilateral frictional contact problem with discontinuous coefficient of friction. The analysis of a class of implicit evolutionary variational inequalities with emphasis on the study of dynamic contact problems for linearly viscoelastic materials may be found in [53, 54]. A mathematical model for a dynamic frictional contact problem with normal compliance and wear was considered in [142]; there, the material was linearly viscoelastic, the coefficient of friction was assumed to be discontinuous, and the existence of a weak solution to the model was obtained by using, again, arguments of multivalued pseudomonotone operators. We refer to [64] for an existence and uniqueness result on the dynamic bilateral contact of a viscoelastic body and a foundation; the model considered there recovers as particular cases both the classical and the modified versions of Coulomb's law of dry friction, and some orthotropic friction laws, as well.

Various other results for dynamic contact problems, considered within the theory of hemivariational inequalities, have been obtained in the literature. For instance, existence and uniqueness results for dynamic viscoelastic contact problems with nonmonotone and multivalued subdifferential boundary conditions describing additional phenomena as wear and adhesion can be found in [25, 26] while contact described by a nonmonotone version of the normal compliance condition, associated to a slip-dependent friction law, was considered in [137]. Also, a dynamic contact problem with nonmonotone skin effects for linearly elastic materials has been studied in [177], by using the vanishing viscosity method for hemivariational inequalities. Finally, existence and uniqueness results for models of a thermoviscoelastic solid in frictional contact with a rigid foundation are presented in [61, 62, 65].

Quasistatic contact problems for nonlinear viscoelastic materials with a constitutive law of a form (8.2) were studied by many authors. References in this topic include [5, 16, 42, 43, 100, 101, 221, 232], see also [102] for a survey. The analysis of two quasistatic frictional contact problems for rate-type viscoplastic materials, including existence results, was performed in [4, 6]. In [4] the contact was modeled with normal compliance and Coulomb's law of dry friction and, in [6], the contact was assumed to be bilateral and associated to the Tresca friction law.

Section 8.2 was written following our paper [181]. A dynamic frictionless contact problem with normal compliance and finite penetration for elastic-visco-plastic materials with a constitutive law similar to (8.30) was considered in [123]. There, an existence result was obtained by using a penalization method. Then the results were extended in [78, 124], in the study of a frictionless contact problem with adhesion and in the study of a frictional contact problem with normal damped response, respectively.

Section 8.3 was written following our paper [185]. The analysis of different mathematical models which describe the contact between an electro-viscoelastic body with a constitutive law of the form (8.46), (8.47) and a conductive foundation can also be found in [20, 21, 146]. In [146] the process was assumed to be quasistatic and the contact was described with normal compliance and the associated friction law. A variational formulation of the problem was derived and the existence of a unique weak solution was obtained, under a smallness assumption on the data. The proof was based on arguments of evolutionary variational inequalities and fixed point. In [21] the process was assumed to be quasistatic, the contact was frictionless and was described with the Signorini condition, and a regularized electrical conductivity condition. The existence of a unique weak solution to the model was proved by using arguments of nonlinear equations with multivalued maximal monotone operators and fixed point. Then a fully discrete scheme was introduced and implemented in a numerical code. Numerical simulations in the study of two-dimensional test problems, together with various comments and interpretations, were also presented. In [20] the process was assumed to be dynamic, the contact was frictionless and described with normal compliance. The influence of the electrical properties of the foundation on the contact process was investigated, both from theoretic and numerical point of view.

The study of dynamic contact problems for electro-viscoelastic bodies within the framework of evolutionary hemivariational inequalities is quite recent. References in this topic include [151] (where a model involving a noncoercive viscosity operator was considered) and [148] (where a frictional contact problem with normal damped response and friction was studied).

It is worth to mention that a global existence results for variational inequalities modeling the dynamic contact with a rigid obstacle were obtained in a number of papers. Thus, a hyperbolic problem with unilateral constraints which models the vibrations of a string in contact with a concave obstacle was treated in [227]. An existence result for a variational inequality with unilateral constraints was obtained in [130] and, as far as we are aware, it represents the only existence result obtained till now in the study of the dynamic contact between an elastic membrane and a rigid obstacle. We also refer to [122] for interesting results on the global solvability and properties of the solutions of linear hyperbolic equations with unilateral constraints. Finally, we mention that a global existence result for a dynamic viscoelastic contact problem with given friction was provided in [120].

The comments above show that the mathematical research related to models arising in Contact Mechanics represents a topic which is continuously evolving. A list of major directions in which it could move in the future is provided in the survey paper [231] as well as in the books [233, 238]. Optimal control of frictional contact problems, study of settings with large coefficient of friction, investigation of thermal effects, and inclusion of additional phenomena such as wear and adhesion into contact models are some of these directions.

References

1. Adams, R.A.: Sobolev Spaces. Academic Press, New York (1975)
2. Adams, R.A., Fournier, J.J.F.: Sobolev Spaces. Elsevier, Oxford (2003)
3. Ahmed, N.U., Kerbal, S.: Optimal control of nonlinear second order evolution equations. J. Appl. Math. Stoch. Anal. **6**, 123–136 (1993)
4. Amassad, A., Fabre, C., Sofonea, M.: A quasistatic viscoplastic contact problem with normal compliance and friction. IMA J. Appl. Math. **69**, 463–482 (2004)
5. Amassad, A., Shillor, M., Sofonea, M.: A quasistatic contact problem with slip dependent coefficient of friction. Math. Methods Appl. Sci. **22**, 267–284 (1999)
6. Amassad, A., Sofonea, M.: Analysis of a quasistatic viscoplastic problem involving Tresca friction law. Discrete Contin. Dyn. Syst. **4**, 55–72 (1998)
7. Andersson, L.-E.: A quasistatic frictional problem with normal compliance. Nonlinear Anal. **16**, 347–370 (1991)
8. Andersson, L.-E.: A global existence result for a quasistatic contact problem with friction. Adv. Math. Sci. Appl. **5**, 249–286 (1995)
9. Andersson, L.-E.: A quasistatic frictional problem with a normal compliance penalization term. Nonlinear Anal. **37**, 689–705 (1999)
10. Andersson, L.-E.: Existence results for quasistatic contact problems with Coulomb friction. Appl. Math. Optim. **42**, 169–202 (2000)
11. Antman, S.S.: Nonlinear Problems of Elasticity, 2nd edn. Springer, New York (2005)
12. Atkinson, K., Han, W.: Theoretical numerical analysis: a functional analysis framework, 3rd edn. Texts in Applied Mathematics, vol. 39, Springer, New York (2009)
13. Aubin, J.P.: L'analyse non linéaire et ses motivations économiques. Masson, Paris, (1984)
14. Aubin, J.-P., Cellina, A.: Differential Inclusions. Set-Valued Maps and Viability Theory. Springer, Berlin, New York, Tokyo (1984)
15. Aubin, J.-P., Frankowska, H.: Set-Valued Analysis. Birkhäuser, Basel, Boston (1990)
16. Awbi, B., Rochdi, M., Sofonea, M.: Abstract evolution equations for viscoelastic frictional contact problems. Z. Angew. Math. Phys. **51**, 218–235 (2000)
17. Baiocchi, C., Capelo, A.: Variational and Quasivariational Inequalities: Applications to Free-Boundary Problems. Wiley, Chichester (1984)
18. Barboteu, M., Fernandez, J.R., Raffat, T.: Numerical analysis of a dynamic piezoelectric contact problem arising in viscoelasticity. Comput. Methods Appl. Mech. Engrg. **197**, 3724–3732 (2008)
19. Barboteu, M., Han, W., Sofonea, M.: Numerical analysis of a bilateral frictional contact problem for linearly elastic materials. IMA J. Numer. Anal. **22**, 407–436 (2002)
20. Barboteu, M., Sofonea, M.: Solvability of a dynamic contact problem between a piezoelectric body and a conductive foundation. Appl. Math. Comput. **215**, 2978–2991 (2009)

S. Migórski et al., *Nonlinear Inclusions and Hemivariational Inequalities*,
Advances in Mechanics and Mathematics 26, DOI 10.1007/978-1-4614-4232-5,

21. Barboteu, M., Sofonea, M.: Modelling and analysis of the unilateral contact of a piezoelectric body with a conductive support. J. Math. Anal. Appl. **358**, 110–124 (2009)
22. Barbu, V.: Nonlinear Semigroups and Differential Equations in Banach Spaces. Editura Academiei, Bucharest-Noordhoff, Leyden (1976)
23. Barbu, V.: Optimal Control of Variational Inequalities. Pitman, Boston (1984)
24. Barbu, V., Precupanu, T.: Convexity and Optimization in Banach Spaces. D. Reidel Publishing Company, Dordrecht (1986)
25. Bartosz, K.: Hemivariational inequality approach to the dynamic viscoelastic sliding contact problem with wear. Nonlinear Anal. **65**, 546–566 (2006)
26. Bartosz, K.: Hemivariational inequalities modeling dynamic contact problems with adhesion. Nonlinear Anal. **71**, 1747–1762 (2009)
27. Batra, R.C., Yang, J.S.: Saint-Venant's principle in linear piezoelectricity. J. Elasticity **38**, 209–218 (1995)
28. Bian, W.: Existence results for second order nonlinear evolution inclusions. Indian J. Pure Appl. Math. **11**, 1177–1193 (1998)
29. Bisegna, P., Lebon, F., Maceri, F.: The unilateral frictional contact of a piezoelectric body with a rigid support. In: Martins, J.A.C., Marques, M.D.P.M. (eds.), Contact Mechanics, pp. 347–354. Kluwer, Dordrecht (2002)
30. Brézis, H.: Equations et inéquations non linéaires dans les espaces vectoriels en dualité. Ann. Inst. Fourier (Grenoble) **18**, 115–175 (1968)
31. Brézis, H.: Problèmes unilatéraux. J. Math. Pures Appl. **51**, 1–168 (1972)
32. Brézis, H.: Opérateurs maximaux monotones et semi-groupes de contractions dans les espaces de Hilbert. Mathematics Studies, North Holland, Amsterdam (1973)
33. Brézis, H.: Analyse fonctionnelle—Théorie et applications. Masson, Paris (1987)
34. Brézis, H.: Functional Analysis, Sobolev Spaces and Partial Differential Equations. Springer, Berlin (2011)
35. Browder, F., Hess, P.: Nonlinear mappings of monotone type in Banach spaces. J. Funct. Anal. **11**, 251–294 (1972)
36. Campillo, M., Dascalu, C., Ionescu, I.R.: Instability of a periodic system of faults. Geophys. J. Int. **159**, 212–222 (2004)
37. Campillo, M., Ionescu, I.R.: Initiation of antiplane shear instability under slip dependent friction. J. Geophys. Res. **102 B9**, 363–371 (1997)
38. Carl, S., Le, V.K., Motreanu, D.: Nonsmooth Variational Problems and Their Inequalities. Springer, New York (2007)
39. Castaing, C., Valadier, M.: Convex Analysis and Measurable Multifunctions. Lecture Notes in Mathematics, vol. 580, Springer, Berlin (1977)
40. Cazenave, T., Haraux, A.: Introduction aux problèmes d'évolution semi-linéaires. Ellipses, Paris (1990).
41. Chang, K.C.: Variational methods for nondifferentiable functionals and applications to partial differential equations. J. Math. Anal. Appl. **80**, 102–129 (1981)
42. Chau, O., Han, W., Sofonea, M.: Analysis and approximation of a viscoelastic contact problem with slip dependent friction. Dyn. Contin. Discrete Impuls. Syst. Ser. B Appl. Algorithms **8**, 153–174 (2001)
43. Chau, O., Motreanu, D., Sofonea, M.: Quasistatic frictional problems for elastic and viscoelastic materials. Appl. Math. **47**, 341–360 (2002)
44. Ciarlet, P.G.: Mathematical Elasticity, vol. I: Three-dimensional elasticity. Studies in Mathematics and its Applications, vol. 20, North-Holland, Amsterdam (1988)
45. Ciulcu, C., Motreanu, D., Sofonea, M.: Analysis of an elastic contact problem with slip dependent coefficient of friction. Math. Inequal. Appl. **4**, 465–479 (2001)
46. Clarke, F.H.: Generalized gradients and applications. Trans. Amer. Math. Soc. **205**, 247–262 (1975)
47. Clarke, F.H.: Generalized gradients of Lipschitz functionals. Adv. Math. **40**, 52–67 (1981)
48. Clarke, F.H.: Optimization and Nonsmooth Analysis. Wiley, Interscience, New York (1983)

49. Clarke, F.H., Ledyaev, Y.S., Stern, R.J., Wolenski, P.R.: Nonsmooth Analysis and Control Theory. Springer, New York (1998)
50. Cocu, M.: Existence of solutions of Signorini problems with friction. Int. J. Eng. Sci. **22**, 567–581 (1984)
51. Cocu, M., Pratt, E., Raous, M.: Existence d'une solution du problème quasistatique de contact unilatéral avec frottement non local. C.R. Acad. Sci. Paris Sér. I Math. **320**, 1413–1417 (1995)
52. Cocu, M., Pratt, E., Raous, M.: Formulation and approximation of quasistatic frictional contact. Int. J. Eng. Sci. **34**, 783–798 (1996)
53. Cocu, M., Ricaud, J.M.: Existence results for a class of implicit evolution inequalities and application to dynamic unilateral contact problems with friction. C.R. Acad. Sci. Paris Sér. I Math. **329**, 839–844 (1999)
54. Cocu, M., Ricaud, J.M.: Analysis of a class of implicit evolution inequalities associated to dynamic contact problems with friction. Int. J. Eng. Sci. **328**, 1534–1549 (2000)
55. Cohn, D.L., Measure Theory. Birkhäuser, Boston (1980)
56. Corneschi, C., Hoarau-Mantel, T.-V., Sofonea, M.: A quasistatic contact problem with slip dependent coefficient of friction for elastic materials. J. Appl. Anal. **8**, 59–80 (2002)
57. Cristescu, N., Suliciu, I.: Viscoplasticity. Martinus Nijhoff Publishers, Editura Tehnică, Bucharest (1982)
58. Deimling, K.: Differential Inclusions. Springer, New York, Heidelberg, Berlin (1980)
59. Deimling, K.: Nonlinear Functional Analysis. Springer, Berlin (1985)
60. Demkowicz, I., Oden, J.T.: On some existence and uniqueness results in contact problems with non local friction. Nonlinear Anal. **6**, 1075–1093 (1982)
61. Denkowski, Z., Migórski S.: A system of evolution hemivariational inequalities modeling thermoviscoelastic frictional contact. Nonlinear Anal. **60**, 1415–1441 (2005)
62. Denkowski, Z., Migórski, S.: Hemivariational inequalities in thermoviscoelasticity. Nonlinear Anal. **63**, e87–e97 (2005)
63. Denkowski, Z., Migórski, S.: On sensitivity of optimal solutions to control problems for hyperbolic hemivariational inequalities. In: Cagnol, J., Zolesio, J.-P. (eds.), Control and Boundary Analysis, pp. 147–156. Marcel Dekker, New York (2005)
64. Denkowski, Z., Migórski, S., Ochal, A.: Existence and uniqueness to a dynamic bilateral frictional contact problem in viscoelasticity. Acta Appl. Math. **94**, 251–276 (2006)
65. Denkowski, Z., Migórski, S., Ochal, A.: Optimal control for a class of mechanical thermoviscoelastic frictional contact problems. Control Cybernet. **36**, 611–632 (2007)
66. Denkowski, Z., Migórski, S., Papageorgiu, N.: An Introduction to Nonlinear Analysis: Theory. Kluwer Academic/Plenum Publishers, Boston (2003)
67. Denkowski, Z., Migórski, S., Papageorgiu, N.S.: An Introduction to Nonlinear Analysis: Applications. Kluwer Academic/Plenum Publishers, Boston (2003)
68. Diestel, J., Uhl, J.: Vector Measures. Mathematical Surveys and Monographs, vol. 15, AMS, Providence, RI (1977)
69. Dinculeanu, N.: Vector Measures. Akademie, Berlin (1966)
70. Doghri, I.: Mechanics of Deformable Solids. Springer, Berlin (2000)
71. Drozdov, A.D.: Finite Elasticity and Viscoelasticity—A Course in the Nonlinear Mechanics of Solids. World Scientific, Singapore (1996)
72. Duvaut, G., Lions, J.-L.: Inequalities in Mechanics and Physics. Springer, Berlin (1976)
73. Dunford, N., Schwartz, J.: Linear Operators. Part I, Wiley, New York (1958)
74. Duvaut, G.: Loi de frottement non locale. J. Méc. Th. et Appl. Special issue, 73–78 (1982)
75. Eck, C., Jarušek, J.: Existence results for the static contact problem with Coulomb friction. Math. Methods Appl. Sci. **8**, 445–468 (1998)
76. Eck, C., Jarušek, J.: Existence results for the semicoercive static contact problem with Coulomb friction. Nonlinear Anal. **42**, 961–976 (2000)
77. Eck, C., Jarušek, J., Krbeč, M.: Unilateral Contact Problems: Variational Methods and Existence Theorems. Pure and Applied Mathematics, vol. 270, Chapman/CRC Press, New York (2005)

78. Eck, C., Jarušek, J., Sofonea, M.: A Dynamic elastic-visco-plastic unilateral contact problem with normal damped response and Coulomb friction. European J. Appl. Math. **21**, 229–251 (2010)
79. Ekeland, I., Temam, R.: Convex Analysis and Variational Problems. North-Holland, Amsterdam (1976)
80. Evans, L.C.: Partial Differential Equations. AMS Press, Providence (1999)
81. Fichera, G.: Problemi elastostatici con vincoli unilaterali. II. Problema di Signorini con ambique condizioni al contorno. Mem. Accad. Naz. Lincei, VIII, vol. 7, 91–140 (1964)
82. Friedman, A.: Variational Principles and Free-boundary Problems. Wiley, New York (1982)
83. Gasinski, L., Papageorgiou, N.S.: Nonlinear Analysis. Series in Mathematical Analysis and Applications, vol. 9, Chapman & Hall/CRC, Boca Raton, FL (2006)
84. Gasinski, L., Smolka, M.: An existence theorem for wave-type hyperbolic hemivariational inequalities. Math. Nachr. **242**, 1–12 (2002)
85. Germain, P., Muller, P.: Introduction à la mécanique des milieux continus, Masson, Paris (1980)
86. Glowinski, R.: Numerical Methods for Nonlinear Variational Problems. Springer, New York (1984)
87. Glowinski, R., Lions, J.-L., Trémolières, R.: Numerical Analysis of Variational Inequalities. North-Holland, Amsterdam (1981)
88. Goeleven, D., Miettinen, M., Panagiotopoulos, P.D.: Dynamic hemivariational inequalities and their applications. J. Optim. Theory Appl. **103**, 567–601 (1999)
89. Goeleven, D., Motreanu, D.: Hyperbolic hemivariational inequality and nonlinear wave equation with discontinuities. In: Gilbert, R.P. et al. (eds.) From Convexity to Nonconvexity, pp. 111–122. Kluwer, Dordrecht (2001)
90. Goeleven, D., Motreanu, D., Dumont, Y., Rochdi, M.: Variational and Hemivariational Inequalities, Theory, Methods and Applications, vol. I: Unilateral Analysis and Unilateral Mechanics. Kluwer Academic Publishers, Boston, Dordrecht, London (2003)
91. Goeleven, D., Motreanu, D.: Variational and Hemivariational Inequalities, Theory, Methods and Applications, vol. II: Unilateral Problems. Kluwer Academic Publishers, Boston, Dordrecht, London (2003)
92. Grisvard, P.: Elliptic Problems in Nonsmooth Domains. Pitman, Boston (1985)
93. Guo, X.: On existence and uniqueness of solution of hyperbolic differential inclusion with discontinuous nonlinearity. J. Math. Anal. Appl. **241**, 198–213 (2000)
94. Guo, X.: The initial boundary value problem of mixed-typed hemivariational inequality. Int. J. Math. Math. Sci. **25**, 43–52 (2001)
95. Guran, A. (ed.): Proceedings of the First International Symposium on Impact and Friction of Solids, Structures and Intelligent Machines. World Scientific, Singapore (2000)
96. Guran, A., Pfeiffer, F., Popp, K. (eds.): Dynamics with friction: modeling, analysis and experiment, Part I. World Scientific, Singapore (1996)
97. Gurtin, M.E.: An Introduction to Continuum Mechanics. Academic Press, New York (1981)
98. Halmos, P.R.: Measure Theory. D. Van Nostrand Company, Princeton, Toronto, London, New York (1950)
99. Han, W., Reddy, B.D.: Plasticity: Mathematical Theory and Numerical Analysis. Springer, New York (1999)
100. Han, W., Sofonea, M.: Evolutionary variational inequalities arising in viscoelastic contact problems. SIAM J. Numer. Anal. **38**, 556–579 (2000)
101. Han W., Sofonea, M.: Time-dependent variational inequalities for viscoelastic contact problems, J. Comput. Appl. Math. **136**, 369–387 (2001)
102. Han, W., Sofonea, M.: Quasistatic Contact Problems in Viscoelasticity and Viscoplasticity. Studies in Advanced Mathematics, vol. 30, American Mathematical Society, Providence, RI-International Press, Somerville, MA (2002)
103. Haslinger, J., Hlaváček, I., Nečas J.: Numerical methods for unilateral problems in solid mechanics. In Handbook of Numerical Analysis, vol. IV, Ciarlet, P.G., and J.-L. Lions (eds.), North-Holland, Amsterdam, pp. 313–485 (1996)

104. Haslinger, J., Miettinen, M., Panagiotopoulos, P.D.: Finite Element Method for Hemivariational Inequalities. Theory, Methods and Applications. Kluwer Academic Publishers, Boston, Dordrecht, London (1999)
105. Hewitt, E., Stromberg, K.: Abstract Analysis. Springer, New York, Inc. (1965)
106. Hild, P.: On finite element uniqueness studies for Coulomb's frictional contact model. Int. J. Appl. Math. Comput. Sci. **12**, 41–50 (2002)
107. Hiriart-Urruty, J.-B., Lemaréchal C.: Convex Analysis and Minimization Algorithms, I, II. Springer, Berlin (1993)
108. Hlaváček, I., Haslinger, J., Nečas, J., Lovíšek, J.: Solution of Variational Inequalities in Mechanics. Springer, New York (1988)
109. Hu, S., Papageorgiou, N.S.: Handbook of Multivalued Analysis, vol. I: theory. Kluwer, Dordrecht, The Netherlands (1997)
110. Ikeda, T.: Fundamentals of Piezoelectricity. Oxford University Press, Oxford (1990)
111. Ionescu, I.R., Dascalu, C., Campillo, M.: Slip-weakening friction on a periodic System of faults: spectral analysis. Z. Angew. Math. Phys. **53**, 980–995 (2002)
112. Ionescu, I.R., Nguyen, Q.-L.: Dynamic contact problems with slip dependent friction in viscoelasticity. Int. J. Appl. Math. Comput. Sci. **12**, 71–80 (2002)
113. Ionescu, I.R., Nguyen, Q.-L., Wolf, S.: Slip displacement dependent friction in dynamic elasticity. Nonlinear Anal. **53**, 375–390 (2003)
114. Ionescu, I.R., Paumier, J.-C.: Friction dynamique avec coefficient dépandant de la vitesse de glissement. C.R. Acad. Sci. Paris Sér. I Math. **316**, 121–125 (1993)
115. Ionescu, I.R., Paumier, J.-C.: On the contact problem with slip rate dependent friction in elastodynamics. Eur. J. Mech. A Solids **13**, 556–568 (1994)
116. Ionescu, I.R., Paumier, J.-C.: On the contact problem with slip displacement dependent friction in elastostatics. Int. J. Eng. Sci. **34**, 471–491 (1996)
117. Ionescu, I.R., Sofonea, M.: Functional and Numerical Methods in Viscoplasticity. Oxford University Press, Oxford (1993)
118. Ionescu, I.R., Wolf, S.: Interaction of faults under slip dependent friction. Nonlinear eigenvalue analysis. Math. Methods Appl. Sci. **28**, 77–100 (2005)
119. Jarušek, J.: Contact problem with given time-dependent friction force in linear viscoelasticity. Comment. Math. Univ. Carolin. **31**, 257–262 (1990)
120. Jarušek, J.: Dynamic contact problems with given friction for viscoelastic bodies. Czechoslovak Math. J. **46**, 475–487 (1996)
121. Jarušek, J., Eck, C.: Dynamic contact problems with small Coulomb friction for viscoelastic bodies. Existence of solutions. Math. Models Methods Appl. Sci. **9**, 11–34 (1999)
122. Jarušek, J., Málek, J., Nečas, J., Šverák, V.: Variational inequality for a viscous drum vibrating in the presence of an obstacle. Rend. Mat. Appl. **12**, 943–958 (1992)
123. Jarušek, J., Sofonea, M.: On the solvability of dynamic elastic-visco-plastic contact problems. ZAMM Z. Angew. Math. Mech. **88**, 3–22 (2008)
124. Jarušek, J., Sofonea, M.: On the solvability of dynamic elastic-visco-plastic contact problems with adhesion. Ann. Acad. Rom. Sci. Ser. Math. Appl. **1**, 191–214 (2009)
125. Johnson, K.L.: Contact Mechanics. Cambridge University Press, Cambridge (1987)
126. Kalita, P.: Decay of energy of a system described by hyperbolic hemivariational inequality. Nonlinear Anal. **74**, 116–1181 (2011)
127. Khludnev, A.M., Sokolowski, J.: Modelling and Control in Solid Mechanics. Birkhäuser-Verlag, Basel (1997)
128. Kikuchi, N., Oden, J.T.: Theory of variational inequalities with applications to problems of flow through porous media. Int. J. Eng. Sci. **18**, 1173–1284 (1980)
129. Kikuchi, N., Oden, J.T.: Contact Problems in Elasticity: A Study of Variational Inequalities and Finite Element Methods. SIAM, Philadelphia (1988)
130. Kim, J.U.: A boundary thin obstacle problem for a wave equation. Comm. Part. Differ. Equat. **14**, 1011–1026 (1989)
131. Kinderlehrer, D., Stampacchia, G.: An introduction to variational inequalities and their applications. Classics in Applied Mathematics, vol. 31, SIAM, Philadelphia (2000)

132. Kisielewicz, M.: Differential Inclusions and Optimal Control. Kluwer, Dordrecht, The Netherlands (1991)
133. Klarbring, A., Mikelič, A., Shillor, M.: Frictional contact problems with normal compliance. Int. J. Eng. Sci. **26**, 811–832 (1988)
134. Klarbring, A., Mikelič, A., Shillor, M.: On friction problems with normal compliance. Nonlinear Anal. **13**, 935–955 (1989)
135. Komura, Y.: Nonlinear semigroups in Hilbert Spaces. J. Math. Soc. Japan **19**, 493–507 (1967)
136. Kufner, A., John, O., Fučik, S.: Function Spaces. Monographs and Textbooks on Mechanics of Solids and Fluids; Mechanics: Analysis. Noordhoff International Publishing, Leyden (1977)
137. Kulig, A.: Hemivariational inequality approach to the dynamic viscoelastic contact problem with nonmonotone normal compliance and slip-dependent friction. Nonlinear Anal. Real World Appl. **9**, 1741–1755 (2008)
138. Kuttler, K.L.: Dynamic friction contact problem with general normal and friction laws. Nonlinear Anal. **28**, 559–575 (1997)
139. Kuttler, K.: Topics in Analysis. Private communication (2006)
140. Kuttler, K.L., Shillor, M.: Set-valued pseudomonotone maps and degenerate evolution inclusions. Commun. Contemp. Math. **1**, 87–123 (1999)
141. Kuttler, K.L., Shillor, M.: Dynamic bilateral contact with discontinuous friction coefficient. Nonlinear Anal. **45**, 309–327 (2001)
142. Kuttler, K.L., Shillor, M.: Dynamic contact with normal compliance, wear and discontinuous friction coefficient. SIAM J. Math. Anal. **34**, 1–27 (2002)
143. Larsen, R.: Functional Analysis. Marcel Dekker, New York (1973)
144. Laursen, T.A.: Computational Contact and Impact Mechanics. Springer, Berlin (2002)
145. Lemaître, J., Chaboche, J.-L.: Mechanics of Solid Materials. Cambridge University Press, Cambridge (1990)
146. Lerguet, Z., Shillor, M., Sofonea, M.: A frictional contact problem for an electro-viscoelastic body. Electron. J. Differential Equations **170**, 1–16 (2007)
147. Li, Y., Liu, Z.: Dynamic contact problem for viscoelastic piezoelectric materials with slip dependent friction. Nonlinear Anal. **71**, 1414–1424 (2009)
148. Li, Y., Liu, Z.: Dynamic contact problem for viscoelastic piezoelectric materials with normal damped response and friction. J. Math. Anal. Appl. **373**, 726–738 (2011)
149. Lions, J.-L.: Quelques méthodes de resolution des problémes aux limites non linéaires. Dunod, Paris (1969)
150. Lions, J.-L., Magenes, E.: Problèmes aux limites non-homogènes I. Dunod, Paris (1968); English translation: Non-homogeneous boundary value problems and applications, vol. I. Springer, New York, Heidelberg (1972)
151. Liu, Z., Migórski, S.: Noncoercive damping in dynamic hemivariational inequality with application to problem of piezoelectricity. Discrete Contin. Dyn. Syst. Ser. B **9**, 129–143 (2008)
152. Liu, Z., Migórski, S., Ochal, A.: Homogenization of boundary hemivariational inequalities in linear elasticity. J. Math. Anal. Appl. **340**, 1347–1361 (2008)
153. Maceri, F., Bisegna, P.: The unilateral frictionless contact of a piezoelectric body with a rigid support. Math. Comput. Model. **28**, 19–28 (1998)
154. Malvern, L.E.: Introduction to the Mechanics of a Continuum Medium. Princeton-Hall, NJ (1969)
155. Martins, J.A.C., Marques, M.D.P.M. (eds.): Contact Mechanics. Kluwer, Dordrecht (2002)
156. Martins, J.A.C., Oden, J.T.: Existence and uniqueness results for dynamic contact problems with nonlinear normal and friction interface laws. Nonlinear Anal. **11**, 407–428 (1987)
157. Matysiak, S.J. (ed.): Contact Mechanics, Special issue of J. Theoret. Appl. Mech. **39**(3) (2001)
158. Maugis, D.: Contact, Adhesion and Rupture of Elastic Solids. Springer, Berlin, Heidelberg (2000)
159. Maz'ja, V.G.: Sobolev Spaces. Springer, Berlin (1985)
160. Migórski, S.: Existence, variational and optimal control problems for nonlinear second order evolution inclusions. Dynam. Systems Appl. **4**, 513–528 (1995)

161. Migórski, S.: Existence and relaxation results for nonlinear second order evolution inclusions. Discuss. Math. Differ. Incl. Control Optim. **15**, 129–148 (1995)
162. Migórski, S.: Evolution hemivariational inequalities in infinite dimension and their control. Nonlinear Anal. **47**, 101–112 (2001)
163. Migórski, S.: Homogenization technique in inverse problems for boundary hemivariational inequalities, Inverse Probl. Sci. Eng. **11**, 229–242 (2003)
164. Migórski, S.: Hemivariational inequalities modeling viscous incompressible fluids. J. Nonlinear Convex Anal. **5**, 217–227 (2004)
165. Migórski, S.: Dynamic hemivariational inequality modeling viscoelastic contact problem with normal damped response and friction. Appl. Anal. **84**, 669–699 (2005)
166. Migórski, S.: Boundary hemivariational inequalities of hyperbolic type and applications. J. Global Optim. **31**, 505–533 (2005)
167. Migórski, S.: Hemivariational inequality for a frictional contact problem in elasto-piezoelectricity. Discrete Contin. Dyn. Syst. Ser. B **6**, 1339–1356 (2006)
168. Migórski, S.: Evolution hemivariational inequality for a class of dynamic viscoelastic nonmonotone frictional contact problems. Comput. Math. Appl. **52**, 677–698 (2006)
169. Migórski, S.: A class of hemivariational inequalities for electroelastic contact problems with slip dependent friction. Discrete Contin. Dyn. Syst. Ser. S **1**, 117–126 (2008)
170. Migórski, S.: Evolution hemivariational inequalities with spplications. In: Gao, D.Y., Motreanu, D. (eds.), Handbook of Nonconvex Analysis and Applications, pp. 409–473. International Press, Boston (2010)
171. Migórski, S., Ochal, A.: Inverse coefficient problem for elliptic hemivariational inequality. In: Gao, D.Y., Ogden, R.W., Stavroulakis, G.E. (eds.), Nonsmooth/Nonconvex Mechanics: Modeling, Analysis and Numerical Methods, pp. 247–262. Kluwer Academic Publishers, Dordrecht, Boston, London (2001)
172. Migórski, S., Ochal, A.: Hemivariational inequality for viscoelastic contact problem with slip-dependent friction. Nonlinear Anal. **61**, 135–161 (2005)
173. Migórski, S., Ochal, A.: Hemivariational inequalities for stationary Navier–Stokes equations. J. Math. Anal. Appl. **306**, 197–217 (2005)
174. Migórski, S., Ochal, A.: A unified approach to dynamic contact problems in viscoelasticity. J. Elasticity **83**, 247–275 (2006)
175. Migórski, S., Ochal, A.: Existence of solutions for second order evolution inclusions with application to mechanical contact problems. Optimization **55**, 101–120 (2006)
176. Migórski, S.,Ochal, A.: Nonlinear impulsive evolution inclusions of second order. Dynam. Systems Appl. **16**, 155–174 (2007)
177. Migórski, S., Ochal, A.: Vanishing viscosity for hemivariational inequality modeling dynamic problems in elasticity. Nonlinear Anal. **66**, 1840–1852 (2007)
178. Migórski, S., Ochal, A.: Dynamic bilateral contact problem for viscoelastic piezoelectric materials with adhesion. Nonlinear Anal. **69**, 495–509 (2008)
179. Migórski, S., Ochal, A.: Quasistatic hemivariational inequality via vanishing acceleration approach. SIAM J. Math. Anal. **41**, 1415–1435 (2009)
180. Migórski, S., Ochal, A.: Nonconvex inequality models for contact problems of nonsmooth mechanics. In: Kuczma, M., Wilmanski, K. (eds.) Computer Methods in Mechanics: Computational Contact Mechanics, Advanced Structured Materials, vol. 1, pp. 43–58. Springer, Berlin, Heidelberg (2010)
181. Migórski, S., Ochal, A., Sofonea, M.: Integrodifferential hemivariational inequalities with applications to viscoelastic frictional contact. Math. Models Methods Appl. Sci. **18**, 271–290 (2008)
182. Migórski, S., Ochal, A., Sofonea, M.: Solvability of dynamic antiplane frictional contact problems for viscoelastic cylinders. Nonlinear Anal. **70**, 3738–3748 (2009)
183. Migórski, S., Ochal, A., Sofonea, M.: Modeling and analysis of an antiplane piezoelectric contact problem. Math. Models Methods Appl. Sci. **19**, 1295–1324 (2009)
184. Migórski, S., Ochal, A., Sofonea, M.: Variational analysis of static frictional contact problems for electro-elastic materials. Math. Nachr. **283**, 1314–1335 (2010)

185. Migórski, S., Ochal, A., Sofonea, M.: A dynamic frictional contact problem for piezoelectric materials. J. Math. Anal. Appl. **361**, 161–176 (2010)
186. Migórski, S., Ochal, A., Sofonea, M.: Weak solvability of antiplane frictional contact problems for elastic cylinders. Nonlinear Anal. Real World Appl. **11**, 172–183 (2010)
187. Migórski, S., Ochal, A., Sofonea, M.: Analysis of a dynamic contact problem for electro-viscoelastic cylinders. Nonlinear Anal. **73**, 1221–1238 (2010)
188. Migórski, S., Ochal, A., Sofonea, M.: Analysis of a frictional contact problem for viscoelastic materials with long memory. Discrete Contin. Dyn. Syst. Ser. B **15**, 687–705 (2011)
189. Migórski, S., Ochal, A., Sofonea, M.: History-dependent subdifferential inclusions and hemivariational inequalities in contact mechanics. Nonlinear Anal. Real World Appl. **12**, 3384–3396 (2011)
190. Mordukhovich, B.: Variational Analysis and Generalized Differentiation I: Basic Theory. Springer, Berlin, Heidelberg (2006)
191. Mordukhovich, B.: Variational Analysis and Generalized Differentiation II: Applications. Springer, Berlin, Heidelberg (2006)
192. Motreanu, D., Panagiotopoulos, P.D.: Minimax Theorems and Qualitative Properties of the Solutions of Hemivariational Inequalities and Applications. Kluwer Academic Publishers, Boston, Dordrecht, London (1999)
193. Motreanu, D., Sofonea, M.: Evolutionary variational inequalities arising in quasistatic frictional contact problems for elastic materials. Abstr. Appl. Anal. **4**, 255–279 (1999)
194. Motreanu, D., Sofonea, M.: Quasivariational inequalities and applications in frictional contact problems with normal compliance. Adv. Math. Sci. Appl. **10**, 103–118 (2000)
195. Naniewicz, Z., Panagiotopoulos, P.D.: Mathematical Theory of Hemivariational Inequalities and Applications. Marcel Dekker, New York (1995)
196. Nečas, J.: Les méthodes directes en théorie des équations elliptiques. Academia, Praha (1967)
197. Nečas, J., Hlaváček, I.: Mathematical Theory of Elastic and Elastico-Plastic Bodies: An Introduction. Elsevier Scientific Publishing Company, Amsterdam, Oxford, New York (1981)
198. Nguyen, Q.S.: Stability and Nonlinear Solid Mechanics. Wiley, Chichester (2000)
199. Ochal, A.: Existence results for evolution hemivariational inequalities of second order. Nonlinear Anal. **60**, 1369–1391 (2005)
200. Oden, J.T., Martins, J.A.C.: Models and computational methods for dynamic friction phenomena. Comput. Methods Appl. Mech. Engrg. **52**, 527–634 (1985)
201. Oden, J.T., Pires, E.B.: Nonlocal and nonlinear friction laws and variational principles for contact problems in elasticity. J. Appl. Mech. **50**, 67–76 (1983)
202. Panagiotopoulos, P.D.: Inequality Problems in Mechanics and Applications. Convex and Nonconvex Energy Functions. Birkhäuser, Boston (1985)
203. Panagiotopoulos, P.D.: Nonconvex problems of semipermeable media and related topics. ZAMM Z. Angew. Math. Mech. **65**, 29–36 (1985)
204. Panagiotopoulos, P.D.: Coercive and semicoercive hemivariational inequalities. Nonlinear Anal. **16**, 209–231 (1991)
205. Panagiotopoulos, P.D.: Hemivariational Inequalities, Applications in Mechanics and Engineering. Springer, Berlin (1993)
206. Panagiotopoulos, P.D.: Modelling of nonconvex nonsmooth energy problems: dynamic hemivariational inequalities with impact effects. J. Comput. Appl. Math. **63**, 123–138 (1995)
207. Panagiotopoulos, P.D.: Hemivariational inequalities and fan-variational inequalities. New applications and results. Atti Semin. Mat. Fis. Univ. Modena Reggio Emilia **43**, 159–191 (1995)
208. Panagiotopoulos, P.D., Pop, G.: On a type of hyperbolic variational-hemivariational inequalities. J. Appl. Anal. **5**, 95–112 (1999)
209. Papageorgiou, N.S.: Existence of solutions for second-order evolution inclusions. Bull. Math. Soc. Sci. Math. Roumanie **37**, 93–107 (1993)
210. Papageorgiou, N.S., Yannakakis, N.: Second order nonlinear evolution inclusions I: existence and relaxation results. Acta Mat. Sin. (Engl. Ser.) **21**, 977–996 (2005)

211. Papageorgiou, N.S., Yannakakis, N.: Second order nonlinear evolution inclusions II: structure of the solution set. Acta Mat. Sin. (Engl. Ser.) **22**, 195–206 (2006)
212. Park, J.Y., Ha, T.G.: Existence of antiperiodic solutions for hemivariational inequalities. Nonlinear Anal. **68**, 747–767 (2008)
213. Park, S.H., Park, J.Y., Jeong, J.M.: Boundary stabilization of hyperbolic hemivariational inequalities Acta Appl. Math. **104**, 139–150 (2008)
214. Pascali, D., Sburlan, S.: Nonlinear Mappings of Monotone Type. Sijthoff and Noordhoff International Publishers, Alpen aan den Rijn (1978)
215. Patron, V.Z., Kudryavtsev, B.A.: Electromagnetoelasticity, Piezoelectrics and Electrically Conductive Solids. Gordon & Breach, London (1988)
216. Pipkin, A.C.: Lectures in Viscoelasticity Theory. Applied Mathematical Sciences, vol. 7, George Allen & Unwin Ltd. London, Springer, New York (1972)
217. Rabinowicz, E.: Friction and Wear of Materials, 2nd edn. Wiley, New York (1995)
218. Raous, M., Jean, M., Moreau, J.J. (eds.): Contact Mechanics. Plenum Press, New York (1995)
219. Rocca, R.: Existence of a solution for a quasistatic contact problem with local friction. C.R. Acad. Sci. Paris Sér. I Math. **328**, 1253–1258 (1999)
220. Rocca, R., Cocu, M.: Existence and approximation of a solution to quasistatic problem with local friction. Int. J. Eng. Sci. **39**, 1233–1255 (2001)
221. Rochdi, M., Shillor, M., Sofonea, M.: Quasistatic viscoelastic contact with normal compliance and friction. J. Elasticity **51**, 105–126 (1998)
222. Rockafellar, T.R.: Convex Analysis. Princeton University Press, Princeton (1970)
223. Rodríguez–Aros, A.D., Sofonea, M., Viaño, J.M.: A class of evolutionary variational inequalities with Volterra-type integral term. Math. Models Methods Appl. Sci. **14**, 555–577 (2004)
224. Rodríguez–Aros, A.D., Sofonea, M., Viaño, J.M.: Numerical approximation of a viscoelastic frictional contact problem. C.R. Acad. Sci. Paris, Sér. II Méc. **334**, 279–284 (2006)
225. Rodríguez–Aros, A.D., Sofonea, M., Viaño, J.M.: Numerical analysis of a frictional contact problem for viscoelastic materials with long-term memory. Numer. Math. **198**, 327–358 (2007)
226. Royden, H.L.: Real Analysis. The Macmillan Company, New York, Collier-Macmillan Ltd, London (1963)
227. Schatzman, M.: A hyperbolic problem of second order with unilateral constraints: the vibrating string with a concave obstacle. J. Math. Anal. Appl. **73**, 138–191 (1980)
228. Scholz, C.H.: The Mechanics of Earthquakes and Faulting. Cambridge University Press, Cambridge (1990)
229. Schwartz, L.: Théorie des Distributions. Hermann, Paris (1951)
230. Shillor, M. (ed.): Recent advances in contact mechanics, Special issue of Math. Comput. Mode. **28** (4–8) (1998)
231. Shillor, M.: Quasistatic problems in contact mechanics. Int. J. Appl. Math. Comput. Sci. **11**, 189–204 (2001)
232. Shillor, M., Sofonea, M.: A quasistatic viscoelastic contact problem with friction. Int. J. Eng. Sci. **38**, 1517–1533 (2000)
233. Shillor, M., Sofonea, M., Telega, J.J.: Models and Analysis of Quasistatic Contact. Lecture Notes in Physics, vol. 655, Springer, Berlin (2004)
234. Showalter, R.: Monotone Operators in Banach Spaces and Nonlinear Partial Differential Equations. vol. 49, Mathematical Surveys and Monographs, American Mathematical Society, Providence, RI (1997)
235. Signorini, A.: Sopra alcune questioni di elastostatica. Atti della Società Italiana per il Progresso delle Scienze (1933)
236. Simon, J.: Compact sets in the space $L^p(0;T;B)$. Ann. Mat. Pura Appl. **146**, 65–96 (1987)
237. Sofonea, M., El Essoufi, H.: A piezoelectric contact problem with slip dependent coefficient of friction. Math. Model. Anal. **9**, 229–242 (2004)
238. Sofonea, M., Han, W., Shillor, M.: Analysis and Approximation of Contact Problems with Adhesion or Damage. Pure and Applied Mathematics, vol. 276, Chapman-Hall/CRC Press, New York (2006)

239. Sofonea, M., Matei, A.: Variational Inequalities with Applications. A Study of Antiplane Frictional Contact Problems. Advances in Mechanics and Mathematics, vol. 18, Springer, New York (2009)
240. Sofonea, M., Rodríguez–Aros, A.D., Viaño J.M.: A class of integro-differential variational inequalities with applications to viscoelastic contact. Math. Comput. Model. **41**, 1355–1369 (2005)
241. Sofonea, M., Viaño, J.M. (eds.), Mathematical Modelling and Numerical Analysis in Solid Mechanics. Special issue of Int. J. Appl. Math. Comput. Sci. **12** (1) (2002)
242. Strömberg, N.: Thermomechanical modelling of tribological systems. Ph.Thesis, D., no. 497, Linköping University, Sweden (1997)
243. Strömberg, N., Johansson, P., A. Klarbring, A.: Generalized standard model for contact friction and wear. In: Raous, M., Jean, M., Moreau, J.J. (eds.) Contact Mechanics, Plenum Press, New York (1995)
244. Strömberg, N., Johansson, L., Klarbring, A.: Derivation and analysis of a generalized standard model for contact friction and wear. Int. J. Solids Structures **33**, 1817–1836 (1996)
245. Tartar, L.: An Introduction to Sobolev Spaces and Interpolation Spaces. Springer, Berlin (2007)
246. Telega, J.J.: Topics on unilateral contact problems of elasticity and inelasticity. In: Moreau, J.J., Panagiotopoulos, P.D. (eds.), Nonsmooth Mechanics and Applications, pp. 340–461. Springer, Wien (1988)
247. Telega, J.J.: Quasi-static Signorini's contact problem with friction and duality. Int. Ser. Numer. Math. **101**, 199–214 (1991)
248. Temam, R., Miranville, A.: Mathematical Modeling in Continuum Mechanics. Cambridge University Press, Cambridge (2001)
249. Triebel, H.: Interpolation Theory, Function Spaces, Differential Operators. North-Holland, Amsterdam (1978)
250. Truesdell, C. (ed.): Mechanics of Solids, vol. III: Theory of Viscoelasticity, Plasticity, Elastic Waves and Elastic Stability. Springer, Berlin (1973)
251. Wilson, W.R.D.: Modeling friction in sheet-metal forming simulation. In: Zabaras et al. (eds.), The Integration of Materials, Process and Product Design, pp. 139–147. Balkema, Rotterdam (1999)
252. Wloka, J.: Partielle Differentialgleichungen, Teubner, B.G., Stuttgart (1982) English translation: Partial Differential Equations. Cambridge University Press, Cambridge (1987)
253. Wriggers, P.: Computational Contact Mechanics. Wiley, Chichester (2002)
254. Wriggers, P., Nackenhorst, U. (eds.): Analysis and simulation of contact problems. Lecture Notes in Applied and Computational Mechanics, vol. 27, Springer, Berlin (2006)
255. Wriggers, P., Panagiotopoulos, P.D. (eds.): New Developments in Contact Problems. Springer, Wien, New York (1999)
256. Xiao, Y., Huang, N.: Browder-Tikhonov regularization for a class of evolution second order hemivariational inequalities. J. Global Optim. **45**, 371–388 (2009)
257. Yang, S.S.: An Introduction to the Theory of Piezoelectricity. Springer, New York (2005)
258. Yang, J.S., Batra, J.M., Liang, X.Q.: The cylindrical bending vibration of a laminated elastic plate due to piezoelectric acutators. Smart Mater. Struct. **3**, 485–493 (1994)
259. Yosida, K.: Functional Analysis, 5th edn. Springer, Berlin (1978)
260. Zeidler, E.: Nonlinear Functional Analysis and its Applications. I: Fixed-point Theorems, Springer, New York (1985)
261. Zeidler, E.: Nonlinear Functional Analysis and its Applications. III: Variational Methods and Optimization. Springer, New York (1986)
262. Zeidler, E.: Nonlinear Functional Analysis and its Applications. IV: Applications to Mathematical Physics. Springer, New York (1988)
263. Zeidler, E.: Nonlinear Functional Analysis and its Applications. II/A: Linear Monotone Operators. Springer, New York (1990)
264. Zeidler, E.: Nonlinear Functional Analysis and its Applications. II/B: Nonlinear Monotone Operators. Springer, New York (1990)

265. Zeidler, E.: Applied Functional Analysis: Main Principles and Their Applications. Springer, New York (1995)
266. Zeidler, E.: Applied Functional Analysis: Applications of Mathematical Physics. Springer, New York (1995)

Index

S. Migórski et al., *Nonlinear Inclusions and Hemivariational Inequalities*,
Advances in Mechanics and Mathematics 26, DOI 10.1007/978-1-4614-4232-5,